FUNDAMENTALS OF NANO-TEXTILE SCIENCE

FUNDAMENTALS OF NANO-TEXTILE SCIENCE

Edited by
Prashansa Sharma, PhD
Devsuni Singh
Vivek Dave, PhD

AAP APPLE
ACADEMIC
PRESS

First edition published 2023

Apple Academic Press Inc.
1265 Goldenrod Circle, NE,
Palm Bay, FL 32905 USA

4164 Lakeshore Road, Burlington,
ON, L7L 1A4 Canada

CRC Press
6000 Broken Sound Parkway NW,
Suite 300, Boca Raton, FL 33487-2742 USA

4 Park Square, Milton Park,
Abingdon, Oxon, OX14 4RN UK

© 2023 by Apple Academic Press, Inc.

Apple Academic Press exclusively co-publishes with CRC Press, an imprint of Taylor & Francis Group, LLC

Library and Archives Canada Cataloguing in Publication

Title: Fundamentals of nano-textile science / edited by Prashansa Sharma, PhD, Devsuni Singh, Vivek Dave, PhD.
Names: Sharma, Prashansa, editor. | Singh, Devsuni, editor. | Dave, Vivek, editor.
Description: First edition. | Includes bibliographical references and index.
Identifiers: Canadiana (print) 20220171165 | Canadiana (ebook) 20220171211 | ISBN 9781774638606 (hardcover) | ISBN 9781774638613 (softcover) | ISBN 9781003277316 (ebook)
Subjects: LCSH: Nanostructured materials. | LCSH: Nanofibers. | LCSH: Textile fabrics. | LCSH: Nanotechnology.
Classification: LCC TA418.9.N35 F86 2023 | DDC 677/.02832—dc23

Library of Congress Cataloging-in-Publication Data

CIP data on file with US Library of Congress

ISBN: 978-1-77463-860-6 (hbk)
ISBN: 978-1-77463-861-3 (pbk)
ISBN: 978-1-00327-731-6 (ebk)

About the Editors

Prashansa Sharma, PhD

Prashansa Sharma, PhD, is Assistant Professor at Department of Home Science, Mahila Maha Vidyalaya, Banaras Hindu University, Varanasi, India. She had more than 12 years of teaching experience at the Banasthali Vidyapith Rajasthan educational institute as well as research experience. She has drafted many research and review papers related to nanotechnology and nanotextile science for publication in national and international journals. She has been recipient of several awards at various conferences, seminars, and conventions. Dr. Sharma is also acting as regular reviewer of journals for the publishing houses of Wiley, Springer, Elsevier, BMC, etc. She completed her MSc and PhD degrees at Banasthali Vidyapith Rajasthan, India, and received a Gold Medal Award from Banasthali Vidyapith during her postgraduate work.

Devsuni Singh

Devsuni Singh is affiliated with the Department of Clothing and Textile, Faculty of Home Science at Banasthali Vidhyapith, Rajasthan, India. Her research interests include development of novel functionality nano-textiles, textile technology, nanotextile science, development of environmentally friendly green nanocomposites, healthcare textiles, textile designing, textile testing, and sustainable textiles. She has experience in nano-textiles and nanoparticles using green herbal synthesis into the textile materials, which helps to promote the emerging fields of nanotechnology and nano-textile science. She received her master's degree in Clothing and Textile with high grades and a merit certificate from Banasthali Vidhyapith, Rajasthan, India.

Vivek Dave

Vivek Dave, PhD, is presently working in the capacity of Associate Professor and Head at Central University South Bihar, Gaya, Bihar, India. His expertise is in working with formulation and development of novel drug delivery systems like PEGylated liposomes, ethosomes, nanoparticles, quantum dots, and nanotechnology. He has been working in the field of novel drug delivery systems and nanotechnology and has published several articles in national and international journals. He has also recorded his expertise in the form of many chapters and books. He has been recipient of several awards at various conferences, seminars, and conventions. Dr. Dave is also acting as regular reviewer of national and international journals for Wiley, Springer, Elsevier, BMC, etc.

Contents

Contributors... *ix*

Abbreviations .... *xiii*

Preface .... *xv*

**PART I: Introduction to Nanofiber Production and Its
Pretreatment Techniques**... 1

1. **Nanofiber and Nanoparticle Production Techniques for Various
Functional Product Developments** 3
   S. Rajesh Kumar, M. Gopalakrishnan, R. Vijayasekar, and A. Ashok Kumar

2. **Flexible Piezoelectric Nanogenerator Composed of Electrospun
Nanofibrous Web**... 31
   S. Wazed Ali and Satyaranjan Bairagi

3. **Nanopretreatments for Textile: Nanoscouring, Nanobleaching,
Nanosoftening, and Nanosurface Activation** 51
   Subhankar Maity, Kunal Singha, and Pintu Pandit

**PART II: Nanoparticle Modification Techniques and Applications of
Nanotechnology in the Textile Industry** 71

4. **Advanced Nanotechnologies for Finishing of Cellulosic Textiles** 73
   Faten Hassan Hassan Abdellatif, Mohamed Mehawed Abdellatif, and
   Hend M. Ahmed

5. **Nano-Finishing on Woolens** 101
   Ajay Kumar, D. B. Shakyawar, Seiko Jose, Vinod Kadam, and N. Shanmugam

6. **Application of Nanotechnology-Based UV Finishes on
Textiles and Their Evaluation** 127
   M. S. Parmar

7. **Nanoparticles Modifications of Textiles Using Plasma Technology**....... 145
   Hend M. Ahmed, Mohamed Mehawed Abdellatif, and
   Faten Hassan Hassan Abdellatif

8. **Nanoparticle Application by Layer-by-Layer Deposition
Technique to Produce Polyelectrolyte Multilayers on Fabrics** 171
   Chet Ram Meena

9.   Nanotechnology in Sports Clothing.. 189
     M. Parthiban and K. Saravanan

10. **Sustainable Physiological Adaptation of Humans to Diverse
     Environment Conditions Using Smart NanoTextiles**............................... 207
     Renu Bala Yadav, Vinay Kumar Yadav, Dharam Pal Pathak, and Rajesh Arora

11. **A New Dimension on Novel Application of Nanotechnology for
     High-Performance Clothing**.. 237
     D. Gopalakrishnan, M. Parthiban, and P. Kandhavadivu

**PART III: Nanotechnology: An Innovative Way for Removal of Textile
     Waste and Nano-Textiles Used for Healthcare Perspective** 275

12. **Nanotechnology Applications for the Management of
     Textile Effluent** 277
     A. S. M. Raja, A. Arputharaj, G. Krishnaprasad, T. Senthil Kumar, and P. G. Patil

13. **Recent Advances of Nanoparticles in the Removal of Textile Dyes**....... 299
     Rekha Sharma, Ankita Dhillon, and Dinesh Kumar

14. **Health Safety and Environment Aspect of Nanotextiles**......................... 317
     Devsuni Singh, Prashansa Sharma, Suman Pant, and Vivek Dave

Index *343*

Contributors

Faten Hassan Hassan Abdellatif
Textile Research Division, Pre-treatment and Finishing of Cellulosic Fabric Department,
National Research Centre, Dokki, Giza, Egypt

Mohamed Mehawed Abdellatif
Chemical Industries Division, Chemistry of Tanning, Materials and Leather Technology,
National Research Centre, Dokki, Giza, Egypt

Hend M. Ahmed
Textile Research Division, Dyeing, Printing and intermediates, National Research Centre, Dokki,
Giza, Egypt

S. Wazed Ali
Department of Textile and Fibre Engineering, Indian Institute of Technology Delhi, Hauz Khas,
New Delhi, India

Rajesh Arora
Defence Institute of Physiology and Allied Sciences (DIPAS), DRDO, Timarpur, Lucknow Road,
Delhi, India

A. Arputharaj
ICAR- Central Institute for Research on Cotton Technology Mumbai-400019, India

Satyaranjan Bairagi
Department of Textile and Fibre Engineering, Indian Institute of Technology Delhi, Hauz Khas,
New Delhi, India

Vivek Dave
Department of Pharmacy, School of Health Sciences, Central University of South Bihar, India

Ankita Dhillon
Department of Chemistry, Banasthali Vidyapith, Rajasthan, India

D. Gopalakrishnan
Department of Fashion Technology, PSG College of Technology, Coimbatore 641004, Tamil Nadu, India

M. Gopalakrishnan
Department of Textile Technology, Bannari Amman Institute of Technology, Sathyamangalam, Erode,
Tamil Nadu, India

Seiko Jose
Textile Manufacturing and Textile Chemistry Division,
ICAR Central Sheep and Wool Research Institute, Avikanagar, India

Vinod Kadam
Textile Manufacturing and Textile Chemistry Division,
ICAR Central Sheep and Wool Research Institute, Avikanagar, India

P. Kandhavadivu
Department of Fashion Technology, PSG College of Technology, Coimbatore 641004, Tamil Nadu, India

G. Krishnaprasad
ICAR- Central Institute for Research on Cotton Technology Mumbai-400019, India

A. Ashok Kumar
Department of Textile Technology, Bannari Amman Institute of Technology, Sathyamangalam, Erode, Tamil Nadu, India

Ajay Kumar
Textile Manufacturing and Textile Chemistry Division,
ICAR Central Sheep and Wool Research Institute, Avikanagar, India

Dinesh Kumar
School of Chemical Sciences, Central University of Gujarat, Gandhinagar, India

S. Rajesh Kumar
Department of Textile Technology, Bannari Amman Institute of Technology, Sathyamangalam, Erode, Tamil Nadu, India

T. Senthil Kumar
ICAR- Central Institute for Research on Cotton Technology Mumbai-400019, India

Subhankar Maity
Department of Textile Technology, Uttar Pradesh Textile Technology Institute, Kanpur, India

Chet Ram Meena
National Institute of Fashion Technology, Jodhpur, Rajasthan, India

Pintu Pandit
National Institute of Fashion Technology, Department of Textile Design, Ministry of Textiles, Govt. of India, NIFT Campus, Mithapur Farms, Patna, India

Suman Pant
Department of Clothing & Textile, Banasthali Vidhyapith, Rajasthan, India

M. S. Parmar
Director (LABS), Northern India Textile Research Association (NITRA), Sector-23, Rajnagar, Ghaziabad 201002, U.P, India

M. Parthiban
Department of Fashion Technology, PSG College of Technology, Coimbatore, Tamil Nadu, India

P. G. Patil
ICAR- Central Institute for Research on Cotton Technology Mumbai-400019, India

Dharam Pal Pathak
Delhi Institute of Pharmaceutical Sciences and Research, M.B. Road, PushpVihar, New Delhi, India

A. S. M. Raja
ICAR- Central Institute for Research on Cotton Technology Mumbai-400019, India

K. Saravanan
Department of Fashion Technology, Bannari Amman Institute of Technology, Sathymangalam, Tamil Nadu, India

Kunal Singha
National Institute of Fashion Technology, Department of Textile Design, Ministry of Textiles, Govt. of India, NIFT Campus, Mithapur Farms, Patna, India

Devsuni Singh
Department of Clothing & Textile, Banasthali Vidhyapith, Rajasthan, India

D. B. Shakyawar
Textile Manufacturing and Textile Chemistry Division, ICAR Central Sheep and Wool Research Institute, Avikanagar, India

N. Shanmugam
Textile Manufacturing and Textile Chemistry Division, ICAR Central Sheep and Wool Research Institute, Avikanagar, India

Rekha Sharma
Department of Chemistry, Banasthali Vidyapith, Rajasthan, India

Prashansa Sharma
Department of Home Science, Mahila Mahavidyalaya, Banaras Hindu University, Varanasi, India

R. Vijayasekar
Department of Textile Technology, Bannari Amman Institute of Technology, Sathyamangalam, Erode, Tamil Nadu, India

Renu Bala Yadav
Defence Institute of Physiology and Allied Sciences (DIPAS), DRDO, Timarpur, Lucknow Road, Delhi, India

Vinay Kumar Yadav
Department of Computer Science and Engineering, Institute of Engineering and Technology (IET), Dr. Rammanohar Lohiya Avadh University, Ayodhya, Uttar Pradesh, India

Abbreviations

[Cnmim]Br	1-alkyl-3-methylimidazolium bromide
2D	two-dimensional
3D	three-dimensional
AATCC	American Association of textile Chemists and Colorists
AFM	atomic force microscopy
Ag–γ-Fe2O3@Cs	Ag coated chitosan capped γ-Fe2O3 binary nanohybrids
AMP clay	amino propyl functionalized magnesium phyllosilicate clay
BET	Brunauer–Emmett–Teller
BNCC	bifunctional nanocellulose
CIO	biocompatible nanohybrids
CMCT	carboxymethylated chitosan
CNns	graphitic carbon nitride (g-C_3N_4) nanosheets
Co_2O_3-NP-AC	cobalt (III) oxide (Co_2O_3) NPs loaded on activated carbon
COD	chemical oxygen demand
CR dye	Congo red dyes
CV	crystal violet
D–R	Dubinin–Radushkevich
E. coli	Escherichia coli
EY	Eosin Y
FTIR	Fourier transform infrared
GA	glutaraldehyde
HCl	hydrochloric acid
HEC	hydroxyethylcellulose
ILs	task-specific ionic liquids
IO	chitosan on γ-Fe_2O_3
LBL	layer-by-layer deposition technique
MB	methylene blue
MG	malachite green
MO dye	methyl orange dye
NPs	nanoparticles
PA	phytic acid
PAA	polyacrylic acid

PEI	poly (ethylene imine)
PAN	electrospun polyacrylonitrile
PANI	polyaniline
PEI	polyethyleneimine
PFO	pseudo-first-order
PS-b-PSAN-capped ZnO precursors	photocatalytic active mesoporous carbon/ZnO hybrid materials from block-copolymer tethered ZnO nanocrystals
PSO	pseudo-second-order
QA	quaternary ammonium
RhB	rhodamine B
SA	sodium alginate
S. Aureus	Staphylococcus Aureus
SEM	scanning electron microscopy
TiO_2	titanium dioxide
TIPT	tetra isopropyl orthotitanate
TEM	transmission electron microscopy
TMAB	tetramethyl ammonium bromide
UPF	ultraviolet protection factor
XRD	X-ray diffraction
ZnO	zinc oxide

Preface

This book, *Fundamentals of Nano-Textile Science,* provides a detailed description of the fundamental principles of nanotechnology in textiles in an easy and effective manner. "Nanotechnology" is becoming an emerging area, growing at an incredible rate in our daily lives and in other ways stimulating the demands of modern living. This book provides an overview of the rapidly growing and innovative developments of nanomaterials in textiles that benefit not only scientists and researchers but also serves as a crucial resource for textile industrialists in developing a new generation of textile fibers as well as products in wide range of new application techniques. The book heightenes the outlook of nanotechnology-based textile enhancement by exploring novel techniques and methods for improving the functionality of textiles.

This book explains the state-of-the-art innovations and developments that overcome the limitations of conventional processes or technology through imparting nanotechnology in the textile industry. The principle of this book is to provide comprehensive knowledge and description in a single volume.

The volume consists of 15 chapters and has been divided into three sections. Each of them covers the main areas of nanotechnology in textile science. The first section of the book introduces the new generation of nano-textile fibers, as it elaborates how the nanofiber production and techniques are used for producing different types of fiber by electrospinning techniques and pretreatments.

The second section deals with nanoparticle modification techniques on the textile for improving the fundamental property of textiles. The book also focuses on nanoparticle modification techniques using a plasma technique or a layer-by-layer deposition technique to develop polyelectrolyte multilayer fabric to enhance the functionality of textile materials.

Various applications of nanotechnology in textile include sports clothing, high-performance clothing or smart nanotextile in military, and smart wearable nanocomposites.

The third and final section of the book provides an informative description of an innovative way for textile wastewater treatment and a healthcare perspective of nanotextile. Our renowned and highly experienced authors provide the readers with the information on latest innovations and novel techniques.

The area of nanotechnology in textiles is wide as the world, which is hard to summarize on the recent innovation of nanotextiles. This book is the conclusion of worldwide research in the aspect of fundamental application of nanotechnology in textile science. It covers almost all the aspects of major researches that is valuable and informative for students and researchers and that should be used in universities and colleges across the globe due to the comprehensive information, updated research, numerous literature references, test methods, subject index, and the abbreviations it provides. It serves as a valuable reference for textile scientists, engineers, technologists, scientific researchers, professors, academics working on the nanotechnological development in textiles.

In the end, the editors would like to express their sincere thanks to all the authors for giving their precious time, effort, and plentiful input to prepare the chapters. We deeply appreciate their contribution, cooperation, and determination to bring out this book. Without their inspiration and the support, the compilation of this book could not have been possible. We would also like to thank the publisher, Apple Academic Press, for recognizing the demand for such a book and for comprehending the increasing importance of the field of nanotechnology in the textile science.

—**Prashansa Sharma**
Devsuni Singh
Vivek Dave

PART I

Introduction to Nanofiber Production and Its Pretreatment Techniques

Nanofiber and Nanoparticle Production Techniques for Various Functional Product Developments

S. RAJESH KUMAR*, M. GOPALAKRISHNAN, R. VIJAYASEKAR, and A. ASHOK KUMAR

Department of Textile Technology, Bannari Amman Institute of Technology, Sathyamangalam, Erode, Tamil Nadu, India

*Corresponding author. E-mail: rajeshkumars@bitsathy.ac.in

ABSTRACT

Nanofibers can be produced from different type of polymers depending on end-use applications like tissue engineering, drug delivery, filtration, etc. The product output characteristics are influenced by so many factors such as from raw material selection to various machinery process conditions. Even though several types of nanofiber production techniques are available in the market, electrospinning technique is quite famous for the production method for its versatility of product developments since 1930s. In the last two decades, it got famous for its effective product outputs, and also various research activities witnessed for its developments like multijet electrospinning, needle-less electrospinning, bubble electrospinning, electroblowing, melt electrospinning, coaxial electrospinning, self-bundling electrospinning, nanospider electrospinning, and so on. The nanoparticle plays a vital role in engineering field for reducing the consumption of the active agent by increasing the surface area. Nano-encapsulations are used in textile and other fields to increase the durability of the finishing. This chapter will highlight some of the important machinery parameters and production-enhancement techniques in various aspects on the fabrication of nanofiber

Fundamentals of Nano-Textile Science. Prashansa Sharma, Devsuni Singh & Vivek Dave (Eds.)
© 2022 Apple Academic Press, Inc. Co-published with CRC Press (Taylor & Francis)

or nanocomposite materials output to meet the end-use applications. This chapter also describes the different methods of nanoparticle development and nano-encapsulation techniques.

1.1   WHY NANOFIBER?

Nanofibers are fibers with diameter in the scale value of nanometer (10^{-9}) in terms of 100 nm or less. Larger surface area due to the smaller fiber size makes the fiber useful in variety of applications like high-performance filtration, battery separation, wound dressing, vascular graft, composite applications, and tissue engineering, and so on. Every nanofibers diameter is different from one to another; it depends on its past history, characteristics of material, and fiber-production techniques. These fibers can be generated from different polymer sources; generally, it can be classified into two major categories as natural polymers (cellulose, collagen, keratin, gelatin, etc.) and synthetic polymers (polyurethanes, polylactic acid [PLA], polycaprolactone, etc.). Development of various polymer materials and its biological activity for the use in medical application field is an important research field for the last two decades. As there is an increase in the knowledge of biological system and materials handling nature, new areas, such as interaction between materials and cells, relationship between molecular structure and macroscopic properties, interaction between the functional groups, signaling between the cells at micro and nanolevels, are also able to control the properties of the biological and artificially intelligent materials very well (Mozafari, 2007). In general, nanoscience can fit in all kind of fields and it brings the hybridization in science to get the benefit out of it by reduction of particle size.

Four major fields of nanotechnology:

1.   **Nanoelectronics** products like semiconductor laser, giant magneto-resistance-based computer hard disks, and so on.
2.   **Nanomaterial** products like nanostructure ceramics, nanocoatings spectacles for scratch resistance, nano UV absorbers for finding the unburnt location.
3.   **Nanotextile** products like nanofibers, nanotubes, nanoclays for textile products, intelligent clothing, and smart textile sensors.
4.   **Nanobiology** products like drugs for cancer fighting, diagnosis, and treatment of aging.

The extent of nanotechnology application is in many areas like potential applications in computers as nanochips' developments, aerospace product development areas like launch vehicles, nanotube manufacturing, various biotechnology applications based on their end-use requirement, and various other fields. Molecular manufacturing is basically related to the manufacturing of small to extremely small circuit boards for mechanical devices. In the engineering field, molecular engineering is also known as nanotechnology. In the atomic and molecular level of changes makes huge impact on products, this understanding makes many scientists and researchers to focus on these nanofiber-manufacturing techniques. Nanofibers are produced by electrical charge to draw very fine fibers from a liquid process is called electrospinning, with smart technology innovations and creative thinking on product end use makes revolution in nanofiber manufacturing in electrospinning technology. Since the focus toward the end product, it makes satisfaction to the industrial needs in filtration areas, medical field, and various other application areas. The nanofibers are strong, tough and also reduce the crack initiation level and propagation rate by means of crack deflection and energy absorption characteristics. Higher value improvement in fracture toughness makes materials as most suitable for filtration applications. This process can further improve with multiple combination of different materials to make nanocomposite materials. In this advanced case-specific nanofillers, such as carbon nanotubes, metal oxides, precious metals, and smart biological agents are incorporated either during electrospinning or in the postprocessing sequences, moreover, to meet the operational requirements, for example, to improve the mechanical stability and decrease of pressure drop, and so on.

1.2 HISTORY OF NANOFIBER DEVELOPMENTS

The term "electrospinning" is derived from "electrostatic spinning." This term got familiarized from the year 1990s onward, but the base idea originated in the 16th century itself, the English scientist Gilbert recorded the magnetic and electrostatic phenomenon on water droplet at the beginning and later stages on change of shape from hemisphere to cone shape. This was the first observation done on liquid body shape changes when the electrostatic force was applied to it (Tucker et al., 2012). Based on this basic phenomenon of liquid body shape changes, various attempts have been made by several researchers and in the 17th century, French scientist Nollet highlighted the fracture caused in water stream due to electric field experimental work (Kleivaitė and Milašius, 2018).

Significant experimental works were carried out by scientist Strutt and Rayleigh in the 18th century; level of critical load was applied on the liquid and the instability level of that concerned liquid medium was calculated and also he commended as electric field cause an increase of liquid medium stability because of charge level on medium at critical level is less and also they commended an amount of charge required for liquid medium droplet formation (Strutt and Lord Rayleigh, 1878). After some years, Charles Vernon and his colleagues described the nanofibers manufacturing process and at the beginning of the 19th century, John Francis patented the process of electrospinning sequence; at the same time, Cooley patented the three-head electrospinning device and also he had done patent on auxiliary electrodes in electrospinning process for the electric spin. Later, he had done research on thin filament formation techniques and done a patent on needle rotation through electrical support, which developed thin strands of filaments (Cooley, 1902).

In the year 1914, Zeleny studied the behavior of fluid droplet in the metal capillaries, this work initiated the mathematical modeling for liquid behavior under electrostatic conditions. Much deeper research works done in this field by Anton Formhalson process conditions and machinery modifications, he had filled nearly 22 patents on electrospinning concepts. In the year 1938, Rozenblum et al., developed elctrospun fiber for filter media application (Tucker et al., 2012). During this time period, several researchers have started investigations on the process possibilities in nanofiber-production technique and patented the experimental setup for the polymer fiber production using electrostatic force.

Nanofiber manufacturing techniques in the electrospinning process is the simplest technique compared to other fiber manufacturing dry spinning, wet spinning, and melt blown processes, but it is a little complex process. After the Second World War, electrospinning was developed in various laboratories due to research interest of scientists but many research findings were not disclosed to the outside world. In 1990s, Geoffrey Ingram Taylor done the theoretical modeling on characteristics of electric field charged fluid droplet shape. Under electric field circumstances, the changed shape of charged fluid called as "Taylor cone," the tip of the capillary tube hemispherical surface of the solution gets elongated to form conical shape due to the increased electric field intensity level in the liquid medium. In the electrospinning fiber manufacturing, when the electric field crosses the critical value, the electric force overcomes the surface tension force and so the solution gets eject from the Taylor cone (Doshi and Reneker, 1995).

In the 20th century, research developments got increased due to the nano-materials application versatility. In 2001, Wang et al. have done a research on fabrication of inorganic materials in electrospinning; in the same year, Bognitzki et al. done a research work on fabrication of porous-natured fabric surface. In the year 2003, multinozzle electrospinning research initiation was done by Gupta et al. and core–shell electrospinning process was done by Sun et al. Due to research interests, various research works started during this time period and so in the year 2005, needle-less electrospinning experiment was done by Jirsak et al. and he also patented his work on the same and Smit et al. reported the first scientific research work on continuous electrospun yarn production concept. In later days, various researchers have reached a remarkable milestone in nanofiber production techniques advancements like 3D-block electrospun fiber work (Teo and Ramakrishna, 2006), electrospun implantable graft work (Nicast, 2008), and nanofiber-based composite works (Kilwell et al., 2012).

1.3 NANOTECHNOLOGY IN FIBER AND TEXTILE MANUFACTURING

Nanotechnology in textile manufacturing field increasing everyday due to versatile product developments in various fields; it makes the research interesting and so many people have focused on new end-use applications and product development. In earlier days, many functional products were developed in specific aspects on various applications, but the product sustainability was the questionable fact; so, the nanofibers and nanocoated products giving high durability and permanent effects led to customer and industry satisfaction. The application of nanotechnology in textile field may increase the durability of the finishing. The increase in durability may be the decrease in particle size; thus, the increase in surface area and increase of the affinity of the chemicals were applied. Fabric coating is one of the famous methods used in textile industry for nanotextile product developments. Coating technique was used to apply nanoparticles, ingredients, or carrier medium on the specific textile fabrics like woven fabrics (or) nonwoven fabrics surfaces to enhance their functionality; at the same time, important characteristic of fabrics like breathability, flexibility, and so on did not get affected due to the coating; so, the interest on new product developments got increased. Out of several application methods, padding technique is most widely used due to their better penetration ability. In padding technique, the nanomaterials are applied in high pressure to get the nanoparticles to fix in the fiber surface permanently. The properties like water repellency, wrinkle resistance, flame retardant, antimicrobial, and soil resistant are the important finishing methods needed for the industry.

1.4 NANOFIBER PRODUCTION TECHNIQUES

Polymeric nanofibers can be obtained from various processing techniques like drawing, template synthesis, self-assembly, phase separation, and electrospinning process. In these all the techniques, electrospinning process is the cost-effective and gives continues fiber production, even though repeatability is possible in all the above cases, there are certain limitations in all the type of production techniques like fiber dimension control is the tedious process in drawing, phase separation and self-assembly processes and the same manner phase separations having limitations on specific polymers (PLLA poly(L-lactic acid) and its blends), self-assembly process having advantages of producing smaller nanofibers (up to 7 nm) as per the need but the problems associated with it as complex process and scaling cannot be done properly as like template synthesis process, in the template synthesis desired shape of material can be obtained from the molten template dimension only. In modern electrospinning processes, all these problems are taken care of and solved with available feasibilities.

1.5 ELECTROSPINNING PROCESS

Development of polymer filaments using electrostatic force is called electrospinning process. In this process, a polymer solution held by its surface tension at the end of the capillary tube is subjected to an electric field. Mechanical properties of the nanofiber depend on the type of manufacturing, material constituent and processing conditions, and so on. Ramakrishna (2005) studied the influence of these parameters on mechanical properties of the nanomaterials. Mutual charge repulsion causes a force directly opposite to the surface tension, when the electric charge gets increased the hemispherical shape of the solution surface gets extent to form a conical shape which is called Taylor cone, as shown in Figure 1.1. At the critical point of charge, the solution gets ejected from the tip of the needle in a continuous form, when the jet travels in the air, the solvents get evaporated and the polymer materials remain leaving the charged solution itself. The instability nature of polymer in air makes waviness structure in fiber formation. Jet ejected speed from needle, distance between the needle tip to collector point, and applied electric charge on fiber also play a major role on jet traveling time in air, depends on these parameters the solvent evaporation rate will get change. Electrically charged polymer solution can be received on the collector end in a web form, this type of fiber production technique, recognized as efficient fabrication technique for nanofibers' production (Taylor, 1969).

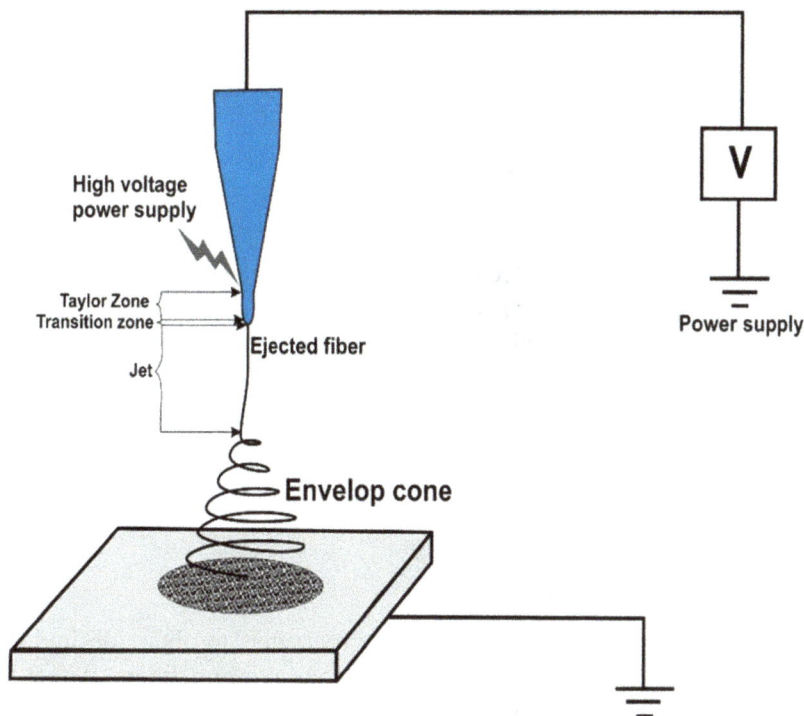

FIGURE 1.1 Electrospinning setup.

Nanofibers can be successfully electrospun by varying one or two of the following parameters, solution properties such as polymer nature, solution viscosity level, and the controlled variables like applied voltage on tip of the needle, hydrostatic pressure on the capillary, distance between the tip to collector point, and the other hand temperature inside the spinning chamber, air pressure, and air velocity. Generally, electrospun nanofiber provides a large ratio of surface area to volume in the small fiber diameters value of 5–0.05 µm. This nature of produced web gives wider opportunities in filtration area as well as wound dressing applications (Doshi and Reneker, 1993).

1.6 ELECTROSPINNING PARAMETERS AND FIBER CHARACTERISTICS

The characteristics of electrospun nanofibers are based on electrospinning parameters like nature of polymer, type of needle, environmental conditions, and so on. In electrospinning nanofiber production stage, several parameters affect the fiber dimension as well as the continuity of the effective fiber length. The parameters which affect the fiber developments are generally categories as the follows:

- Processing parameters
  - Type of needle
  - Needle length
  - Needle tip to collector distance
  - Type of collector and its angle
- Raw material properties
  - Type of polymer
  - Solvent material
  - Solution concentration
- Ambient parameters
  - Temperature
  - Relative humidity
  - Atmospheric pressure

It is very important to consider these parameters to obtain desired fiber length and fiber morphologies like continuous uniform fibers, defect-free fiber surface (free from beads and pores on the structure), and controlled single fiber diameter collected in the collector zone.

1.6.1   PROCESS PARAMETER

The first study on electrospinning process variables was done by Deitzel et al. (2001); they identified that the increasing voltage in electrospinning process results the changes in surface of the polymer solution droplet from which the electrospinning jet originated. Megelski et al. (2002) found in their study that fiber diameter decreases with increasing applied voltage.

1.6.2   TIP TO COLLECTOR DISTANCE

The effect of needle diameter on thermal properties of fluid nanofiber diameter was done by Macossay et al. (2007). The SEM image evident that the diameter of needle and the average diameter of the nanofiber have no influence. Solution fluid coming out from the syringe directly gets charged from high-voltage electric field; at this time, the important point need to be considered is the distance between the tip of the needle to collector point; when the distance changes, the morphological structure of nanofiber also gets changed by elongating. The distance between these two points need to adjust according to the type of solution and type of solvent evaporation nature for to get the solid nanofiber from the ejected jet. Solvent material should get evaporated from the solution before it gets deposited into the collector point. The material coming out from syringe already got charged from the high-voltage electric field and so if the solvent does not get evaporated properly due to short distance, it results beads formation (Ramakrishnan et al., 2005).

In some cases, it is based on the type of solution properties, and distance has no influence on the produced fiber. The first study on electrospinning process variables was done by Deitzel et al. (2001), they identified that the increase in voltage of electrospinning process will change the droplets of polymer solution surface; Megelski et al. (2002) also found that the fiber diameter decreases with increasing applied voltage.

The effect of needle diameter on thermal properties of fluid nanofiber diameter was done by Macossay et al. (2007). He had done the analysis in scanning electron microscope (SEM); it showed there was no correlation between needle diameter and the average nanofiber diameter. Solution fluid coming out from the syringe directly gets charged from high-voltage electric field. At this time, the important point that needs to be considered is the distance between the tip of the needle to the collector point; the morpho-logical structure of the nanofiber also got changed when the distance changes because of elongating nature of the material. The distance between these two points need to adjust according to the type of solution and type of solvent evaporation nature to get the solid nanofiber from the ejected jet. Solvent material should get evaporated from the solution before it gets deposited into the collector point. The material coming out from syringe already got charged from the high-voltage electric field and so if the solvent does not get evaporated properly due to short distance, it results beads formation (Ramakrishnan et al., 2005).

In some cases, it is based on the type of solution properties; the effect of varying distance may or may not have a significant effect on the produced fiber diameter. However, Megelski et al. (2002) observed beads formation during their research work, when the distance was too low. In another case, increasing the distance results in decreasing of fiber diameter was observed (Ayutsede et al., 2005). The results of beads may be due to the instable nature of jet and its use when the highest field strength between the two points happens (Deitzel et al., 2001). However, in another case, longer distance between the two points increased the fiber diameter due to less electrostatic field strength (Lee et al., 2004). The excess amount of solvents will merge the fibers when the distance is low, and it will form junctions and bonding in inter- and intra-layer segments. The additional strength of the resultant scaffold can be achieved by this interconnected fiber mesh (Buchko et al., 1999). These experiments clearly show that optimum electrostatic field strength is required to control the stretching property of the jet and also it support for staple fiber formation; these can be controlled by adjusting the tip to collector distance and the type of solvent nature.

1.6.3 SOLUTION PROPERTIES

Complexity level in nanofiber production through electrospinning process is mainly based on the type of polymer characteristics, such as the crystalline polymers, polylactic acid, polyacrylic nitrile, and the polyethylene oxide can easily produce in fiber form. However, it is not easy to make nanofibers from elastomers whose polymer chain is flexible in the amorphous state (Yamashita et al., 2007). Nanofiber-producing polymers are self-induced by electrical connection, when the solid polymer gets dissolved in a solvent, the solution viscosity gets modified proportionally with respect to polymer concentration. Higher polymer concentration and higher fluid viscosity level leads to larger nanofiber diameter. At low concentration of fluid leads to low viscosity in the solution and so electrospray will happen instead of electrospinning (Deitzel et al., 2001).

At the low solution concentrations, a mixture of fibers/bead formation and branching of fibers takes place, shown in Figure 1.2. When the solution concentration increases, the shape of the beads gets changed from spherical form to spindle-like fiber form, when the level of molecular weight increases the fibers get changed into the ribbon form (Bhardwaj and Kundu, 2010). Change of polymer concentration, feed rate, and addition of salt on polycaprolactone nanofiber diameter and length were experimented by Vaseashta (2007) and Beachley (2009). Megelski et al. (2002) done research on feed rate changes and showed results in change of diameter; they have concluded that reducing polymer solutions cause reduction in fiber length and diameter values and so increasing the values in a proper way is the only solution to increase the fiber dimensions.

Fiber-forming solution temperature stability is one of the important parameters that needs to focus on staple fiber formation. Low heat stability materials, such as PLA, showed instable nature when the temperature exceeds beyond the range (180°C–230°C) (Ya et al., 2010). In this research work, they have observed partial blocks in spinnerate holes at the temperature of 200°C; when the temperature rises to 210°C, many small crystal particles at the surface of web was observed. In the same condition of raising temperature, at 230°C showed very brittle fiber formations; at this stage, even the gentle touch also makes the great impact on web breakage. Beyond the 230°C, decomposition of PLA was observed in their research work and also below the temperature of 210°C also not supports for fiber formation. So, the value of finding spinning temperature finding also taking important role on fiber formation, in this case for PLA spinning temperature falling at 220°C.

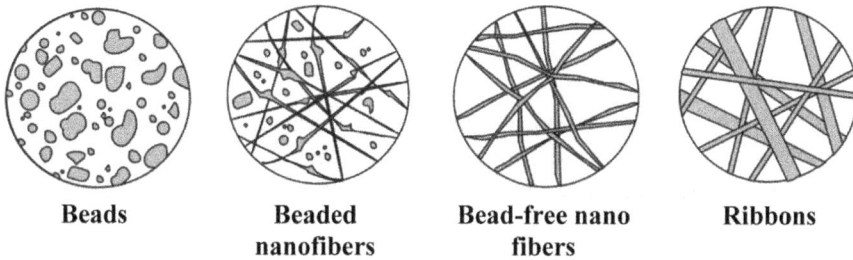

| Beads | Beaded nanofibers | Bead-free nano fibers | Ribbons |

Molecular weight increases

FIGURE 1.2 Nanofiber formation.

1.6.4 ELECTRO HYDRODYNAMICS

Electrostatic spinning is a spontaneous process; it involves the ejection of a viscous solution from an electrically charged drop. Subsequent drawing and solidification of the jet lead to the formation of a uniform, thin fiber. Unlike the conventional fiber manufacturing processes, the driving forces for the electrospinning are based on electrostatic interactions. Electrostatic field generated in the electrically driven polymer fluid stream called as jet, it is a self-induced process. Each jet ejected from the source solidifies and produces infinite solid filament with less than 1 µm diameter (nanofiber). Dynamics of electrically charged liquid called as electrohydrodynamics (EHD). Electrospinning physical phenomenon can be included within the EHD; this process is based on the electrostatic repulsion of charges ruled by the Coulomb law shown in Figure 1.3. It states that like-for-like charges repel; it shows cylindrical electrode charges will be more at the sharpest edge, so it is important to consider basic laws of physics during designing of cylindrically shaped electrodes.

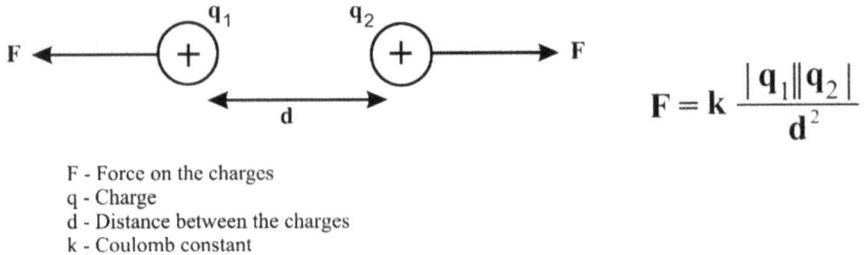

$$F = k \frac{|q_1||q_2|}{d^2}$$

F - Force on the charges
q - Charge
d - Distance between the charges
k - Coulomb constant

FIGURE 1.3   Coulombic repulsive forces.

One point of researchers describing the electrospinning is the process of dynamic balance between the surface tension force and Coulombic repulsive forces. As the result of electric field increased, the repulsive force overcomes the liquid surface tension and the droplets splits into several smaller drops, called electrospraying. When the polymer viscoelastic behavior and sufficient entanglements between molecules in the liquid happen, fluid drops get transformed into charged jets and solidifies in polymer filaments under a dynamic balance of air drag, electric field, and the gravitational field forces (Vaseashta, 2007). The electrospinning is mainly dependent on the level of electrostatic field applied. It is significant to eliminate these hotspots which are not necessary on spinning electrode; without considering the electrostatic field, for designing electrodes results unbalanced conditions like more production on the ends of the electrodes instead of center of electrode (Niu et al., 2009).

1.6.5   AMBIENT CONDITIONS

Environmental conditions like humidity, atmospheric pressure, and temperatures show greater impact on fiber formation in electrospinning process. Any interaction between the surrounding and the polymer solution may have an effect on the electrospun fiber morphology. Mixed gases in the air also show a greater impact on fiber morphology as electrostatic interactions with gas make changes in fiber morphology. For example, when the helium gas interacts with electrostatic field, fiber gets affect by splitting due to high energy level in the fluid jet. In another case, Baumgarten observed in his research twice the diameter of electrospun fiber when the fiber formation took place in Freon-12 environment (Baumgarten et al., 1971).

The high level of electrostatic charge present in the jet always tries to interact with atmospheric conditions. When the humidity level increases, surface nature of the nanofiber gets changed to porous form as water get condensed on the surface on the fiber. Since electrospinning is influenced by external high voltage electric supply, any changes happening in the electrospinning surroundings will also affect the process sequence of nanofiber development (Megelski et al., 2002). Another research work was carried out by Casper in the field of porous structure changes and the depth level of porous nature. During his experiment, polysulfone dissolved in tetrahydrofuran showed less than 50% humidity level and so the results on smooth fiber surface was recorded and also the increase of humidity level showed the porous nature changes in the terms of shape as well as in the depth level in the fiber physical structure (Casper et al., 2004). The pores present in the surface may influence the breath figures of the thin film (Srnivasarao et al., 2001). When the humidity level increases, the water deposition on the jet causes the volatile solvents to evaporate quickly. The pores of the breath figure are even and uniform in size by the changing from dynamic condition to static condition of nanofiber (Megelski et al., 2002).

1.7 MULTIJET ELECTROSPINNING TECHNIQUES

Various multijet electrospinning techniques were reported in the past. Cui et al. (2007) reported quantitative study on the orthogonal design using electrospun has been reported. The researchers focus on response surface methodology for the easiness. Sukigara analyzed the fiber diameter by changing the electric field, concentration, and spinning distances of the *Bombyx mori* in the electrospinning. To establish a relationship between the variables, they have used factorial experimental design, in that they assigned three parameters with nine factors as a quadratic polynomial method. As per their understanding from the result of experiments carried out, fiber diameter increases with increasing of concentration and also they have observed various impacts for the electric field due to solution concentration. Based on these experimental results, they suggested and concluded that the interactions and coupling effects are always present between the different parameters (Sukigara et al., 2004).

In general, the production of nanofibers in electrospinning process is very low, due to single delivery output. To overcome the production issues, research works on multiple-output delivery systems were focused on many researchers in R&D Laboratories. These modernized techniques can be categorized in three major groups: (a) single-needle electrospinning (SNE) for multijets; (b) multijets from multiple needle electrospinning (MNE); (c) needle-less electrospinning for multijets (Nayak et al., 2012).

1.7.1 SINGLE NEEDLE FOR MULTIJETS

Single-needle electrospinning setup, generally, ejects only one jet from the tip of needle. Yamashita et al. (2007) first time found the multijets formation in SS needle electrospinning setup with grooved tip. The glassy syringe was mounted on the tip with polybutadiene (PB) material without beads formation. This setup has significant effect in electric field distribution and some degree of solution blockage at the needle tip is shown in Figure 1.4. In the other hand, multiple Taylor cone formation experiment done by Vaseashta for the multijet formation from the single needle was collected in curved plate (Vaseashta, 2007). Although the mechanism of controlled high production technique from SNE jet splitting is yet to be analyzed.

1.7.2 MULTIPLE NEEDLES FOR MULTIJETS

In the research interest of people in electrospinning production enhancement attempted in various ways like increasing the number of needles and change of angles, and so on. In the MNE system, the needle configuration and the gage between the needles place an important role of fiber formation, due to the repulsive forces acting between the needle makes fiber morphological changes. Needles can be arranged in linear configuration, such as triangular, circular, hexagonal, and elliptical forms. For example, a linear configuration of four needles can be arranged for nanoweb fabrication technique. Several investigations have been reported in the 2D configurations. For instance, the elliptical arrangements and the concentric arrangements of the multijet spinning heads provide higher productivity and stability of jet development process (Tomaszewski and Szadkowski, 2005). Yamashita et al. investigated and designed various types of multi-nozzle heads for the issues associated with production aspects and nozzle clogging of electrospinning. These modifications increased production of nanofibers (Yamashita et al., 2007).

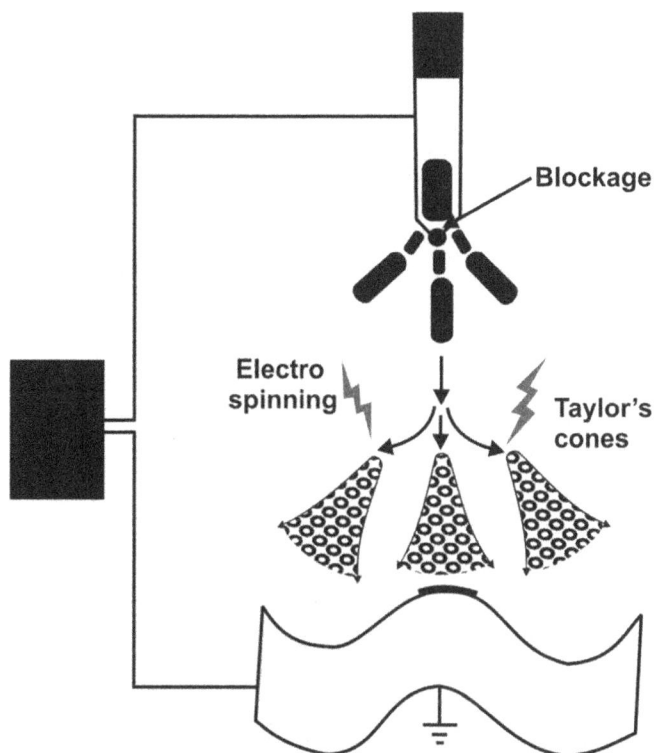

FIGURE 1.4 Single needle for multijet formation.

1.7.3 NEEDLE-LESS ELECTROSPINNING FOR MULTIJETS

The application suitability of any nanofiber material is based on the characteristics of fiber-forming polymer and morphological structure of polymer, such as crystallinity nature of the material. Scale-up solution is needed for commercial application of nanofibers. Normally, the needle electrospinning has power supply unit with high voltage, spinnerate, and a collector unit. It helps to control the process variables, but the problem is the production rate which is very slow (<300 mg/h/spl) and in some cases the added risk that blockages in needle may also happen based on polymer characteristics (Teo and Ramakrishna, 2006). The advancements on multineedle spinnerate setup need a large working space and careful designing between the needles required so that the strong repulsion and fiber-structure damages can be avoided. Invention on new methods of needle-less electrospinning setup, the needle blocks, throughput and higher density of jets, need to be considered (Niu et al., 2009). In the modern technique, multiple jets can be produced from open liquid surface found by Yarin and Zussman, they reported the new method of nanofiber production with the combination of two-layer system as such as normal magnetic field in the bottom and the electric field on the top to achieve steady vertical spikes as like shown in Figure 1.5. Nanofibers were electrospun at 10^5 V/m from numerous cones located in the free surface, and the lower layer magnetic field supported to punch the polymer solution for fiber formation. They have also highlighted that jetting from the free surface of a polymer melt could also be possible at higher electric field of the order of 10^8 V/m (Yarin and Zussman, 2004).

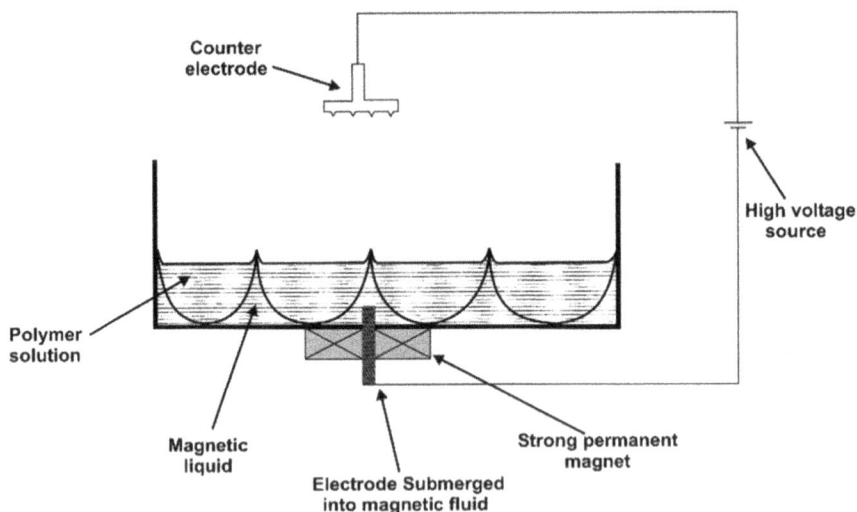

FIGURE 1.5   Needle-less electrospinning.

1.8   NANO-ENCAPSULATION

Nanoparticles are solid in nature, diameter ranging from 1 to 100 nm in size. The nanoparticle is the generic term for nanospheres and nanocapsules. In nanosphere, the drugs are adsorbed at the surface of the sphere or it may be encapsulated within the sphere. In nanocapsules, the drug is limited in vesicular systems covered by polymeric membrane (Couvreur et al., 1995).

The method of production of nanoparticles is categorized into two main classes:

1. Polymerization method
2. Direct from macromolecules

1.8.1 POLYMERIZATION METHOD

Emulsion polymerization and interfacial polymerization methods used to produce nanoparticles. In emulsion polymerization, the organic and aqueous medium is used as a solvent.

1.8.1.1 EMULSION POLYMERIZATION

Emulsion is one of the methods to produce the nanoparticles in a faster manner (Kreuter, 1990). There are two methodologies: continuous organic phase and an aqueous phase. In the former method, the monomers are dispersed into an emulsion or into the nonsoluble material for monomer or inverse microemulsion. Nanospheres of polyacrylamide were first developed by this emulsion method (Ekman and Sjöholm, 1978; Lowe and Temple, 1994). The aggregation of nanoparticles is protected by using surfactants or soluble protective polymer (Kreuter, 1991). This procedure requires toxic organic solvents, monomers, initiators, and surfactants, and moreover, the process is difficult to produce nanoparticles. The nanoparticles, polyethyl cyanoacrylate, polymethyl methacrylate, and polybutyl cyanoacrylate were developed by dispersion of monomers into solvents such as *n*-pentane, toluene, and cyclohexane. Fluorescein, triamcinolone, timolol, and pilocarpine are encapsulated by this method (El-Samaligy et al., 1986). In the aqueous phase method, the monomers are dissolved in an aqueous continuous phase solution with an initiator may be ionic or a free radical. Alternatively, using the high energy radiation, ultraviolet or strong visible light, or *g* radiation, the monomer molecules were transformed into an initiating free radical (Vauthier et al., 2003).

1.8.1.2 INTERFACIAL POLYMERIZATION

The nanoencapsule of poly(alkyl cyanoacrylate) is prepared with this method. The method produces the nanoencapsules at a faster rate due to the rapid polymerization within seconds (Couvreur et al., 2002). The ion initiates the polymerization medium (Al Khouri Fallouh et al., 1986). Cyanoacrylate and drug dispersed in a mixture of ethanol and oil. The mixture is then transferred into the aqueous solution by slowly with or without surfactants. Nanocapsules are developed spontaneously by polymerization of cyanoacrylates. Polymerization initiates once the cyanoacrylates contact with ions present in water. Then, the nanocapsules are purified by evaporating the solution under vacuum. The nanoencapsules, poly(isohexyl cyanoacrylate) and poly(isobutyl cyanoacrylate), are prepared by this method (Watnasirichaikul et al., 2000).

1.8.1.3   INTERFACIAL POLYCONDENSATION

Interfacial polycondensation is another method to prepare polymeric nanoparticles by interfacial polycondensation using of lipophilic monomer and hydrophilic monomer in the presence of or without surfactants. The nanocapsules produced by this method are smaller than 500 nm. The polyurethane and poly(ether urethane) copolymers are synthesized by a modified interfacial polycondensation method (Bouchemal et al., 2004; Montasser et al., 2002).

1.8.1.4   NANOPARTICLES FROM PREFORMED POLYMERS

The non-biodegradation or slow biodegradation of monomers used in micellar polymerization and toxic nature of the residual molecules in the polymerization medium, the preformed polymers used instead of monomers.

1.8.1.5   EMULSIFICATION/SOLVENT EVAPORATION

Two steps are involved in emulsification–solvent evaporation method: first, emulsification in an aqueous phase, and in the second step, evaporation of polymer solvent and inducing polymer precipitation as nanospheres. The dissolved drug-containing polymer solution is dispersed into nanodroplets with dispersing agents and chloroform, a nonsolvent. The drug of nanosphere precipitated out in the form of a fine dispersion. The residual solvents are removed subsequently by raising the temperature. The size of the nanosphere is dependent on stir rate, viscosity of the polymer solution, type and quantity of dispersing agent, and temperature (Tice and Gilley, 1985).

1.8.1.6 SOLVENT DISPLACEMENT AND INTERFACIAL DEPOSITION

It is a similar method based on spontaneous emulsification of the organic internal phase containing the dissolved polymer. This method produces only nanocapsules, whereas the former method produces both nanosphere and nanocapsules (Fessi et al., 1989; Quintanar-Guerrero et al., 1998).

1.8.1.7 EMULSIFICATION/SOLVENT DIFFUSION

The polymer which is encapsulating the core is dissolved in a partially water-soluble solvent. For the production of nanoparticles, solvents which are used to produce nanoparticles are diffused by feeding the excess amount of water when the organic solvent is partially miscible with water. Then, emulsify the polymer, water, and saturated solvent in the presence of a stabilizer in an aqueous solution. The resultant nanospheres or nanocapsules are obtained after removing the organic solvent (Quintanar-Guerrero et al., 1998).

1.8.1.8 SALTING OUT WITH SYNTHETIC POLYMERS

In this technique, the solvent which is miscible in the aqueous medium is separated out by the salting-out technique. This technique is a modified emulsification/solvent diffusion method. In this method, the solvent, acetone is used to dissolve the polymer and drug, and then it is subsequently emulsified in the aqueous solution containing an electrolyte such as calcium chloride, magnesium chloride, and magnesium acetate. As the diffusion of the acetone is enhanced, the oil in water emulsion is diluted out with an aqueous solution. This diffusion of acetone induces the formation of nanospheres (Lambert et al., 2001; Quintanar-Guerrero et al., 1998).

1.8.1.8.1 *Albumin Nanoparticles Produced in an External-Oily Emulsion*

Preparation of albumin nanospheres involves two methods: the first one is thermal treatment at elevated temperatures normally at 95°C–170°C and another one is chemical treatment with isooctane, vegetable oils, or aqueous medium. The other methods are either of these two with slight modifications (Patil, 2003). In the thermal method, the albumin droplets are stabilized at

175°C–180°C for 10 min. To reduce the viscosity of the oil phase, the mixture is diluted with ethyl ether after cooling. This method is suitable only for the material that is not heat sensitive. The heat-sensitive materials, the serum albumin nanoparticles produced by emulsifying in cottonseed oil at 25°C. Then the particles are suspended in cross-linking agents like 2,3-butadiene or formaldehyde. The resultant nanoparticles are centrifuged and dried by lyophilization. The elimination of cottonseed oil is still problematic in purification process (Widder et al., 1979).

1.8.2 METHODS FOR PRODUCTION OF NANOPARTICLES

Several methods are available for production of nanoparticles. Here, some of those are listed:

1. Electrochemical deposition
2. Sol–gel process
3. Gas-phase condensation
4. Plasma-enhanced vapor decomposition
5. Sputtering technique

1.8.2.1 ELECTROCHEMICAL DEPOSITION

Electrochemical deposition is a metallic coating on a base material by electro-chemical reduction of metal ions. It is mainly to reduce the friction and wear, desired electrical properties, reduce corrosion, and to increase heat resistance. It is sometimes called electroplating and sometimes electro-crystallization, due to the crystalline nature of the coating. In electrodeposition, the cathode and an anode are dipped in the electrolyte. Mostly, the electrolytes are metal-containing liquid. When electric current passes through the electrodes, the metal ions are deposited on the metal (Shinomiya et al., 2006).

1.8.2.2 SOL–GEL METHOD

The wet-chemical technique of production of nanomaterial is also called chemical solution deposition. The precursors, carbonates or acetates or nitrates, are dissolved in deionized water. The preparation of colloidal substances using the starting materials is called a gel. Polyvinyl alcohol, the gelling agent, is added to the starting material to produce a gel. A coating is made on a substrate with controlled temperature, pH, and viscosity. Finally, the coated product is annealed with suitable temperature (Callister and Rethwisch, 2004; Lu and Jagannathan, 2002).

1.8.2.3 GAS-PHASE CONDENSATION METHODS

In this method, the particle size of the nanoparticle is smaller than 10 nm. In inert gas environmental conditions, the metals are introduced at lower pressure (0.5–4 Torr) in a temperature-controlled alumina crucible. Due to the metal collision, the metal vapors are cooled rapidly in the presence of inert gas and the supersaturated vapor undergoes homogeneous nucleation. The size of the particle is influenced by the pressure of inert gas and higher atomic weight (Eilers and Tissue, 1995).

1.8.2.4 PLASMA-ENHANCED CHEMICAL VAPOR DEPOSITION

This method is widely adopted for making the thin-film production industry. In conventional chemical vapor deposition (CVD), thermally decompose the inert gas at 500°C–1000°C, whereas the PECVD method, even at room temperature, the inert gases are dissociated in the presence of plasma. For sensitive materials, damaged by temperature can use this method. Plasma can modify the properties of the film due to the concurrent "ion bombardment" of plasma (Malesevic et al., 2008).

Direct and remote methods are commonly practiced in PECVD. In the direct method, the substrate, the inert gas, and the gas dilutant are directly fed into the reacting chamber, whereas, in the remote method, the plasma is generated separate from the region of deposition and then the inert gases are introduced into plasma chamber (Teo et al., 2003).

1.8.2.5 SPUTTERING

The efficient method to produce nanoparticles is sputtering at high pressure. In this method, the standard magnetron, DC/RF used to sputter the metal at high pressure (0.23–1.5 Torr) in an inert gas atmosphere. At this high pressure, the gas-phase nucleation is formed. The nanoparticles are formed on the gold finger are collected using the scraper. The advantages of this method are no chemicals or solvents are used and improved control of the process (Heo et al., 2005; Seol et al., 2003). The basic theories involved with special reference to the behavior of various acoustical parameters have also been presented.

1.9 NANOFIBER APPLICATIONS

Different types of electrospun fibers produced from the various solutions create the opportunity in many application areas like filter membranes, composite applications, tissue engineering, biosensors, drug-delivery applications, and so on. Nonwoven composite fabrics can be produced by changing the composition of the spinning solutions successively to obtain the different layers of polymers one above the other layer in deposited form. Some of the major commercial applications of electrospun fibers are listed below:

1. Thin polymeric membranes for support
2. Composite development by reinforcing nanofibers
3. Textile surface characteristic changes from hydrophilic to hydro-phobic form
4. Wound dressing application
5. Thin layer of nonwoven fabric production, and so on

The electrospun nonwoven nanofiber mats are known for their 3D porous structures with interconnected. These electrospun nanofiber mats are relatively larger surface area. They provide unique outputs with high class of different combinations of ideal material as a matrix form used in tissue engineering. Numerous scaffolds, which are used for tissue engineering, were manufactured by electrospun using biopolymers (Bhattarai et al., 2004).

Nanocomposites are the materials of the 21st century having an annual growth rate of more than 25% because of their unique property-changing features creating opportunity in various application fields. For aerospace applications, properties like mechanical, thermal, electrical, chemical, and biodegradable properties are the essential one to focus on the material selections, due to the provisions in electrospinning process for combining various essential properties expanding its potential features in aerospace applications. Selection of nanofillers and the matrix enhances certain proper-ties that are desired (Rathod et al., 2017). The effect of high temperatures on polymer nanocomposites was investigated by Koo and Pilato (2005) by the use of phenol, epoxy, nylon, and so on and demonstrated the feasibility of these polymeric materials for flame retardant coatings, insulation for rocket propulsion, ablative materials for rocket nozzles, and so on. In the production process of filtration membranes, the time of electrospun fiber production process takes the key role on changing the efficiency level in filtration application. When the longer time is taken in the fiber production, it results more deposition of nanofibers on the single layer mat and decreases the filtration performance. On the other hand, reduction on fiber deposition time (compare with single-layer timing) in multilayer mat makes a significant filtration efficiency level, more than 99%. The fibrous filters are with the fiber diameters in a few dozen micrometers, and the filter porosity is usually around 80–90%. Fibrous structures maintain low resistance to the air flow with the high-efficiency indication by removing the micrometer and submicrometer particles. Especially for the submicrometer aerosol particles, nanfibrous media give a favorable solution compared with other filter media (Podgorski et al., 2006).

Gulrajani (2006) reviewed the various nanofinishes applied on the textile materials, such as hydrophobic finish, self-cleaning textile material characteristics, antimicrobial fabric finish, and effect of TiO in self-cleaning materials. Various research developments in nanotechnology finishes are clearly seen in improving textile properties. In synthetic fiber production, to pursue a nanoscale emulsification through more accurate results on fabric finishes can be possible to achieve. It can be emulsified into nanomicelles made into nanosols/nanocapsules form to apply evenly on the textile substrate; it changes the performance characteristics of textile materials in various finishing applications like stain-resistant, wrinkle resistance, change of hydrophilic and hydrophobic natures, and shrink-proof abilities (Qian and Hinestroza, 2004).

1.10 CONCLUSION

Electrospinning is not a complicated one as other fiber manufacturing techniques, it's a simple and versatile product-development technique; it supports to produce nanofibers with added value as per requirement in the laboratory as well as in the industry. The versatile applications, such as textiles, filtration, medicine, and so on, make this technique the most versatile technique to produce nanofibers. Recent times in the last two decades, there has been an explosive growth in electrospinning processes and the research findings also support for the various application opportunities because of its unique inherent characteristics. Despite process versatility and various advantages offering by electrospinning, the throughput of nanofibers has been a serious bottleneck problem that limits its real-time applications. Increasing the production rate makes the series issues like mutual repulsion between the jets, high voltage (safety concerns), fiber reproducibility, and environmental-related issues, and so on. In the cost-effectiveness of electrospinning setup, it is insufficient to examine only at the fiber production rate. It is very difficult to compare other types of fiber production systems with electrospinning system due to its lesser production rate and essentially required shorter distance for fiber production and also the type of equipment which is used for nanofiber productions; generally, it's in the range of less than 10 cm, whereas in other systems, 6–10 m range essentially required to stretch the produced filaments.

In the fiber development stage, process parameters such as the needle and the collector distance not significantly changing the fiber diameter, whereas it will control the beads formation in the process sequence. However, the large shearing force overcomes the surface tension acts on the solution, which supports for nanofiber nonwoven development without droplet formation. In modern-disk electrospinning, fiber developments are from the disk edge area, and the starting voltage for the fiber development process is playing an important role on fiber production, when the voltage increases the disk spun nanofiber became finer with a narrower diameter distribution, whereas nanofibers from the cylinder nozzle mainly produced based on the polymer concentration as well as the level of applied voltage. A staple-jet production from polymer solution ensures the proper output nanofiber; it can be obtained by changing the controlled variables. These are important considerations on the nanofiber production areas to increase the potential market growth and also to create new opportunities for the researchers to focus on versatile applications in electrospun nanofibers.

Nano-encapsulation is another area to coat or finish the textile fibers in nanoscale level. In this chapter, various methods to produce the nano-encapsulation and nanoparticles are described, out of which, very few methods were used or suitable to produce the nano-encapsulation and nanoparticles for textile applications.

KEYWORDS

- **nanofibers**
- **nanocomposites**
- **electrospinning**
- **electrospraying**
- **nanofiber production**

REFERENCES

Al Khouri Fallouh, N.; Roblot-Treupel, L.; Fessi, H; Devissaguet, J. P.; Puisieux, F. Development of a New Process for the Manufacture of Polyisobutylcyanoacrylate Nanocapsules. *Int. J. Pharm.* **1986,** *28* (2–3), 125–132.

Bhardwaj, N.; Kundu, S. C. Electrospinning: A Fascinating Fiber Fabrication Technique. *Biotechnol. Adv.* **2010,** *28* (3), 325–347.

Bhattarai, S. R.; Bhattarai, N.; Yi, H. K.; Hwang, P. H.; Cha, D. I.; Kim, H. Y. Novel Biodegradable Electrospun Membrane: Scaffold for Tissue Engineering. *Biomaterials* **2004,** *25* (13), 2595–2602.

Bouchemal, K.; Briançon, S.; Perrier, E.; Fessi, H.; Bonnet, I.; Zydowicz, N. Synthesis and Characterization of Polyurethane and Poly(Ether Urethane) Nanocapsules Using a New Technique of Interfacial Polycondensation Combined to Spontaneous Emulsification. *Int. J. Pharm.* **2004,** *269* (1), 89–100.

Callister, W. D.; Rethwisch, D. G. 2004, Edition 6.

Cooley, J. F. P1_US692631.pdf. US Patent, 1902.

Couvreur, P.; Dubernet, C.; Puisieux, F. Controlled Drug Delivery with Nanoparticles: Current Possibilities and Future Trends. *Eur. J. Pharm. Biopharm.* **1995,** *41*, 2–13.

Couvreur, P.; Barratt, G.; Fattal, E.; Vauthier, C. Nanocapsule Technology: A Review. *Crit. Rev. Ther. Drug Carrier Syst.* **2002,** *19* (2), 99–134.

Cui, W.; Li, X.; Zhou, S.; Weng, J. Investigation on Process Parameters of Electrospinning System through Orthogonal Experimental Design. *J. Appl. Polym. Sci.* **2007,** *103* (5), 3105–3112.

Deitzel, J. M.; Kleinmeyer, J.; Harris, D.; Beck Tan, N. C. The Effect of Processing Variables on the Morphology of Electrospun Nanofibers and Textiles. *Polymer* **2001,** *42* (1), 261–272.

Doshi, J.; Reneker, D. H. Electrospinning Process and Applications of Electrospun Fibers. *J. Electrostat.* **1995,** *35*, 151–160.

Eilers, H.; Tissue, B. M. Synthesis of Nanophase ZnO, Eu_2O_3, and ZrO_2 by Gas-Phase Condensation with Cw-CO_2 Laser Heating. *Mater. Lett.* **1995,** *24* (4), 261–265.

Ekman, B.; Sjöholm, I. Improved Stability of Proteins Immobilized in Microparticles Prepared by a Modified Emulsion Polymerization Technique. *J. Pharm. Sci.* **1978,** *67* (5), 693–696.

El-Samaligy, M. S.; Rohdewald, P.; Mahmoud, H. A. Polyalkyl Cyanoacrylate Nanocapsules. *J. Pharm. Pharmacol.* **1986,** *38* (3), 216–218.

Fessi, H.; Puisieux, F.; Devissaguet, J. P.; Ammoury, N.; Benita, S. Nanocapsule Formation by Interfacial Polymer Deposition Following Solvent Displacement. *Int. J. Pharm.* **1989,** *55* (1), R1–R4.

Gulrajani, M. L. Nano Finishes. *Indian J. Fibre Text. Res.* **2006,** *31*, 187–201.

Heo, C. H.; Lee, S. B.; Boo, J. H. Deposition of TiO_2 Thin Films Using RF Magnetron Sputtering Method and Study of Their Surface Characteristics. *Thin Solid Films* **2005,** *475* (1–2), 183–188.

Kleivaitė, V.; Milašius, R. Electrospinning—100 Years of Investigations and Still Open Questions of Web Structure Estimation. *Autex Res. J.* **2018,** *18* (4), 398–404.

Koo, J. H.; Pilato, L. Polymer Nanostructured Materials for High Temperature Applications. *SAMPE J.* **2005,** *41* (2), 7–19.

Kreuter, J. Large-Scale Production Problems and Manufacturing of Nanoparticles. In *Specialized Drug Delivery System*; Tyle, P., Ed.; Marcel Dekker: New York, 1990; pp 257–266.

Kreuter, J. Nanoparticle-Based Drug Delivery Systems. *J. Controlled Release* **1991,** *16,* 169–176.

Lambert, G.; Fattal, E.; Couvreur, P. Nanoparticulate System for the Delivery of Antisense Oligonucleotides. *Adv. Drug Delivery Rev.* **2001,** *47,* 99–112.

Lowe, P. J.; Temple, C. S. Calcitonin and Insulin in Isobutylcyanoacrylate Nanocapsules: Protection Against Proteases and Effect on Intestinal Absorption in Rats. *J. Pharm. Pharmacol.* **1994,** *46* (7), 547–552.

Lu, C. H.; Jagannathan, R. Cerium-Ion-Doped Yttrium Aluminum Garnet Nanophosphors Prepared Through Sol–Gel Pyrolysis for Luminescent Lighting. *Appl. Phys. Lett.* **2002,** *80* (19), 3608–3610.

Macossay, J.; Marruffo, A.; Rincon, R.; Eubanks, T.; Kuang, A. Effect of Needle Diameter on Nanofiber Diameter and Thermal Properties of Electrospun Poly(Methyl Methacrylate). *Polym. Adv. Technol.* **2007,** *18* (3), 180–183.

Malesevic, A.; Vitchev, R.; Schouteden, K.; Volodin, A.; Zhang, L.; Tendeloo, G. V.; Vanhulsel, A.; Haesendonck, C. V. Synthesis of Few-Layer Graphene Via Microwave Plasma-Enhanced Chemical Vapour Deposition. *Nanotechnology* **2008,** *19* (30), 305604.

Megelski, S.; Stephens, J. S.; Chase, D. B.; Rabolt, J. F. Micro- and Nanostructured Surface Morphology on Electrospun Polymer Fibers. *Macromolecules* **2002,** *35* (22), 8456–8466.

Montasser, I.; Fessi, H.; Coleman, A. W. Atomic Force Microscopy Imaging of Novel Type of Polymeric Colloidal Nanostructures. *Eur. J. Pharm. Biopharm.* **2002,** *54* (3), 281–284.

Mozafari, M. R. (Ed.). *Nanomaterials and Nanosystems for Biomedical Applications*; Springer Netherlands: Amsterdam, 2007.

Nayak, R.; Padhye, R.; Kyratzis, I. L.; Truong, Y. B.; Arnold, L. Recent Advances in Nanofibre Fabrication Techniques. *Text. Res. J.* **2012,** *82* (2), 129–147.

Niu, H.; Lin, T.; Wang, X. Needleless Electrospinning. I. A Comparison of Cylinder and Disk Nozzles. *J. Appl. Polym. Sci.* **2009,** *114* (6), 3524–3530.

Patanaik, A.; Anandjiwala, R. D.; Rengasamy, R. S.; Ghosh, A.; Pal, H. Nanotechnology in Fibrous Materials—A New Perspective. *Text. Progress* **2007,** *39* (2), 67–120.

Patil, G. V. Biopolymer Albumin for Diagnosis and in Drug Delivery. *Drug Dev. Res.* **2003,** *58* (3), 219–247.

Qian, L.; Hinestroza, J. P. Application of Nanotechnology for High Performance Textiles. *J. Text. Apparel, Technol. Manage.* **2004,** *4* (1), 1–7.

Quintanar-Guerrero, D.; Allémann, E.; Fessi, H.; Doelker, E. Preparation Techniques and Mechanisms of Formation of Biodegradable Nanoparticles from Preformed Polymers. *Drug Dev. Ind. Pharm.* **1998,** *24* (12), 1113–1128.

Ramakrishna, S. (Ed.). *An Introduction to Electrospinning and Nanofibers*; World Scientific, 2005.

Rathod, V. T.; Kumar, J. S.; Jain, A. Polymer and Ceramic Nanocomposites for Aerospace Applications. *Appl. Nanosci.* **2017,** *7* (8), 519–548.

Seol, J.; Lee, S.; Lee, J.; Nam, H.; Kim, K. Electrical and Optical Properties of CuZnSnS Thin Films Prepared by RF Magnetron Sputtering Process. *Solar Energy Mater. Solar Cells* **2003,** *75* (1–2), 155–162.

Shinomiya, T.; Gupta, V.; Miura, N. Effects of Electrochemical-Deposition Method and Microstructure on the Capacitive Characteristics of Nano-Sized Manganese Oxide. *Electrochim. Acta* **2006,** *51* (21), 4412–4419.

Taylor, G. Electrically Driven Jets. *Proc. R. Soc. A: Math., Phys. Eng. Sci.* **1969,** *313* (1515), 453–475.

Teo, K. B. K.; Lee, S. B.; Chhowalla, M.; Semet, V.; Binh, V. T.; Groening, O.; Castignolles, M.; Loiseau, A.; Pirio, G.; Legagneux, P.; Pribat, D.; Hasko, D. G.; Ahmed, H.; Amaratunga, G. A. J.; Milne, W. I. Plasma Enhanced Chemical Vapour Deposition Carbon Nanotubes/ Nanofibres How Uniform do They Grow? *Nanotechnology* **2003**, *14* (2), 204–211.

Teo, W. E.; Ramakrishna, S. A Review on Electrospinning Design and Nanofibre Assemblies. *Nanotechnology* **2006**, *17* (14), R89–R106.

Tice, T. R.; Gilley, R. M. Preparation of Injectable Controlled-Release Microcapsules by a Solvent-Evaporation Process. *J. Control Release* **1985**, *2*, 343–352.

Tomaszewski, W.; Szadkowski, M. Investigation of Electrospinning with the Use of a Multi-Jet Electrospinning Head. *Fibres Textiles East. Europe* **2005**, *13* (4), 22–26.

Sukigara, S.; Gandhi, M.; Ayutsede, J.; Micklus, M.; Ko, F. Regeneration of Bombyx Mori Silk by Electrospinning. Part 2. Process Optimization and Empirical Modeling Using Response Surface Methodology. *Polymer* **2004**, *45* (11), 3701–3708.

Tucker, N.; Stanger, J. J.; Staiger, M. P.; Razzaq, H.; Hofman, K. The History of the Science and Technology of Electrospinning from 1600 to 1995. *J. Eng. Fibers Fabrics* **2012**, *7* (2), 63–73.

Vaseashta, A. Controlled Formation of Multiple Taylor Cones in Electrospinning Process. *Appl. Phys. Lett.* **2007**, *90* (9), 093115.

Vauthier, C.; Dubernet, C.; Fattal, E.; Pinto-Alphandary, H.; Couvreur, P. Poly (alkylcyanoac-rylates) as Biodegradable Materials for Biomedical Applications. *Adv. Drug Delivery Rev.* **2003**, *55* (4), 519–548.

Watnasirichaikul, S.; Davies, N. M.; Rades, T.; Tucker, I. G. Preparation of Biodegradable Insulin Nanocapsules from Biocompatible Microemulsions. *Pharm. Res.* **2000**, *17* (6), 684–689.

Widder, K. J.; Flouret, G.; Senye, A. Magnetic Microspheres: Synthesis of a Novel Parental Drug Carrier. *J. Pharm. Sci.* **1979**, *68* (1), 79–82.

Ya, L.; Bowen, C.; Guoxiang, C. Development and Filtration Performance of Polylactic Acid Meltblowns. *Text. Res. J.* **2010**, *80* (9), 771–779.

Yamashita, Y.; Ko, F.; Tanaka, A.; Miyake, H. Characteristics of Elastomeric Nanofiber Membranes Produced by Electrospinning. *J. Text. Eng.* **2007**, *53* (4), 137–142.

Yarin, A. L.; Zussman, E. Upward Needleless Electrospinning of Multiple Nanofibers. *Polymer* **2004**, *45* (9), 2977–2980.

CHAPTER 2

Flexible Piezoelectric Nanogenerator Composed of Electrospun Nanofibrous Web

S. WAZED ALI* and SATYARANJAN BAIRAGI

Department of Textile and Fibre Engineering, Indian Institute of Technology Delhi, Hauz Khas, New Delhi, India

*Corresponding author. E-mail: wazed@textile.iitd.ac.in

ABSTRACT

In recent days, nanotechnology has drawn a great interest in various fields such as filtration, medical, functional textiles, sensors, actuators, and so forth. In this respect, electrospun nanofiber has become one of the most promising materials to fulfill the advanced requirements of the above-said applications. Electrospinning is one of the attractive technologies where nanodimensional fibrous structure is developed by application of high electric field in the respective polymer solutions. Among the different applications, electrospinning has pinched a tremendous role on the fabrication of textile-based highly efficient piezoelectric sensors and actuators. The electrospun webs in various forms like pure polymer, nanocomposite, and hybrid based have been utilized in the piezoelectric end uses. In this chapter, details of electrospinning method and recent development of electrospun web-based piezoelectric nanogenerators have been discussed systematically.

2.1 INTRODUCTION

Nanotechnology is the current buzzword to the scientific community. This technology has recently been exploited in the textile community in a much

Fundamentals of Nano-Textile Science. Prashansa Sharma, Devsuni Singh & Vivek Dave (Eds.)
© 2022 Apple Academic Press, Inc. Co-published with CRC Press (Taylor & Francis)

significant way. There are many lucrative functional properties which have been imbued in textile structure sector. Antimicrobial, UV protective, fire retardant, filtration, packaging, ballistic, and so on are the few functionalities to mention in this respect. Among different nanomaterials, nanofibrous structure is one of the magic materials, which has found numerous applications in technical textiles. Electrospinning is the most advantageous route to prepare nanofibrous structure in terms of controlling the fiber diameter and it's easy to design the fiber structure than other methods. Nowadays, a surge of research has also been done on the nanofiber-based piezoelectric nanogenerator.

Piezoelectricity is a property of the materials which can be used for generating electric charge when the respective materials are subjected to a mechanical stress or vice-versa (Kim et al., 2011). In the year 1880, Curie brothers first established this piezoelectric property in Rochelle salt (Curie and Curie, 1880). Thereafter, many researches have been carried out in the field of piezoelectricity. This approach is one of the most important findings to reduce the environmental pollution made by fuel energy, by harvesting various mechanical energies available in our surroundings. Mechanical energy is one of the most abundant energies among the other energies (thermal and solar energy) which can be available in different forms such as wind energy, vibration energy, energy due to the human body movement, and so on (Fan et al., 2016; Wang and Wu, 2012). These wastage energies can be harvested in terms of useful electrical energy through the piezoelectric materials. This also needs to be mentioned that piezoelectric technology has enhanced capability to generate high energy density as compared to other energy harvesting technologies (triboelectric, pyroelectric, electromagnetic, etc.) (Khalifa et al., 2019).

There are verities of piezoelectric materials such as lead zirconate titanate (PZT), sodium niobate ($NaNbO_3$), zinc stannate ($ZnSnO_3$), potassium sodium niobate (KNN), poly(vinylidene difluoride) (PVDF), and its copolymer poly(vinylidene difluoride trifluoroethylene) (PVDF-TrFE), and different bio-based materials (like onion skin, eggshell, fish scale, fish bladder, etc.) (Bairagi and Ali, 2019a, 2019b; Ghosh and Mandal, 2016; Karan et al., 2018; Lee et al., 2014; Maiti et al., 2017; Niu et al., 2015; Nour et al., 2014). Thus, piezoelectric materials are broadly classified into three subgroups, that is, inorganic, organic, and composite (polymer plus filler). Among these piezoelectric materials, most promising materials are PZT and PVDF. But, they have some disadvantages too. In the case of PZT, it has very high piezoelectric properties (piezoelectric constant ~ 400–700 pC/N, Curie temperature ~ >400°C), but it contains more than 60% lead which is not eco-friendly. On the other hand, PVDF has enough flexibility and easy availability but its piezoelectric properties are low (piezoelectric constant ~ −30 pC/N, dielectric constant ~ 6–12). To resolve the individual problems of the inorganic and polymer-based piezoelectric materials, research work has turned its path into the composite-based piezometric materials, which can demonstrate the combined effect of both the polymer and inorganic filler. PVDF is one of the most usable piezoelectric polymers due its higher electroactive property. It contains five different crystal phases among which β-phase is mainly responsible for the piezoelectric property of PVDF. One important challenge is that to obtain piezoelectric effect from PVDF polymer, its more stable α-crystal phase must be transferred into the β-crystal phase. There are different techniques to convert α to β phase, such as mechanical stretching, incorporation of filler into PVDF matrix, electrical poling of the polymer, or by electrospinng technique. Electrospinning technique is one of the lucrative techniques to impart piezoelectric properties in the PVDF polymer in a simple manner (Chang et al., 2007; Salimi and Yousefi, 2004).

In this chapter, a detailed description of the electrospinning method and different piezoelectric nanogenerators based on electrospinning nanofibrous webs have been represented systematically. First, electrospinning method has been discussed in details in Section 2.2 and then various research works on the electrospun web-based piezometric nanogenerator have been elaborated in Section 2.3.

2.2 ELECTROSPINNING

2.2.1 WHAT IS ELECTROSPINNING?

In recent days, "nanofiber" has drawn a great interest for its promise in wide spectrum application areas due to small pores, high surface area, and consistency of the fiber-forming process unlike commercial spinning procedures, such as biomedical (tissue engineering, scaffolds), flexible electronics, protective clothing, nanofiltration, healthcare, defense, biotechnology, and environmental engineering (Bhardwaj and Kundu, 2010). There are different techniques by which nanofibers can be developed like electrospinning, melt blown using molecular dies, bi-component fibers, phase separation, or self-assembly. Among these, electrospinning is most promising technique, since it is most controllable as compared to others. Electrospinning was first discovered by Rayleigh in 1987. "Electrospinning" term is derived from the "electrostatic spinning," where nanofiber gets developed due to the electrostatic repulsive force. In electrospinning process, the polymer solution is held on the tip of the needle because of surface tension. The nano to submicron fibers get generated in the electrospinning due to the potential difference between the needle tip and the collector. Once an electric field is applied to the polymer solution, a mutual repulsive static force is developed. When this induced repulsive force in the polymer solution overcomes the surface tension, a jet is ejected from the needle tip. Furthermore, a hemispherical shape like jet is developed at the needle tip when electric field is further increased, which is called the "Taylor Cone." The total assembly of the electrospinning is depicted in Figure 2.1. Working parameters are very important for developing desirable fiber quality like uniform diameter, beads-free, high-strength fiber. All the working parameters are subdivided into three groups such as (i) solution parameters (solution viscosity, solution concentration, solution conductivity, vapor pressure of solution, solution temperature, and surface tension), (ii) process parameters (applied voltage, solution flow rate, type of collector, type of needle tip orifice, and needle tip to collector distance), and (iii) ambient parameters (temperature, relative humidity, and air velocity in the electrospinning chamber). The effects of different working parameters are explained in the next sections.

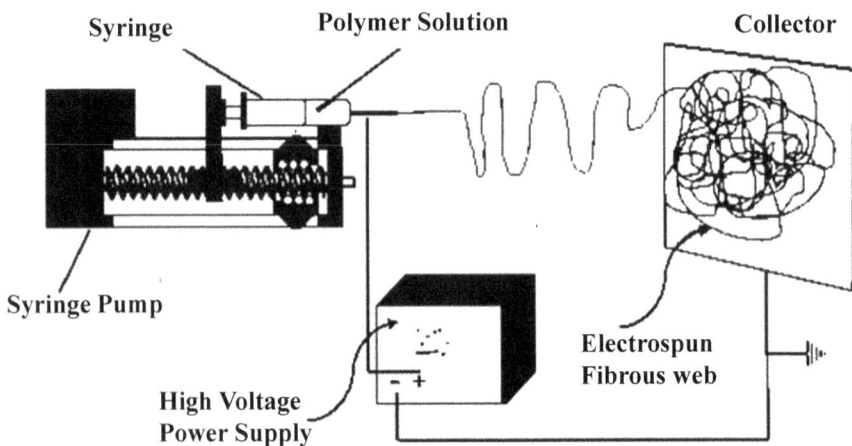

FIGURE 2.1   Electrospinning set-up.

2.2.2   DIFFERENT PARTS IN ELECTROSPINNING SET-UP

The schematic line diagram of a typical electrospinning unit is shown in Figure 2.1.

For the electrospinning process, polymer should be soluble or meltable. Polymer can be dissolved into organic solvent or water. The electrospinning machine is consisting of the following main parts:

1. **Metallic needle:** A metallic needle must be attached with the plastic syringe with different solution reservoir capacity (2, 5 mL, or so). This metallic needle acts as both function electrode and spinneret. The production of the electrospinning process can be enhanced by using more than one metallic needle as spinneret but for this situation, maintaining a uniform diameter fiber production could be a difficult task. The needle diameter (inner) can get varied and is generally attached with the syringe pump.

2. **Feeding pump:** The plastic syringe with different capacities can be used as a feeding unit where the solution or melt is reserved. The feeding rate can be varied from a few microliters to milliliter per hour. The feeding rate is controlled by using a pump or piston and the viscosity of the solution.

3. **Collector:** The role of collector in the electrospinning process is very sensitive to control fiber diameter and to obtain beads-free fibers. The collector can be of different types such as drum type, flat-plate type, or disc type. The drum-type collector can be used to prepare nanofibers of different diameters by changing the drum rpm. To get aligned fiber, rotating collectors (drum or disc) can be used, whereas randomly oriented fibers will be developed for the stationary collector (flat plate).

4. **High-voltage supply:** Nanofibers are developed in the process of electrospinning by application of an external electric field between the metallic needle tip and the collector.

2.2.3 DIFFERENT PARAMETERS IN ELECTROSPINNING TO CONTROL THE FIBER PROPERTIES

2.2.3.1 SOLUTION PARAMETERS

2.2.3.1.1 Viscosity of the Solution

Fiber morphology is extremely dependent on the solution viscosity. At very low viscosity, smooth and continuous fibers cannot be obtained. The very high viscosity of the solution is also problematic for developing continuous defect-free fibers because at the higher viscosity, the solution becomes hard and its ejection out of the needle tip becomes very difficult. However, there is less tendency to form beaded fibers at the higher viscosity. Fong et al. studied on beads formation in the electrospinning technique by using poly(ethylene) oxide (PEO) solution and they observed that bead formation gradually gets decreased with increase in the solution viscosity. The SEM images of their work have been depicted in Figure 2.2 (Fong et al., 1999). Therefore, an optimum viscosity is required for producing smooth and continuous fibers. The viscosity mainly depends on the concentration of the polymer and the molecular mass of the polymer.

2.2.3.1.2   *Concentration of Polymer Solution*

The concentration of the polymer is also another important solution parameter for developing beads-free smooth electrospun fibers. Viscosity of the solution can be maintained by changing the amount of polymer. For beads-free smooth continuous fibers, an optimum concentration of the polymer must be maintained. At the very low concentration (say, $\eta < 1$ poise), droplets will be generated rather than fibers. However, at very high concentration of the polymer (say, $\eta > 20$ poise), spinning will be hindered by decontrolling the flow of solution through the needle tip, since at higher concentration polymer becomes too viscous due to higher cohesive nature of the solution. The effect of concentration on the fiber morphology is shown in Figure 2.3.

2.2.3.1.3   *Molecular Weight of the Polymer*

Molecular mass of the polymer also has effect on the fiber structure. The entanglement of the polymer chain strongly depends upon the molecular weight of the polymer. Higher molecular weight provides higher number of polymer chain entanglement and vice-versa. The viscosity of the solution also depends on the molecular weight of the polymer. Therefore, an optimum molecular weight of the polymer is required to develop smooth and continuous fibers. Koski et al. (2004) investigated the effect of molecular weight of polymer on the fiber morphology by using polyvinyl alcohol (PVA) as a polymer. They concluded that beaded fibers are developed with the low molecular weight polymer. However, for the very high molecular weight polymer, fiber diameter also gets increased (Fig. 2.4).

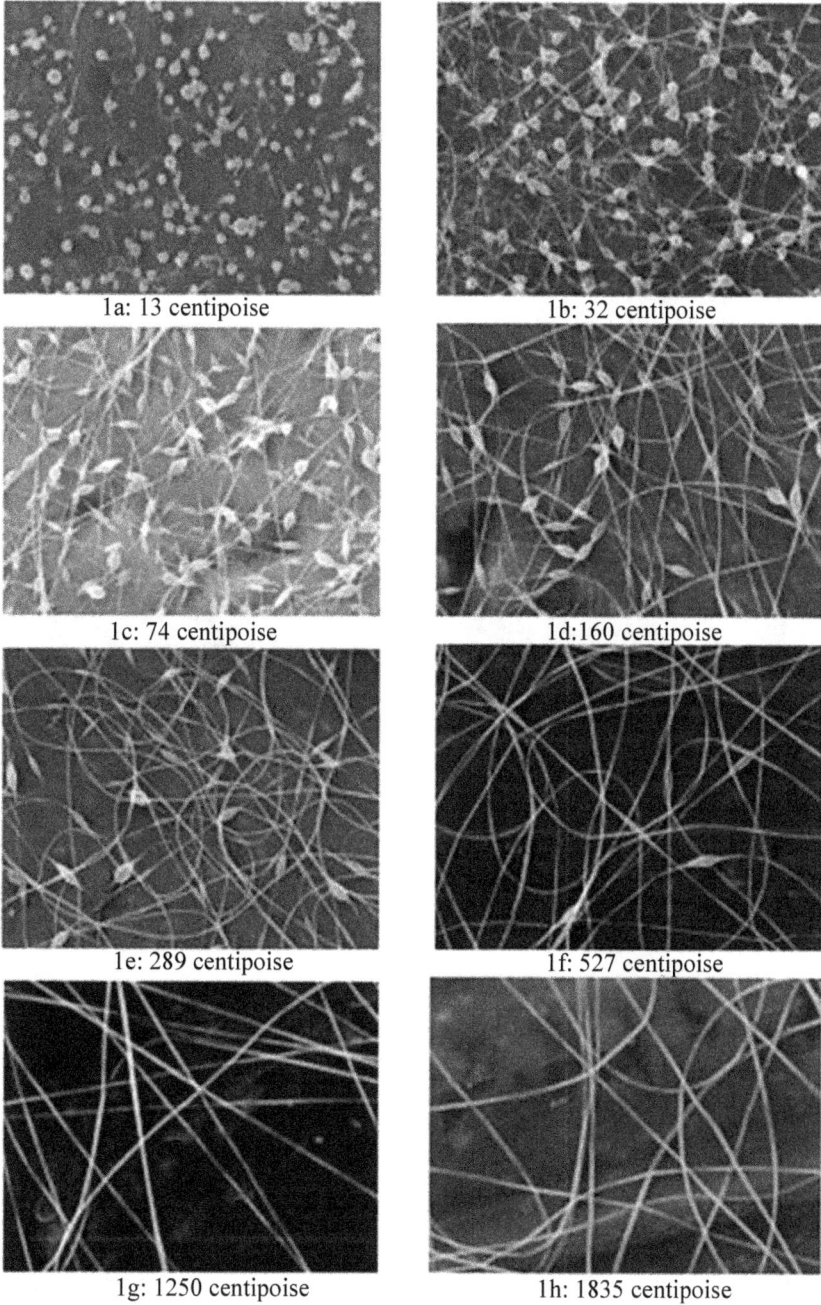

FIGURE 2.2 Effects of solution viscosity on the morphology of the fiber (Reprinted with permission from Fong et al., 1999. © Elsevier.).

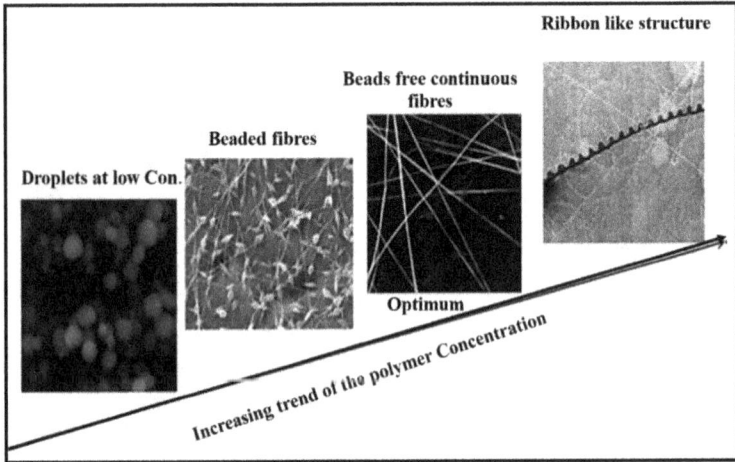

FIGURE 2.3   Morphology of the fiber at different concentrations of polymer (Reprinted with permission from Deitzel et al., 2001. © Elsevier.).

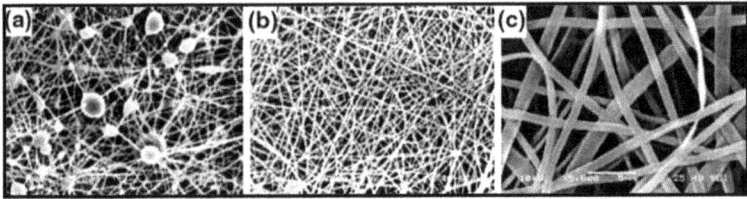

FIGURE 2.4   Effect of molecular weight [(a) 9000–10,000 g/mol; (b) 13,000–23,000 g/mol; and (c) 31,000–50,000 g/mol] on the fiber morphology (Reprinted with permission from Koski et al., 2004. © Elsevier.).

2.2.3.1.4   *Surface Tension of the Solution*

Surface tension of the polymer solution is also another important factor, which absolutely depends on the viscosity and the concentration of the polymer solution. In case of higher surface tension of the polymer solution, beaded fibers will be prominent and vice-versa.

2.2.3.2   *PROCESS PARAMETERS*

2.2.3.2.1   *Applied Voltage*

Supplied external voltage has a decisive role in the electrospinning process. Once, applied voltage is crossed the brink limit, the jet can be ejected from the needle tip. The applied voltage also controls the fiber diameter in the electrospinning process. Zhang et al. (2005) studied on the effect of applied voltage on the fiber diameter distribution and morphology of the fiber in electrospinning process by using PVA solution and they concluded that fiber diameter gets increased with increase in the applied voltage, which can be seen from Figure 2.5.

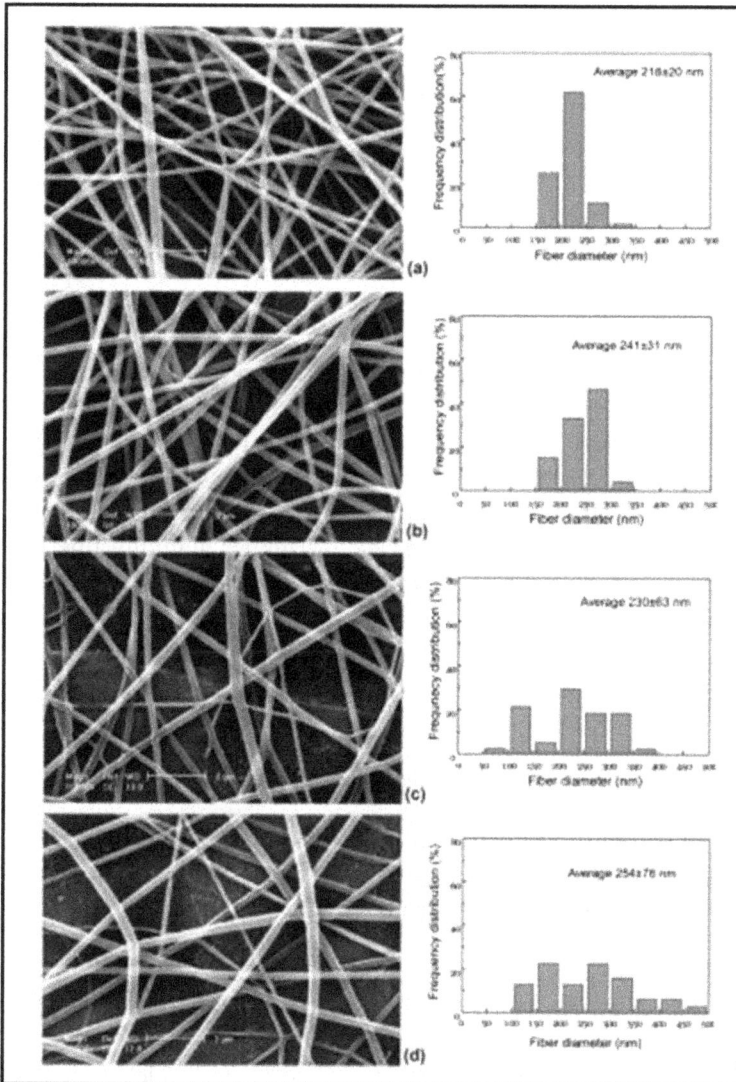

FIGURE 2.5   The effect of applied voltage [(a) 5 kV, (b) 8 kV, (c) 10 kV, and (d) 13 kV] on the surface morphology and diameter of the fibers (Reprinted with permission from Zhang et al., 2005. © Elsevier.).

2.2.3.2.2   Flow Rate of the Solution

Flow rate of the polymer solution inside the plastic syringe is another important parameter which can control the fiber diameter. If flow rate is low, the
polymer solution can be charged sufficiently, and finally, beads-free smooth
fiber would be developed.

2.2.3.2.3   Tip to Collector Distance

The distance between the collector and the tip has effect on the fiber diameter and morphology. Higher distance will provide enough time for solvent
evaporation from the fiber and thus enough stretching can take place. In the
case of lower distance, fiber with bigger diameter will be developed.

2.2.3.2.4   Ambient Parameters

Ambient parameters (temperature and humidity) also have immense effect
on the fiber diameter and morphology. At the lower humidity, evaporation
rate of the solvent from the fiber surface will be improved. As a result, thinner
diameter fiber can be produced and for the higher humidity fiber diameter
will be increased. At the higher temperature, evaporation rate will be higher
and vice-versa.

Effect of the all working parameters on the fiber diameter during the
electrospinning process is given below:

$$dt = \left[\gamma \varepsilon \frac{Q^2}{I^2} \frac{2}{\pi(2\ln x - 3)} \right]^{1/3} \tag{2.1}$$

where d is the fiber diameter; γ and ε are the surface tension and dielectric
constant of the solution; Q and I are the flow rate and the current carried by
the fiber; and χ is the ratio of the initial jet length to the nozzle diameter.

2.3   ELECTROSPUN WEB-BASED PIEZOELECTRIC GENERATORS

Electrospinning is one of the most feasible ways to develop highly efficient
piezoelectric nanogenerator as compared to other fiber-formation techniques
like solution spinning or melt spinning. This is due to the opportunity of
spun nanofiber to get poled during electrospinning process itself under high

electrical field. In addition, mechanical stretching of the nanofiber also takes place. Therefore, extra arrangement of poling process could be eliminated unlike the solution cast and melt-extruded fiber or film to obtain reasonable piezometric performance from the same. In case of composite-based materials, there is a chance of poling both the polymer and filler materials at the same time (depends on the materials used and the electric filed applied) in electrospinning process, resulting in higher piezoelectric effect. For instance, KNN nanostructures have been incorporated into the PVDF polymer to develop a flexible electrospun web-based piezoelectric nanogenerator by Teka et al. (2018). It has been observed that KNN nanostructures incorporated PVDF polymer-based nanogenerator can generate ~ 1.9 V output voltage by simple hand tapping on the device. In another study, Bairagi and Ali (2019c) have also reported an electrospun nanocomposite web-based piezometric nanogenerator, where they have investigated the effect of aspect ratio of the filler (KNN nanorods [NR]) material in the piezometric performance of the PVDF polymer. Finally, they have concluded that the aspect ratio of the filler has an immense effect on the piezometric properties of the materials due to (i) higher dipole moment provided by the higher aspect ratio KNN NRs, (ii) in-situ poling of both the KNN filler and PVDF polymer at the same time, (iii) mechanical stretching of the fiber because of the higher electric field in the electrospinning method, and (iv) higher deformation tendency of the PVDF polymer chains since larger contact area between the KNN NRs and PVDF polymer matrix during energy tapping. The developed KNN/PVDF polymer-based nanogenerator showed an output voltage of ~17.5 V and current of ~1.81 μA, under the application of 1.1 kPa pressure by the finger tapping on the device. The KNN in the form of nanoparticles (NPs) has also been incorporated into the PVDF-TrFE polymer matrix to form a flexible lead-free piezoelectric nanogenerator by using electrospinning method (Kang et al., 2015). The developed nanogenerator showed an output voltage ~0.98 V and current ~78 nA, with the loading of 10% KNN NPs as depicted in Figure 2.6.

Research has also been carried out by using the ZnO as a filler material in the form of particles, NRs, and nanowires into the PVDF or its copolymer (PVDF-TrFE) to develop piezoelectric nanogenerator. ZnO is a semiconducting metal oxide material, which can show piezoelectric effect without further electrical poling. Therefore, by using ZnO as a filler material into the piezoelectric polymer, higher piezoelectric performance is expected from the nanocomposite (ZnO/polymer)-based nanogenerator. Li et al. (2019) has developed two different types of piezoelectric nanogenerators, one can be

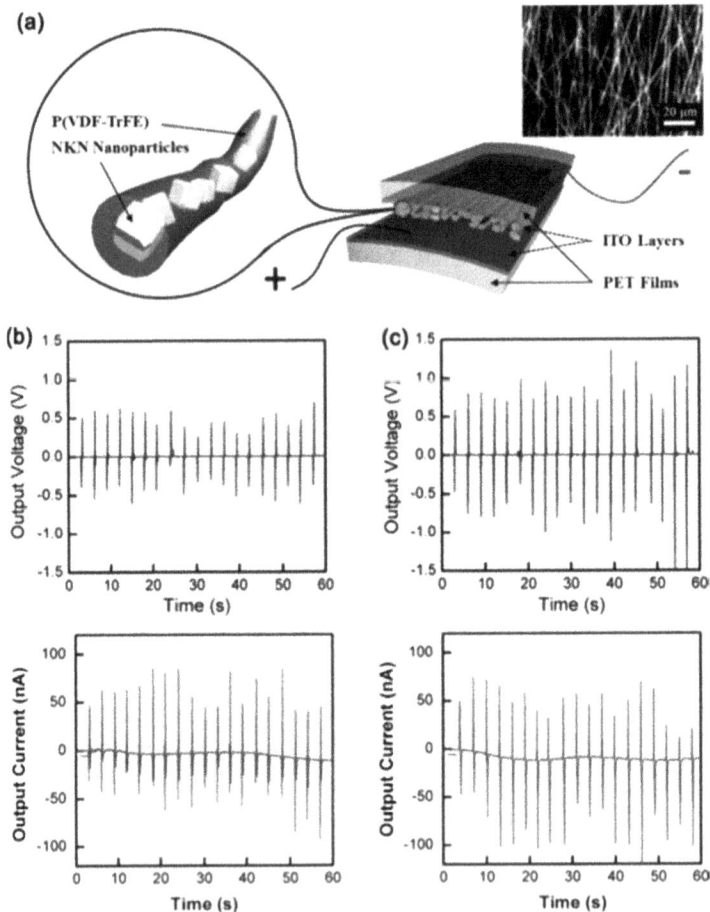

FIGURE 2.6   (a) Image of the nanogenerator, (b) output signal of the pristine polymer-based nanogenerator, and (c) output signal of the nanocomposite-based nanogenerator (Reprinted with permisison from Kang et al., 2015. © Elsevier).

classified as PVDF/ZnO NP-based and the other as PVDF/ZnO NR-based electrospun nanocomposites. They have evaluated that the piezoelectric performance of both the nanogenerators by application of same level of pressure. It has been found that PVDF/ZnO NRs nanocomposite-based nanogenerator showed the higher output voltage (~85 V) and current (~2.2 µA) in contrast to the PVDF/ZnO NP-based nanogenerator (output voltage ~60 V and current 1.7 µA). It has been concluded that NRs having higher aspect ratio have higher β-nucleating effect in the PVDF polymer matrix, whereas for the NPs of ZnO would have lower β-nucleating effect due to the lower aspect ratio of the particles. Furthermore, in the case of NRs, combined piezoelectric effect would also be higher than the particles form of the ZnO. The doping of ZnO has also been evaluated to confirm its effect in piezoelectric properties of the polymer. Parangusan et al. (2019) developed pure PVDF-HFP electrospun fiber-based nanogenerator and Co-ZnO/PVDF-HFP nanocomposite electrospun fiber-based nanogenerator. It has been found that Co-doped ZnO incorporated to PVDF-HFP-based nanogenerator can generate ~2.8 V output voltage, whereas for pure PVDF-HFP-based nanogenerator, the output voltage is ~150 mV. The authors have concluded that doped ZnO has higher β-nucleation effect than the pure PVDF-HFP polymer, for which Co-ZnO/PVDF-HFP nanogenerator showed higher piezoelectric performance. In another study, Ponnamma et al. (2019) has evaluated the piezoelectric performance of the Fe-doped ZnO and cellulose nanocrystal (CNC)-loaded PVDF-HFP polymer-based nanogenerator. This hybrid composite material has been developed by using electrospinning method by optimizing the process parameters. Finally, they have compared the piezoelectric performance of the net polymer (PVDF-HFP)-based nano-generator and hybrid (CNC/Fe-ZnO/PVDF-HFP) electrospun fiber-based nanogenerator. They observed that 12 V output voltage and 1.9 μ/cm^2 current density generated by this hybrid nanogenerator is 60 times higher than the pure PVDF-HFP-based nanogenerator as shown in Figure 2.7.

Barium titanate (BT) ($BaTiO_3$) also is an emerging ceramic-based piezoelectric material which can easily be used in the polymer as a filler material to improve the resultant performance of the polymer. BT has higher dielectric and piezometric constant for which it has been used in the different polymer to form composite-based piezoelectric materials. Hu et al. (2019) investigated the piezoelectric performance of the $BaTiO_3$ NPs/PVDF-TrFE electrospun fiber-based nanogenerator and $BaTiO_3$ NPs/PVDF-TrFE electrospun fibers with penetrated electrode-based nanogenerator, respectively. It has been found that the deposition of penetrated electrode material on the $BaTiO_3$ NPs/PVDF-TrFE electrospun fibers can improve the piezoelectric performance of the nanocomposite-based nanogenerator. This is because of the conducting path created by the penetrated electrode material inside the nanocomposite for which charges can easily flow towards the electrodes and will show the higher output signal as compared to simple $BaTiO_3$ NPs/PVDF-TrFE electrospun fiber-based nanogenerator. In various research works, it can also be found that the different conductive fillers (CNT, Ag, PANI (polyaniline), Cu, etc.) have been used in combination with piezoelectric filler in the polymer matrix to improve the energy-harvesting efficacy by providing a conductive path inside the piezoelectric materials.

For instance, Shi et al. (2018) has fabricated a CNT/BT/PVDF electrospun nanocomposite-based piezoelectric nanogenerator and evaluated the synergistic effect of the conductive carbon nanotube (CNT) and BT in the PVDF polymer matrix. The eletrospun nanocomposite is composed of 0.15% CNT and 15% BT incorporated into the PVDF polymer. They observed that this CNT/BT/PVDF-based nanogenerator can generate an output voltage of ~11 V and power of ~ 4.1 µW under the frequency of 2 Hz and a strain of 4 mm as shown in Figure 2.8.

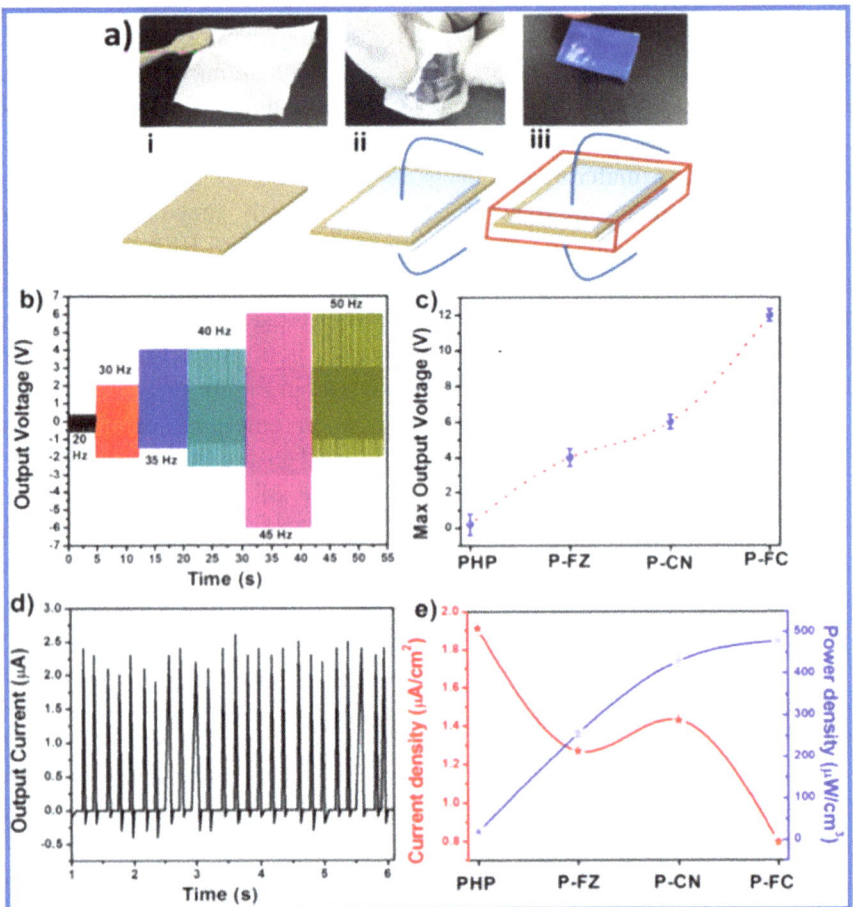

FIGURE 2.7 (a) Schematic diagram of the nanogenerator fabrication, (b) output voltage with varying frequencies, (c) output signals of the various nanogenerators, (d) output current of the nanogenerator at the frequency of 45 Hz, and (e) current and power density of the nanogenerator (Reprinted with permission from Ponnamma et al., 2019. © Elsevier).

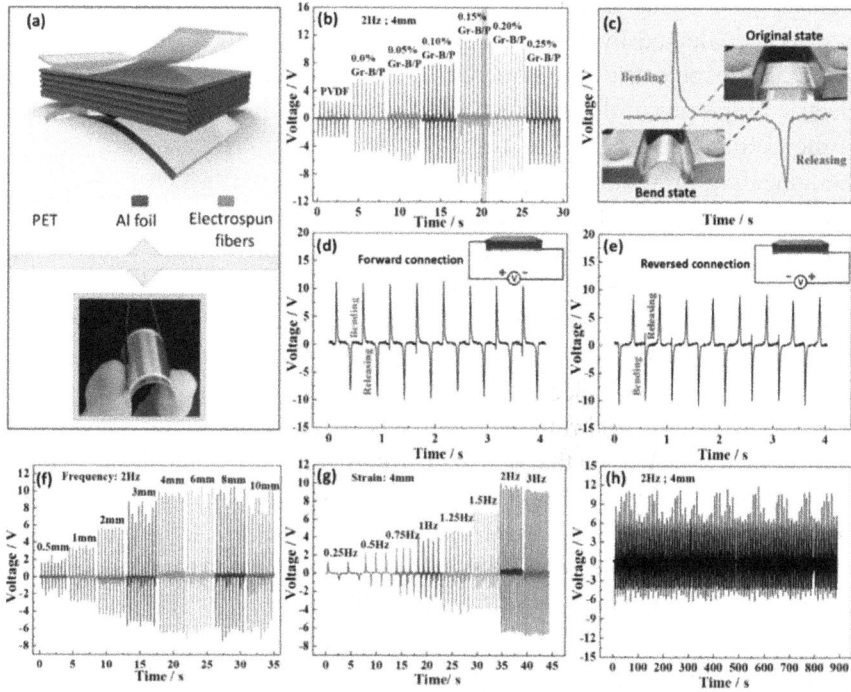

FIGURE 2.8 (a) Image of the nanogenerator, (b) changes of the output voltage with the time, (c) zoomed image of a single signal peak, (d) output signal of the nanogenerator with the forward connection, (e) output signal of the nanogenerator with the reverse connection, (f) output signal of the nanogenerator with different strain values, (g) output signal with different frequencies, and (h) durability of the nanogenerator with the cyclic experiments (Reprinted with permision from Shi et al., 2018. © Elsevier).

PVDF-TrFE polymer is the most promising candidate in the piezoelectric applications due to its self-poled nature with compare to other polymeric piezoelectric materials (PP, Nylon 11, PLA, and PVDF). Research is also growing extremely with the PVDF-TrFE-based polymer in recent days. Kim et al. (2018) has developed a multilayer-based piezoelectric nanogenerator by using PVDF-TrFE electrospun web in which they have used microbead-based electrode material for enhancing the piezoelectric performance of the PVDF-TrFE polymer. The microbeads in the electrode material provide a critical compression and tension force to the electrospun nanofibers so that nanofibers get deformed largely and respond to a larger piezoelectric effect from the multilayer piezoelectric nanogenerator. The PVDF-TrFE polymer-based aligned nanofiber has been generated by electrospinning method where drum-type collector has been used to collect the nanofiber on its surface

(Jiang et al., 2018). The effect of annealing of this electrospun nanofiber has been evaluated by measuring the piezoelectric effect of the nanofiber by using piezoelectric force microscopy. It has been found that the piezo-electric properties of the annealed fiber show maximum value with compare to the normal nanofibers due to the higher amount of β-phase nucleated in the annealing process. With connection of this research, Kim et al. (2019) have also developed the aligned electrospun nanofiber based on PVDF-TrFE polymer and investigated the piezoelectric performance with various annealing temperature of the PVDF-TrFE electrospun nanofibers. They have observed that the aligned and thermally annealed nanofiber showed the power density value of 1.3 mW/cm^3. Moreover, the piezoelectric performance of the PVDF-TrFE-based electrospun nanofiber also can be improved by using parallel electrode-based rotating drum as a collector during the electrospin-ning method. It is known that due to the rotating motion of the drum, orienta-tion of the dipole can be improved for which resultant output voltage also get increased. The parallel electrode electrospun nanofiber-based piezoelectric nanogenerator composed of PVDF-TrFE polymer showed the output voltage of ~12 V and current ~150 nA (You et al., 2019).

Besides, KNN BT filler, lead zirconium titanate (PZT) has also been used in the piezoelectric applications as the most usable filler material because of its higher piezoelectric property as mentioned in the introduc-tion part. Due to higher rigidity of the ceramic-based (PZT) piezoelectric materials, development of composite-based piezoelectric nanogenerators and their application have drawn a great interest in recent days. Lee et al. (2019) has reported a piezoelectric nanogenerator based on PZT electrospun fibers. The nanogenerator is developed by deposition of PZT nanofiber on the metal mesh for controlling the composite structure. Then this whole assembly has been covered by using the PDMS (polydimethylsiloxane) polymer to protect it from environmental issues. This nanogenerator showed an output voltage ~1.1 V and current ~1.4 μA under a bending strain on the nanogenerator.

2.4   CONCLUSIONS

In this chapter, the details of electrospinning method and electrospun nanofiber-based piezoelectric nanogenerators have been discussed. One can understand the basics of electrospinning method in terms of different parameters (solution, process, and ambient parameters) which have very important role on the fiber production and its properties. In addition, a detailed study on the different electrospun nanocomposite-based piezoelectric materials has been represented. It has been highlighted that electrospinning method is an emerging technology to develop flexible, light-weight, and highly efficient piezoelectric nanogenerator without extra-arrangements of required poling process to prepare a piezoelectric material in general unlike as demanded in case of solution cast and melt extrusion method to prepare either film or fiber. The output signal generated by the electrospun fiber-based nanogenerator shows better value which can be enough to power up different small-scale portable and wearable electronic gadgets that can reduce the problem of battery which gets exhaust very quickly and needs an external power source to recharge. Finally, this composition provides a brief idea about the electrospinning process and its use to produce nanofiber-based piezoelectric nanogenerator for energy harvesting.

ACKNOWLEDGMENTS

The authors are grateful to Science and Engineering Research Board (SERB), The Govt. of India for funding (File No. YSS/2014/000964) a research work (on nanocomposite-based flexible piezoelectric materials) from which some findings are reported here with proper references.

KEYWORDS

- **electrospinning**
- **nanocomposite**
- **piezoelectricity**
- **energy harvesting**
- **textiles**

REFERENCES

Bairagi, S.; Ali, S. W. A Unique Piezoelectric Nanogenerator Composed of Melt-Spun PVDF/ KNN Nanorod-Based Nanocomposite Fibre. *Eur. Polym.* **2019a,** *116,* 554–561.
Bairagi, S.; Ali, S. W. Effects of Surface Modification on Electrical Properties of KNN Nanorod-Incorporated PVDF Composites. *J. Mater. Sci.* **2019b,** *54,* 11462–11484.

Bairagi, S.; Ali, S. W. Influence of High Aspect Ratio Lead-Free Piezoelectric Filler in Designing Flexible Fibrous Nanogenerator: Demonstration of Significant High Output Voltage. *Energy Technol.* **2019c,** *7,* 1900538.

Bhardwaj, N.; Kundu, S. C. Electrospinning: A Fascinating Fiber Fabrication Technique. *Biotechnol. Adv.* **2010,** *28,* 325–347.

Kang, H.B.; Han, C.S.; Pyun, J.C.; Ryu, W.H.; Kang, C-Y; Cho, Y.S. (Na, K) NbO$_3$ Nanoparticle-Embedded Piezoelectric Nanofiber Composites for Flexible Nanogenerators. *Compos. Sci. Technol.* **2015,** *111,* 1–8.

Chang, Y. M.; Lee, J. S.; Kim, K. J. Heartbeat Monitoring Technique Based on Corona-Poled PVDF Film Sensor for Smart Apparel Application. *Solid State Phenom.* **2007,** *124–126,* 299–302.

Curie, J.; Curie, P. Développement par Compression de l'Électricité Polaire Dans les Cristaux Hémièdres à Faces Inclinées. *Bull. Soc. Minéral. France* **1880,** *3,* 90–93.

Deitzel, J. M.; Kleinmeyer, J.; Harris, D.; Beck Tan, N. C. The Effect of Processing Variables on the Morphology of Electrospun. *Polymer* **2001,** *42,* 261–272.

Fan, F. R.; Tang, W.; Wang, Z. L. Flexible Nanogenerators for Energy Harvesting and Self-Powered Electronics. *Adv. Mater.* **2016,** *28,* 4283–4305.

Fong, H.; Chun, I.; Reneker, D. H. Beaded Nanofibers Formed During Electrospinning. *Polymer* **1999,** *40*(16), 4585–4592.

Ghosh, S. K.; Mandal, D. High-Performance Bio-piezoelectric Nanogenerator Made with Fish Scale. *Appl. Phys. Lett.* **2016,** *109.*

Hu, X.; Yan, X.; Gong, L.; Wang, F.; Xu, Y.; Feng, L. Improved Piezoelectric Sensing Performance of P(VDF-TrFE) Nanofibers by Utilizing BTO Nanoparticles and Penetrated Electrodes. *ACS Appl. Mater. Interfaces* **2019,** *11,* 7379–7386.

Jiang, Y.; Gong, L.; Hu, X.; Zhao, Y.; Chen, H.; Feng, L.; Zhang, D. Aligned P(VDF-TrFE) Nanofibers for Enhanced Piezoelectric Directional Strain Sensing. *Polymer* **2018,** *10,* 1–12.

Karan, S. K.; Maiti, S.; Paria, S.; Maitra, A.; Si, S. K.; Kim, J. K.; Khatua, B. B. A New Insight towards Eggshell Membrane as High Energy Conversion Efficient Bio-piezoelectric Energy Harvester. *Mater. Today Energy* **2018,** *9,* 114–125.

Khalifa, M.; Mahendran, A.; Anandhan, S. Durable, Efficient, and Flexible Piezoelectric Nanogenerator from Electrospun PANi/HNT/PVDF Blend Nanocomposite. *Polym. Compos.* **2019,** *40,* 1663–1675.

Kim, H. S.; Kim, J. H.; Kim, J. A Review of Piezoelectric Energy Harvesting Based on Vibration. *Int. J. Precis. Eng. Manage.* **2011,** *12,* 1129–1141.

Kim, Y. W.; Lee, H. B.; Yeon, S. M.; Park, J.; Lee, H. J.; Yoon, J.; Park, S. H. Enhanced Piezoelectricity in a Robust and Harmonious Multilayer Assembly of Electrospun Nanofiber Mats and Microbead-Based Electrodes. *ACS Appl. Mater. Interfaces* **2018,** *10,* 5723–5730.

Kim, S.; Yoo, J.; Cho, Y. S. Flexible Piezoelectric Energy Generators Based on P(VDF-TrFE). *Mater. Res. Express.* **2019,** *6,* 086311.

Koski, A.; Yim, K.; Shivkumar, S. Effect of Molecular Weight on Fibrous PVA Produced by Electrospinning. *Mater. Lett.* **2004,** *58,* 493–497.

Lee, J. H.; Lee, K. Y.; Gupta, M. K.; Kim, T. Y.; Lee, D. Y.; Oh, J. Highly Stretchable Piezoelectric-Pyroelectric Hybrid Nanogenerator. *Adv. Mater.* **2014,** *26,* 765–769.

Lee, H.; Kim, H.; Kim, D. Y.; Seo, Y. Pure Piezoelectricity Generation by a Flexible Nano-generator Based on Lead Zirconate Titanate Nanofibers. *ACS Omega* **2019,** *4,* 2610–2617.

Li, J.; Chen, S.; Liu, W.; Fu, R.; Tu, S.; Zhao, Y. High Performance Piezoelectric Nanogenerators Based on Electrospun ZnO Nanorods/Poly(vinylidene fluoride) Composite Membranes. *J. Phys. Chem. C* **2019**, *123*, 11378–11387.

Maiti, S.; Kumar Karan, S.; Lee, J.; Kumar Mishra, A.; Bhusan Khatua, B.; Kon Kim, J. Bio-waste Onion Skin as an Innovative Nature-Driven Piezoelectric Material with High Energy Conversion Efficiency. *Nano Energy* **2017**, *42*, 282–293.

Niu, Y.; Yu, K.; Bai, Y.; Wang, H. Enhanced Dielectric Performance of $BaTiO_3$/PVDF Composites Prepared by Modified Process for Energy Storage Applications. *IEEE Trans. Ultrason. Ferr.* **2015**, *62*, 108–115.

Nour, E. S.; Khan, A.; Nur, O.; Willander, M. A Flexible Sandwich Nanogenerator for Harvesting Piezoelectric Potential from Single Crystalline Zinc Oxide Nanowires. *Nanomater. Nanotechnol.* **2014**, *4*, 24.

Parangusan, H.; Ponnamma, D.; Al Ali Almaadeed, M. Toward High Power Generating Piezoelectric Nanofibers: Influence of Particle Size and Surface Electrostatic Interaction of $Ce-Fe_2O_3$ and $Ce-Co_3O_4$ on PVDF. *ACS Omega* **2019**, *4*, 6312–6323.

Ponnamma, D.; Parangusan, H.; Tanvir, A.; AlMa'adeed, M. A. A. Smart and Robust Electrospun Fabrics of Piezoelectric Polymer Nanocomposite for Self-powering Electronic Textiles. *Mater. Design* **2019**, *184*, 108176.

Salimi, A.; Yousefi, A. A. Conformational Changes and Phase Transformation Mechanisms in PVDF Solution-Cast Films. *J. Polym. Sci. Pol. Phys.* **2004**, *42*, 3487–3495.

Shi, K.; Sun, B.; Huang, X.; Jiang, P. Synergistic Effect of Graphene Nanosheet and $BaTiO_3$ Nanoparticles on Performance Enhancement of Electrospun PVDF Nanofiber Mat for Flexible Piezoelectric Nanogenerators. *Nano Energy* **2018**, *52*, 153–162.

Teka, A.; Bairagi, S.; Shahadat, M.; Joshi, M.; Ziauddin Ahammad, S.; Wazed Ali, S. Poly(vinylidene fluoride) (PVDF)/Potassium Sodium Niobate (KNN)-Based Nanofibrous Web: A Unique Nanogenerator for Renewable Energy Harvesting and Investigating the Role of KNN Nanostructures. *Polym. Advan. Technol.* **2018**, *29*, 2537–2544.

Wang, Z. L.; Wu, W. Nanotechnology-Enabled Energy Harvesting for Self-Powered Micro-/Nanosystems. *Angew. Chem. Int.* **2012**, *51*, 11700–11721.

You, S.; Zhang, L.; Gui, J.; Cui, H.; Guo, S. A Flexible Piezoelectric Nanogenerator Based on Aligned P(VDF-TrFE) Nanofibers. *Micromachines* **2019**, *10*, 302.

Zhang, C. Yuan, X.; Wu, L.; Han, Y.; Sheng, J. Study on Morphology of Electrospun Poly (vinyl alcohol) Mats. *Eur. Polym.* **2005**, *41*, 423–432.

CHAPTER 3

Nanopretreatments for Textile: Nanoscouring, Nanobleaching, Nanosoftening, and Nanosurface Activation

SUBHANKAR MAITY[1*], KUNAL SINGHA[2], and PINTU PANDIT[2]

[1]*Department of Textile Technology, Uttar Pradesh Textile Technology Institute, Kanpur, India*

[2]*National Institute of Fashion Technology, Department of Textile Design, Ministry of Textiles, Govt. of India, NIFT Campus, Mithapur Farms, Patna, India*

Corresponding author. E-mail: maity.textile@gmail.com

ABSTRACT

With the advent of nanoscience nanotechnology, the emergence of new field of research and innovations in textile wet processing is well known; these processes are called nanopretreatment and nanofinishing. Pretreatment processes of textiles, such as scouring, bleaching, softening, and surface activation, finish, such as like self-cleaning, antimicrobial, water repellence, flame retardant, etc. can be achieved by the application of various nanomaterials and processes. Applications of nano-TiO_2 and nano-ZnO in textiles by various means are some of the common nanofinishing methods established to develop many such qualities and performance of textiles. Nanosilver, nanosililca, carbon nanotubes, maghemite iron oxide nanoparticles, etc. are potential nanomaterials explored for acquiring various functional properties of the treated textiles without significant deterioration of mechanical properties. This chapter deals with various nanopretreatment processes, such as

Fundamentals of Nano-Textile Science. Prashansa Sharma, Devsuni Singh & Vivek Dave (Eds.)

nanoscouring, nanobleaching, nanosoftening, and some nanosurface activation processes, characterization, and performance of the treated textiles.

3.1 INTRODUCTION

Nanoscience and nanotechnology are the new fields of study and research in materials science where various nanomaterials are employed for the development of new materials with much better functionality, durability, and performance. In the field of textile wet processing, there are pretreatment and finishing methods of textiles where prominent scope of nanotechnology has been established. Scouring is the most important pretreatment process of textile where surface impurities, oil, and wax are removed from fiber surface so that they become suitable for absorbing moisture, dye, and other finishing agents. Bleaching is another process during which process the textile fibers become whiter. There are many textile-finishing processes by which special properties are imparted on textile cloths, such as antimicrobial, self-cleaning, UV protection, fire retardant, easy care, softening, etc. Deviating from the conventional pretreatment and finishing processes, nanotechnology routes can show us alternative methods by which we can obtain the treated textiles with much better functionality, performance, and durability. This chapter deals with the scouring and bleaching processes of textile materials in nanotechnology route that is called nanopretreatment, and some textile nanofinishing properties, such as UV protection, antimicrobial, self-cleaning, easy care and softening treatment.

Natural fibers, such as cotton, wool, silk, etc. have impurities, such as natural waxes, pectin, minerals, proteins, ashes, seed coats, etc. Synthetic fibers have added minerals and oils during their manufacturing. All these surface impurities need to be cleaned and removed from the fiber surface for improving hydrophilicity and surface adhesion. It is really difficult to remove surface wax of cotton fiber completely though it is desirable. It is well practiced to remove other impurities from cotton by conventional scouring. The traditional process of scouring involves the treatment of fibers in certain alkali solution at high temperature for long duration. During the process, fats and waxes are either converted into water-soluble matters or emulsified into water emulsifiable matters, oils are saponified, pectins are converted into soluble salts of pectic acid, proteins and amines are converted to soluble amino acids and ammonia, and minerals become water-soluble. Dedicated machines are available for textile industry for scouring. Kier and Jiggers are the suitable machines for batch scouring. More recently known machine for scouring is J-box which is used for continuous scouring operation in the industry. Regarding the chemicals used for scouring, soda lime was the cheapest alkali used for cotton scouring that has insignificant detrimental effect on fiber properties. Now, sodium hydroxide and sodium carbonate are used in mixtures for both batch and continuous process. Along with these alkalis, various anionic or nonionic detergents,such as sodium alkyl sulfate, polyethoxylated compounds, etc. and auxiliaries are added to improve the wax removal process. The auxiliaries are used for various purposes, such as preventing hard water problems, reducing surface tension, increasing disso-lution of insoluble fats and waxes. Even though the alkali scouring can effi-ciently remove almost all surface waxes and impurities, the process requires high energy, water and generates undesirable effluents that are harmful to the environment. Also, alkali in such adverse treatment condition, oxidizes cellulose to produce oxycellulose, thus is deteriorating the tensile strength of cotton. Therefore, alternative eco-friendly processes are adopted and proposed by various researchers. Bioscouring is an alternative process that is conducted at cold temperature with minimal effluent generation (Choe et al., 2004). Bioscouring is conducted with enzymes that have specific action on target molecules of the impurities existing on fiber surface. However, the process requires either long time or requires elevated temperature which challenges the merits of the process and hence is commercially not accepted. Recently, this enzymatic scouring is performed with detergents for better effect and it demands the replacement of alkali scouring (Hasanbeigi and Price, 2015). However, the use of detergents is not desirable from an envi-ronmental point of view. All these demerits and limitations can be resolved by the nano-pretreatment route.

3.2 NANOSCOURING

The quality and durability of dyeing, printing, and surface finishing of textile fibers depend on proper pretreatment so that the textile surface becomes hydrophilic. In this respect, scouring is a vital pretreatment process in which various natural impurities, such as wax, oil, fat, gum, foreign matters, addi-tives, etc. are removed from the surface of the fibers to improve the hydro-philicity of the surface. As a result, absorption or affixation of dye molecules and other functional finishes become convenient and durable. The scouring process is an essential pretreatment process for all natural fibers, such as cotton, wool, silk, jute, flax, etc. where there is natural waxy coating on the fiber surface, and removal of same is difficult. In the case of man-made fibers, oils are manually applied on the surface of man-made fibers as spin finishes to improve the spinning process. Removal of such applied oils is easier. Nanotechnology is the suitable alternative process for scouring of textile fibers as discussed in the following section.

3.2.1   NANOSCOURING OF COTTON

Nanosized TiO_2 particles are the best nanoparticles explored in materials science due to their low bandgap energy. When TiO_2 nanoparticles are irradiated with light rays possessing energy higher than their bandgap, electrons and hole pairs are liberated from their surface. This phenomenon is called photocatalytic effect. Attributing to this photocatalytic effect, TiO_2 has been explored for various applications in textiles, such as water purification, UV protection, self-cleaning, antibacterial effect, etc. The same photocatalytic event of TiO_2 nanoparticles can be used for the decomposition of natural oil and wax impurities and coloring substances from gray cotton fabrics to improve whiteness and hydrophilicity. The process is termed as nanophotoscouring. The fabrics loaded with nano-TiO_2 also have added benefits like self-cleaning, antimicrobial, and UV protection (Montazer and Morshedi, 2012). In this process, an aqueous solution of nano-TiO_2 powder (0.5–10g/L) is prepared by adding 10% citric acid (CA), 6% sodium hypophosphite (SHP), and 0.05% dispersing agent and by sonication for 5min. Desized cotton fabrics are impregnated with this dispersion solution at 40: 1 M: L ratio and padded which is followed by drying and curing at high temperature. The cured cotton fabrics are then either irradiated under UV-A lamps for 10min or exposed in direct sunlight for 7days. During the exposure to UV rays, negative electron and positive holes' pairs are generated on nano-TiO_2 particles which can react with atmospheric oxygen and water molecules to produce superoxide and hydroxyl radicals. These active radicals on the surface of nano-TiO_2 particles are able to decompose the surface wax of cotton fibers and thereby enhance the hydrophilicity. In this manner, cotton fibers become efficient in absorbing water more quickly, as evidenced by contact angle meter. Sunlight-irradiated fabrics are found to be more hydrophilic than the UV-A lamp-irradiated fabrics (Montazer and Morshedi, 2012). Because, the broad range of UV-B, UV-A, and visible wavelengths are present in the sunlight that may result in higher activity of nano-TiO_2 particles as compared with UV-A alone. This hydrophilicity of cotton fabrics is attributed to the decomposition of the surface wax through photocatalytic action of the TiO_2 nanoparticles under sunlight.

3.2.2 NANOSCOURING OF WOOL AND SILK

Nanophotoscouring is also applicable for protein fibers, such as wool and silk, because the same technique is also useful for the decomposition of natural wax and other hydrophobic substances from the surface of these protein fibers. Wool has natural waxes which consist of esters of various long-chain fatty acids having long-chain alcohols and steroids. Esters of 1- and 2- alkanols and 1,2-diols, triterpene alcohols, sterol esters, and free acids and sterols are available in the wool wax. The photocatalytic decomposition of the esters occurs during sunlight irradiation and as a result, esters are converted into aldehyde and alcohols first and further decomposed until removed from the wool surface. Various oxidizing radicals, such as OH^*, RO^*, and RO_2^* which are generated on the surface of TiO_2 during sunlight irradiation are responsible for the photocatalytic decomposition (Gauthier et al., 2006).

ZnO nanoparticles are also explored for their use in scouring of wool fabrics where these are in situ synthesized on wool surface. During this process, zinc acetate is dissolved in water and water/ethanol media in various concentrations (10%–30%) with constant stirring at 30°C and then wool fabrics are immersed in the solutions followed by the addition of ammonia and NaOH solution one by one drop wise at a temperature of 90°C till 1h. As a result, zinc acetate is converted to $Zn(OH)_2$, and thereafter, wool fabrics are taken out of the solution and heated at 80°C for 10h to convert $Zn(OH)_2$ to ZnO by dehydration. In this manner, ZnO nanoparticles are in situ synthesized over raw wool surface and treated wool fabrics are kept at sunlight for 7days for obtaining scoured and hydrophilic wool fabrics. The water contact angle measured in a contact angle tester and water absorption time of various ZnO-treated samples are given in Table 3.1 (Montazer et al., 2013). The results depict that the nanophotoscouring process is very effective for scouring of wool fabric, and due to which wool surface becomes hydrophilic. Water contact angle of untreated fabric was 85 degree and that is reduced to 0 degree in the case of samples prepared in water/ethanol media. Water absorption time of untreated wool fabric was 7200s and that is reduced to 0s in the case of all treated fabrics.

The nano-ZnO particles generate positive holes and electrons under UV irradiation present in the sunlight. The electrons are responsible for the reduction reactions of Zn^{2+} to Zn^{1+} and generation of the anions of oxygen (O^{2-}). Whereas, the holes are responsible for the oxidation reaction of water molecules and the anions of oxygen (O^{2-}) and hydroxyl (-OH) groups. The generation of hydrophilic hydroxyl groups improves the hydrophilic properties of the wool fabrics as shown in Figure 3.1. Therefore, the photocatalytic activity generates abundant hydroxyl groups on the wool surface resulting in improved hydrophilicity and smaller water contact angle.

TABLE 3.1   Effects of Nanophotoscouring of Wool by In Situ Synthesized ZnO Nanoparticles.

Sample	Contact angle (°)	Water absorption time (s)
Untreated	85	7200
Wool fabric treated with 10% zinc acetate in water solution	8 6	0
Wool fabric treated with 20% zinc acetate in water solution	7.5	0
Wool fabric treated with 30% zinc acetate in water solution	7.2	0
Wool fabric treated with 30% zinc acetate in water/ethanol (50:50) solution	0	0
Wool fabric treated with 30% zinc acetate/ethanol (25:75) in water solution	0	0

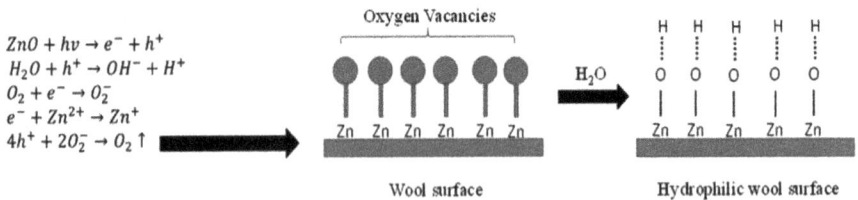

FIGURE 3.1   Generation of hydrophilic surface of wool by ZnO nanoparticle-induced nanophotoscouring.

Hydrophilicity study is conducted for the wool fabrics after the nanoscouring process and the results obtained are shown in Figure 3.2. It is observed that water absorption time becomes shorter after all types of treatment processes, and among them, best result is obtained in the case of sample number 6 which is the wool fabric treated with 5g/L TiO_2 dispersion solution prepared in the presence of 10% citric acid (CA) which is used as a binder and 6% sodium hypophosphite (SHP) which is used as a catalyst. For this sample, water absorption time decreases to 192s from 7200s for raw wool fabric (Montazer et al., 2013, Hashemikiaa and Montazer, 2012).

3.3 NANOBLEACHING

Though surface wax, oil, and impurities can be removed by scouring process, there are coloring matters still remain on the fiber surface contributing a yellow to brown appearance of the fibers. These coloring matters need to be destroyed to obtain whiter appearance without causing any detrimental effect to the fiber structure and its properties. Strong oxidizing or reducing agents are used conventionally for destroying coloring matters from the surface of cotton fiber to obtain bleached white fibers. Combining of scouring and bleaching processin strong alkaline solution is also practiced. Hypochlorite bleaching which is the most common in the industry, and is suffering from few limitations, such as the disability of obtaining desirable whiteness, demand of fully scoured fiber as starting material, loss of tensile strength, corrosion to the equipment, unpleasant chlorine odor, and generation of harmful chlorine bi-products. Peroxide bleaching is costly and is not popular for cotton, though it is suitable for wool. Peroxide bleaching is conducted in alkaline pH and it causes partial degradation to wool. Sulfur dioxide and sodium bisulfite bleaching of wool produce a temporary bleaching effect. Use of strong pH solution, oxidizing/reducing agent makes textile fiber harsh in feel and deteriorates the mechanical properties. Bleaching in the route of nanotechnology can be an effective alternative for mitigating those limitations of the existing bleaching methods.

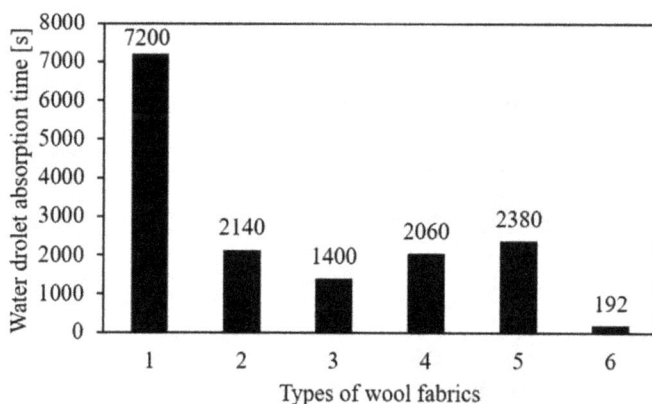

FIGURE 3.2   Water adsorption time of wool fabrics: (1) raw wool, (2) 6g/L H_2O_2 treatment in acidic media, (3) 6g/L H_2O_2 treatment in alkaline media, (4) treated with 0.25g/L TiO_2/10% citric acid (CA)/6% sodium hypophosphite (SHP), (5) 0.25g/L TiO_2/10% CA/6% SHP, (6) treated with 5g/L TiO_2/10% CA/6% SHP. (Reprinted from Montazer, M.; Morshedi, S. Photo Bleaching of Wool Using Nano TiO_2 under Daylight Irradiation. *J. Ind. Eng. Chem.* **2014**, *20*, 83–90).

3.3.1 NANOBLEACHING OF COTTON

Scoured cotton fabrics need to be coated with nano-TiO_2 particles and are kept under sunlight or UV irradiation in the laboratory. As a result, oxidizing radicals produced by the photocatalytic effect of TiO_2 in the presence of UV rays attack the coloring matters, pectin, and hemicellulose possessing polysaccharide and cause the breakdown of polymers. The bleaching mechanism by discoloring the natural pigments on natural fibers is attributed to the generation of oxidizing radicals, such as OH^*, RO^*, and RO_2^* by the TiO_2 nanoparticles. The scheme of reactions is given in Equations 1–15. This mechanism of bleaching is called nanophotobleaching and is shown in Figure 3.3. It has been reported that the whiteness of the desized cotton fabrics improved significantly after the nanophotobleaching. Because, the natural tints, coloring matters of cotton fibers, including pectin and other impurities are destroyed from the fiber surface to make fiber cleaner and whiter under UV radiation and daylight. The photocatalytic effect is proved to be harmless on the cellulosic chain of the cotton fiber, because the oxidizing radicals cannot oxidize high molecular weight cellulosic molecules. It is also reported that the daylight irradiation produced a whiter cotton fabric than the artificial UV-A irradiation. This is because sunlight contains a broad range of UV-B, UV-A, and visible wavelength of lights which may play a role for better activation of nano-TiO_2 particles as compared with UV-A alone. While comparing the bleaching efficiency by this route with the conventional peroxide bleaching, it is observed that photobleached cotton fabrics with 7g/L TiO_2 loading have similar whiteness index as compared with the H_2O_2 (0.5g/L) bleached cotton fabrics (Montazer and Morshedi, 2012).

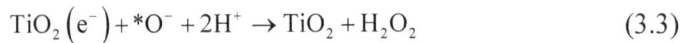

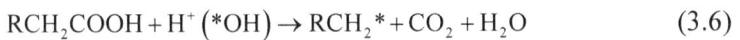

$$TiO_2 + hv \rightarrow TiO_2\left(e^- + h^+\right) \tag{3.1}$$

$$TiO_2\left(e^-\right) + O_2 \rightarrow TiO_2 + {}^*O^- \tag{3.2}$$

$$TiO_2\left(e^-\right) + {}^*O^- + 2H^+ \rightarrow TiO_2 + H_2O_2 \tag{3.3}$$

$$TiO_2\left(e^-\right) + H_2O_2 \rightarrow TiO_2 + {}^*OH + OH^- \tag{3.4}$$

$$TiO_2\left(h^+\right) + H_2O_{ads} \rightarrow TiO_2 + {}^*OH_{ads} + H^+ \tag{3.5}$$

$$RCH_2COOH + H^+\left({}^*OH\right) \rightarrow RCH_2{}^* + CO_2 + H_2O \tag{3.6}$$

$$RCH_2{}^* + O_2 \rightarrow RCH_2OO^* \tag{3.7}$$

$${}^*OH + H_2O_2 \rightarrow H_2O + {}^*OOH \tag{3.8}$$

$$RCH_2O_2* + *OOH \rightarrow RCH_2OOOH \qquad (3.9)$$

$$RCH_2OOOH + RCH_2COOH \rightarrow RCH_2OH + RCHO + O_2 + CO_2 \qquad (3.10)$$

$$RCOOCH_3 + h^+ (*OH) \rightarrow RCH_3 + CO_2 + H_2O \qquad (3.11)$$

$$RCH_2OH + *OH \rightarrow RCHOH* + H_2O \qquad (3.12)$$

$$RCHOH* + H_2O \rightarrow RCH(OH)_2 + *H \qquad (3.13)$$

$$RCH(OH)_2 \rightarrow H_2O + RCHO \qquad (3.14)$$

$$RCHO + *OH \rightarrow RCO + H_2O \rightarrow RCOOH + *H \qquad (3.15)$$

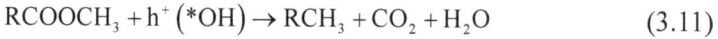

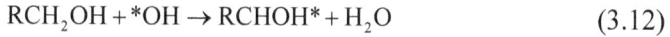

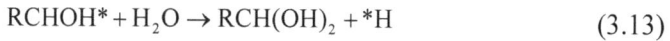

In fact, this nanopretreatment process gives us a combined effect of scouring and bleaching. In addition, other characteristics, such as self-cleaning, UV protection, and antibacterial effects are also obtained simultaneously. But, the demerits of the process include (1) deterioration of the tensile strength of cotton fabric due to the treatment in acidic condition and (2) it is a time-consuming process. In another process, cotton fabrics are treated with nanotitanium tetraisopropoxide and acetic acid in an ultrasonic bath (50kHz, 50W) for 4h at 75°C as shown in Figure 3.4. This sonochemical process is reported to be effective for the formation of covalent bonds between the –OH groups of cotton fiber and TiO_2. The advantage of this process is that no deterioration of tensile strength of cotton occurs due to the treatment condition (Akhavan and Montazer, 2014).

3.3.2 NANOBLEACHING OF WOOL AND SILK

Wool fabrics are used for photobleaching. Photobleaching is done by treatment of wool fabrics with nanocrystallized TiO_2 particles (Montazer and Pakdel, 2010). In this process, aqueous dispersion solutions of TiO_2 nanoparticles of various concentrations from 0.001 to 0.075g/L are prepared by mixing sodium hypophosphite (SHP), and citric acid (CA) as cross-linking agents with sonication of the mixture for several minutes in an ultrasonicator. Wool fabrics are treated with the solution for 32min and then cured at 120°C for 2min. The nanophotobleaching process is conducted by exposing the fabric samples under UV-C rays for 3h. A decreasing phenomenon of photoyellowing of wool fabrics is observed due to this nanopretreatment process. TiO_2 particles absorb UV-C rays and prevent photooxidation of wool which is responsible for the yellowing. Figure 3.5 shows the decrease of photoyellowing of wool fabric due to increase of TiO_2 concentration (Montazer and Pakdel, 2010).

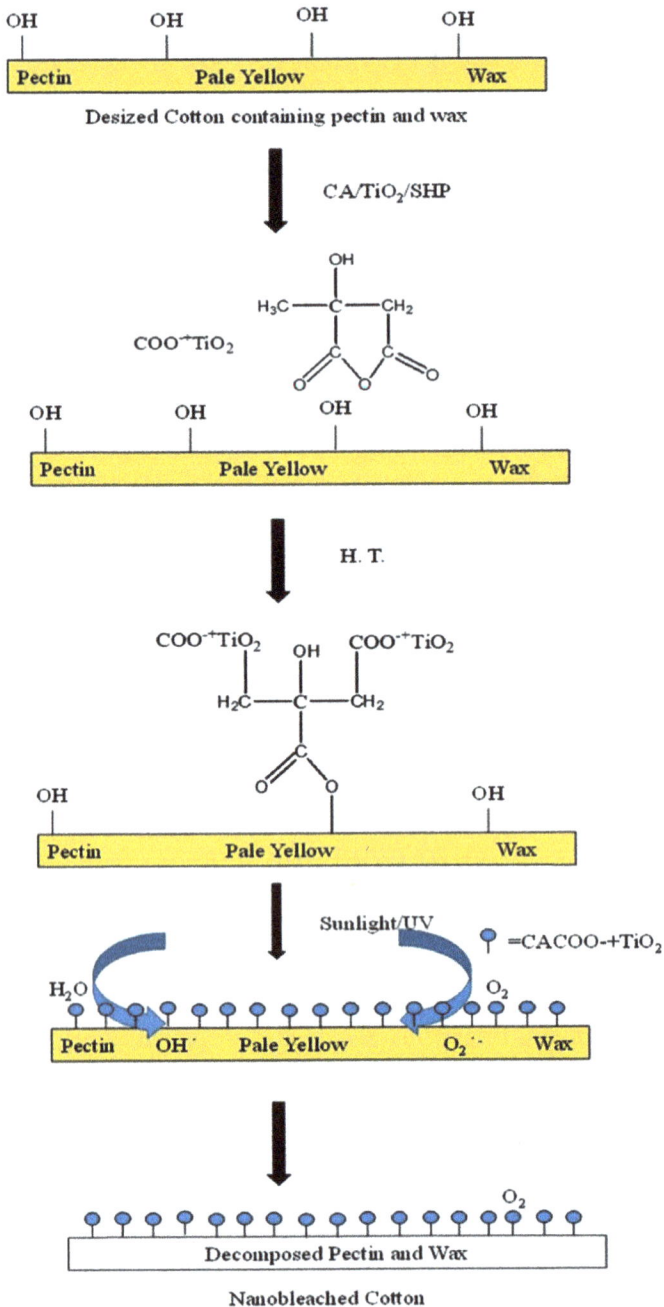

FIGURE 3.3   The schematic presentation of nanophotoscouring and nanophotobleaching of cotton fiber by TiO$_2$ nanoparticles.

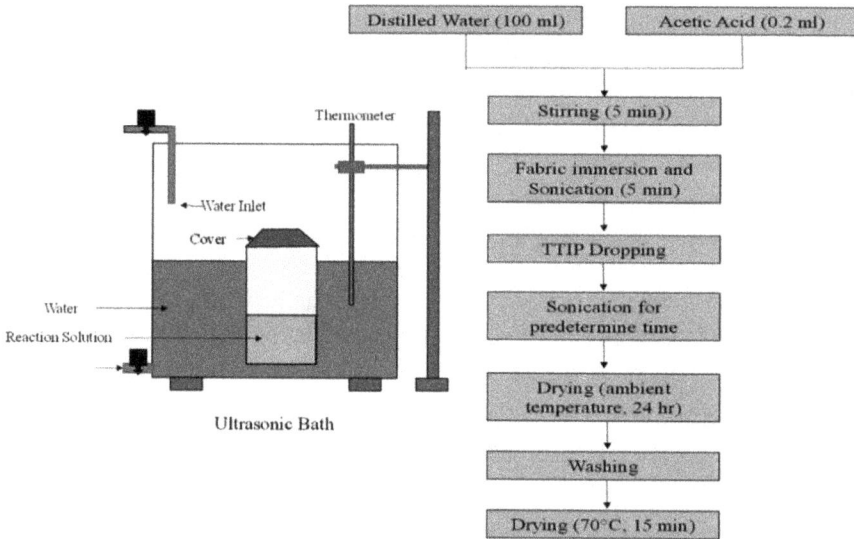

FIGURE 3.4 Ultrasonic bath sonication set-up for in situ sonosynthesis of TiO$_2$ nanoparticles on cotton fabric. (Reprinted with permission from Akhavan, F.; Montazer, M. In Situ Sonosynthesis of Nano TiO$_2$ on Cotton Fabric. *Ultrason. Sonochem.* **2014**, *21*, 681–691)

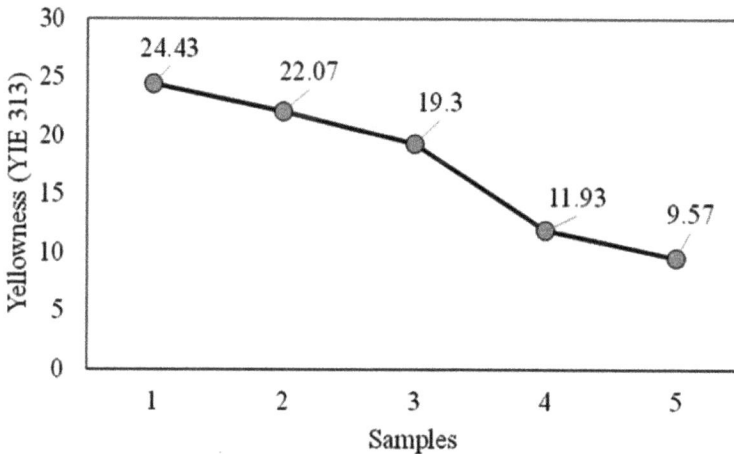

FIGURE 3.5 Yellowness Index of various wool fabrics after 3h UV irradiation by artificial UV-C sources: (1) raw wool; (2) oxidized wool; (3) wool treated with 0.001g/L) TiO$_2$; (4) wool treated with 0.038g/L TiO$_2$; and (5) wool treated with 0.075g/L TiO$_2$.

Nanophotobleaching of wool fabrics is demonstrated by ZnO nanoparticles which are in situ synthesized on fabric surface from the precursor of zinc acetate in water and ethanol media (Montazer et al., 2013). In wool fibers, the natural pigments are eumelanin and pheomelanin which cause the wool to have cream or pale yellow color. When UV rays of energy higher than the bandgap of ZnO incident on the ZnO particles, they generate electrons in their conduction band and holes in their valence band. Holes possess strong oxidation potentials and oxidize water molecules to produce hydroxyl radicals (*OH). Whereas, electrons do have strong reduction potential and reduce oxygen molecules into the anions of superoxide radicals. These hydroxyl radicals and anions of superoxide radicals destroy various coloring substances present on the wool surface. The scheme of reaction is given in Equations 16–27. In this way, photobleaching occurs and yellowness index of wool fabric decreases. The yellowness index of the untreated fabric is 35.32 and that is found reduced to 23.84 after the nanophotobleaching by ZnO in situ synthesized in water/ethanol media (Montazer et al., 2013).

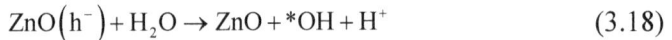

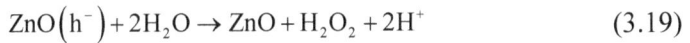

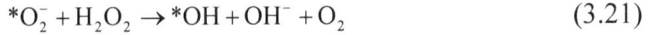

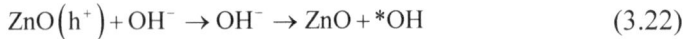

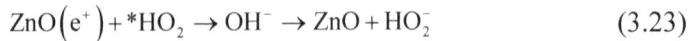

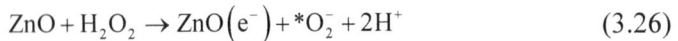

$$ZnO + hv\left(e^- + h^-\right) \tag{3.16}$$

$$ZnO\left(e^-\right) + O_2 \rightarrow ZnO + {}^*O_2^- \tag{3.17}$$

$$ZnO\left(h^-\right) + H_2O \rightarrow ZnO + {}^*OH + H^+ \tag{3.18}$$

$$ZnO\left(h^-\right) + 2H_2O \rightarrow ZnO + H_2O_2 + 2H^+ \tag{3.19}$$

$${}^*O_2^- + H^+ \rightarrow {}^*H_2O \tag{3.20}$$

$${}^*O_2^- + H_2O_2 \rightarrow {}^*OH + OH^- + O_2 \tag{3.21}$$

$$ZnO\left(h^+\right) + OH^- \rightarrow OH^- \rightarrow ZnO + {}^*OH \tag{3.22}$$

$$ZnO\left(e^+\right) + {}^*HO_2 \rightarrow OH^- \rightarrow ZnO + HO_2^- \tag{3.23}$$

$$HO_2^- + H^+ \rightarrow H_2O_2 \tag{3.24}$$

$$2HO_2 \rightarrow O_2 + H_2O_2 \tag{3.25}$$

$$ZnO + H_2O_2 \rightarrow ZnO\left(e^-\right) + {}^*O_2^- + 2H^+ \tag{3.26}$$

$$ZnO\left(e^-\right) + H^2O_2 \rightarrow ZnO + {}^*OH + OH^- \tag{3.27}$$

Melamine which is thought to be responsible for the pale yellow color of animal fibers are also disintegrated and decomposed by TiO_2 nanoparticles in the presence of sunlight (Montazer and Morshedi, 2014). The decomposition mechanism of melamine is shown in Figure 3.6. The decomposition of melamine occurs due to ring opening reaction which is caused due to the neucleophilic attack by OOH⁻ ions generated in the course of photocatalytic reaction (Montazer and Morshedi, 2014). As a result, the whiteness index of the textile material improves and bleaching effect is achieved. It is observed that the photobleached wool fabrics treated with nano-TiO_2 particles are whiter than the bleached fabrics obtained by hydrogen peroxide treatment.

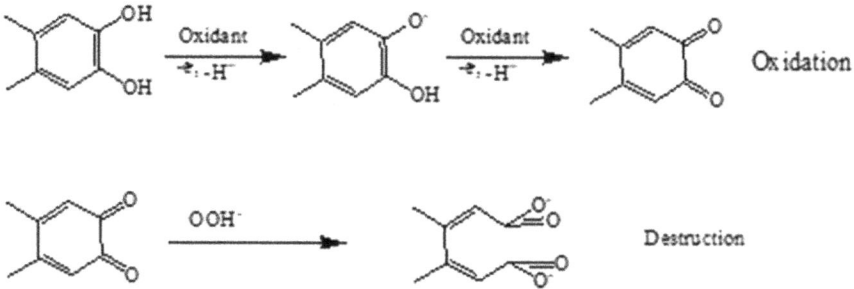

FIGURE 3.6 Decomposition reaction of melamine by the attack of TiO_2 in the presence of sunlight.

3.4 NANOBIOPHOTOSCOURING

Some biocatalysts like enzymes are explored for the enhancement of photo-catalytic activity of weak photocatalyst like α-Fe_2O_3. As a result, the hybrid enzyme system becomes active in visible light illumination and produces the required photocatalytic cleaning of surface impurities. One such attempt has been made by Kamad and Sohb (2015), they used α-Fe_2O_3 in conjunction with horseradish peroxidase (HRP, MW 42,000), a biocatalyst which is acting as redox enzyme also. This hybrid catalyst system was experimented for the oxidation of luminol as a model reaction. Under visible light illumination, bare α-Fe_2O_3 exhibits poor photocatalytic oxidation of the same. However, the incorporation of immobilized HRP in the catalyst system improves the catalytic effect induced by visible light, and it was found that α-Fe_2O_3 is efficient in breaking down of luminol (Kamad and Sohb, 2015). In view of the above, there is scope for exploring various enzymes for enhancing the photocatalytic effects of nanoparticles and the method is called nanobiophotoscouring.

3.5 NANOFINISHING AND NANOSURAFCE ACTIVATION

With the advent of nanoscience and nanotechnology, a new field of research has taken birth in the area of textile finishing which is called as "nanofinishing." The nanofinishing methods have opened up various new approaches to improve the existing finishing processes. In the course of time, new functional textile products have been developed with active surface actions that can serve one or few specific properties with much better efficiency and durability. Textile surface can be activated by embedding various nanomaterials, such as nanosized ZnO, TiO_2, SiO_2 Al_2O_3 Ag, and carbon nanotubes, etc. By embedding these nanomaterials on textile surface, various functional effects can be achieved. Depending upon the types of nanomaterials and their degree of loading, such activated textiles can possess various properties, such as self-cleaning, softening, water repellency, UV resistance, antimicrobial, etc.

3.5.1   NANOCLEANING

It is observed that the cotton fabrics coated with the nanocrystalline TiO_2 can remove various stains and dirts, and this action is attributed to the consequences of the photocatalytic effect of TiO_2 under sunlight. It is reported that TiO_2-loaded cotton textiles eliminate the stains of wine, coffee, tea, methylene blue, other dye molecules, etc. by destroying the chromophore(s) of the stains under daylight irradiation (Meilert et al., 2005; Montazer and Seifollahzadeh, 2011; Hashemikiaa and Montazer, 2012; Akhavan and Montazer, 2014). The self-cleaning ability of the nano-TiO_2-coated textiles is attributed to the highly oxidative intermediates produced on the surface of the textiles. The mechanism of dismissing of the stains and release of CO_2 due to light irradiation is given in Equation (28).

$$C_xH_yN_yS_W + hv + H_2O_w + O_2 \rightarrow CO_2 + H_2O + SO_p + NO_q \qquad (3.28)$$

The visible light-induced self-cleaning ability of cotton textiles has been evidenced by a simultaneous coating of the film of TiO_2 nanoparticles and AgI particles on their surface (Wu and Long, 2011). In comparison with the solo TiO_2- treated cotton, it is observed that the performance of AgI–Nano-TiO_2-coated cotton fabrics is significantly better under visible light exposure experiment. The superior performance is attributed to the synergistic effect of AgI and TiO_2 nanoparticles. This photocatalytic activity and performance by AgI/TiO_2 nanoparticles for cotton textiles are found to be maintained upon several numbers of photocatalytic cycles (Wu and Long, 2011).

3.5.2 NANOSOFTENING

Softness is the most vital quality of the textile cloths from consumers' point of view. The comfort of the clothing is mostly associated with the softness of the same and purchase decision of the garment by the user solely depends on the softness of the cloth. There are household softeners available in the retail market in the form of detergents, soaps, or softener itself, which are added to the fabric during washing or after wash to get the gentle feel of the fabric and make them more comfortable to wear by the consumer. These treatments are temporary and require to be applied during each washing cycle. Semipermanent and permanent softeners are used in textile industry to obtain cost-effective softening treatment, which are wash-durable, resistant to high temperatures, nonyellowing, and nondetrimental for dye-shade. Like nanopretreatments and nanofinishing methods, nanosoftening methods are also available through the science of nanotechnology. The advantage of nanosoftening agents is that their nanometer range of size enables them to easily penetrate into fiber structure and resulting in a plasticizing action by enhancing the molecular mobility and brings down the glass transition temperature of the fiber. As a result, the mechanical response of the fiber becomes easier at relatively low temperature and the material behaves softer. In this regard, nanoemulsions of silicone-based softeners have been successfully applied on textile materials for achieving high levels of softness. By another approach, softness of the textile materials can be achieved by reducing fiber to fiber friction. In this approach, nanoplatelets, such as nanolayers of silicate (nanoclay) or nanographene platelets are applied on the textile surface which form their nanolayers and reduce the friction at the fiber surface.

3.5.2.1 SOFTENING WITH NANOCLAY

Clays are available in nature as phyllosilicates. Montmorillonite (MMT) is one such clay available in nature with a chemical structure $(Na, Ca)_{0.33}(Al, Mg)_2(Si_4O_{10})(OH)_2nH_2O$. It has a 2:1 layered sheet-like structure and the sheets contain negative charges in the interlayer spaces as well as on the surface which can be neutralized by cations, such as Na^+, Ca^{2+} etc. Thus, such clays possess ion-exchanging property. This ion-exchange property can be tailored as per the requirement of improving ion absorption capacity and catalytic ability. These clays are also superior antimicrobial agents due to their nontoxic and environment-friendly characteristics. Nowadays, clays are manufactured in nanosized particles for enhancing the functionality and incorporating the same in textile materials by various means. The popular incorporation process is either adding the nanoclay particles in melt spinning dope with textile polymers or preparing a dispersion solution of the same and coating textiles by dip-dry-cure process (Asadi and Montazer, 2013). These nanoclay-treated fabrics are found suitable for various functional applications for their characteristics, such as flame retardant, UV protection, and importantly antimicrobial property of the denim fabrics without back-staining. The treatment can also improve the handle of the garment, making the garment softer (Sadeghian et al., 2013a, 2013b, 2015).

3.5.2.2   SOFTENING WITH NANOSILICONES

Silicones are the conventional softening agents for garments, and recently, nanosilicons which are the nanosized particles are explored extensively for the softening treatment. Depending upon the particle size, various silicone emulsion solutions can be prepared. Milky solution of macro particles of size 150–300nm, hazy solution of semimacro particles of size 80–120nm, and transparent solution of microparticles of size less than 40nm are prepared. The microemulsion of silicones is prepared by grafting hydrophilic amino groups in the primary silicone chain, and as a result, very good shear stability can be achieved. Nanoemulsion of silicones is produced with the particle size less than 10nm and which is the most advanced application in the softening treatment of garments. These nanosilicone particles can easily penetrate in textile substrates and provide excellent softness (Roy Choudhury et al., 2012). The nanosilicone particles enhance lubrication at the fiber level of the garments and improve softness and resist pill formation (Celik et al., 2010).

3.5.3   UV PROTECTION

The UV blockers used for the surface activation of textiles are oxides of some metals and semiconductors like ZnO, TiO_2, SiO_2 and Al_2O_3, etc. TiO_2 and ZnO nanomaterials are mostly explored for this purpose. These nanoparticles have large specific surface area and are capable of absorbing UV rays and scattering the same in the form of visible wavelength. Thus, textile materials are able to protect us from harmful UV radiation (Wong et al., 2006). The ability of UV blocking of cotton fabrics is brought about by coating the fabrics with a nanofilm of TiO_2 by employing a sol–gel method. This treatment is found durable and can bear up to 50 laundering cycles without loss (Viswanath and Ramachandran, 2010). In another study, ZnO nanorods of 10–50nm in length are assembled on the surface of cotton fabrics and as a result, the fabric is able to restrict UV rays. These cotton fabrics are observed as excellent performer in terms of various functional properties including UV protection. The presence of ZnO nanorods over cotton surface did not deteriorate physical and mechanical properties of the fabric. It is reported that only 2% loading of such ZnO nanorods is sufficient to block 75% of the UV radiation. But, these ZnO nanorods have poor chemical stability and gradually dissolved in water. In comparison, TiO_2 particles have superior chemical stability and are insoluble in water. The major limitation of the TiO_2 nanoparticles is their narrower range of UV absorption capability. However, as compared to conventional methods, such nanofinishes have better durability attributing to large specific surface area of the nanoparticles resulting in their high surface energy that is responsible for their enhanced affinity with textile substrates (Mankodi and Agarwal, 2011).

3.5.4 WATER REPELLENT AND EASY CARE FINISH

Easy care hydrophobic nanofinish of cellulosic fabrics can be possible by nanocoating of TiO_2 particles. If the contact angle subtended by a fluid droplet on the solid surface is less than 90 degree, the solid surface is termed as a hydrophilic surface corresponding to the same fluid. When the contact angle is >90 degree, the surface is termed as a hydrophobic surface corresponding to the fluid. When the contact angle is >150 degree, the surface becomes superhydrophobic. A superhydrophobic surface creates the water or oil repellency leading to self-cleaning effect. The self-cleaning textiles can be prepared by coating with some functional nanomaterials by the route of nanotechnology. A superhydrophobic surface is created by this process with a contact angle of 120 degree–160 degree. As a result, water molecule could not wet the surface and a peach fuzz effect is achieved. This finish does not deteriorate the mechanical properties of the treated fabric. In this process, Ag nanoparticles are applied on the textile surface. The silver particles destroy various organic compounds, such as dirt, contaminants, as well as microorganisms resulting in minimal washing of the cloths. The Ag nanoparticles exhibit water repellency effect by creating nanowhiskers over textile surface which are made of hydrocarbons and have about 1/1000th of the size of a typical cotton fiber. Ag nanoparticles are also used in conjunction with TiO_2 particles to coat over textile surface either in colloidal form or in particular form for improved functionality. The fixation of the particles on textile surface can be improved by high temperature curing. The high temperature curing treatment on cotton or polyester fibers produces activated surface induced by oxygen containing diverse polar groups. These polar groups increase the synergy of Ag blending with TiO_2 on textile surface. High frequency plasma treatment in the presence of oxygen and vacuum UV lead plasma treatments are also tried for increasing the adhesion of Ag and TiO_2 on textile surface (Qi et al., 2011; Kiwi Pulgarin, 2010).

3.5.5   ANTIMICROBIAL NANOFINISH

Natural textile fibers, such as cotton, wool, silk, etc. are liable to bacterial attack for prolong unwashed use in day today life. Synthetic fibers are also helping the growth of the microbes in cloths. As preventive measures, antibacterial disinfection and finishing techniques have been developed by treating various natural and synthetic textiles with silver nanoparticles, TiO_2 nanoparticles, and zinc oxide (ZnO) nanoparticles, etc. In most of the cases, the treatment process involves the preparation of their colloidal solutions (25–50 ppm) followed by dip-dry-cure process. These nanosized metallic ions and metallic compounds trigger a photocatalysis reaction in the presence of UV or visible light, presence of atmospheric oxygen, thus generating O^{2-} radical ions which are responsible for the destruction or killing of microorganisms which are in contact with the fiber. The advantage of using these nanosized particles is that huge numbers of particles can be accumulated per unit area over textile surface. As a benefit, the antibacterial efficiency can be maximized (Fouda, 2012).

3.6   CONCLUSION

Overviewing the history of textile pretreatment and finishing processes during the few last decades, there are considerable concern about complexity, energy consumption, wastage of water and chemicals, and effluent load to the environment. Tremendous efforts have been made to minimize them for the development of sustainable processes. In the advent of nanotechnology, suitable nanotreatment processes show the path for sustainable scouring, bleaching, cleaning, softening, UV blocking, antimicrobial activity, and water-repellent finishing of textiles. These nanotreatment processes guarantee minimal chemical wastage, energy-saving, low toxicity, maximum efficiency, and better performance. The photocatalytic activity of nanoparticles, such as TiO_2, Zno, Ag, etc. is the phenomenon responsible for decomposing selective unwanted organic materials from the textile surface under UV and/or sunlight irradiation to achieve scouring, bleaching, and self-cleaning? In the same line, the nanofinising processes like softening, water repellence, and UV protection are also successful. These processes avoid harsh chemical treatment conditions, and as a result, the mechanical and comfort properties of the textiles do not deteriorate. The nanopretreatment and nanofinishing of textile treatment have already proved to be successful alternative methods to existing processes and have a bright future ahead.

KEYWORDS

- **nanoscouring**
- **nanobleaching**
- **nanosoftening**
- **nano-TiO$_2$**
- **nano-ZnO**
- **nanosilver**
- **nanosililca**
- **iron oxide nanoparticles**

REFERENCES

Akhavan, F.; Montazer, M. In Situ Sonosynthesis of Nano TiO$_2$ on Cotton Fabric. *Ultrason. Sonochem.* **2014**, *21*, 681–691.

Asadi, M.; Montazer, M. Multi-Functional Polyester Hollow Fiber Nonwoven Fabric with Using Nano Clay/Nano TiO$_2$/Polysiloxane Composites. *J. Inorg. Organomet. Polym. Mater.* **2013**, *23*, 1358–1367.

Celik, N.; Deg¯irmenci, Z.; Kaynak, H. K. Effect of Nano-Silicone Softener on Abrasion and Pilling Resistance and Color Fastness of Knitted Fabrics. *Tekstil ve Konfeksiyon.* **2010**, *1*, 41–47.

Choe, E. K.; Nam, C. W.; Kook, S. R.; Chung, C.; Cavaco-Paulo, A. Implementation of Batchwise Bioscouring of Cotton Knits. *Biocatal. Biotransform.* **2004**, *22*, 375–382.

Fouda, M. M. Antibacterial Modification of Textiles Using Nanotechnology. In *de A Search for Antibacterial Agents*; Bobbarala, D. V., Ed.; INTECH Open Access Publisher, 2012; pp 47–72.

Gauthier, E.; Boyer, C.; Thivel, P. X.; Delpech, F.; Rou, J. C. Experimental Study of Odorous Ester Photocatalysis. *Environ. Eng. Manag. J.* **2006**, *5*, 1001–1010.

Hasanbeigi, A.; Price, L. A Technical Review of Emerging Technologies for Energy and Water Efficiency and Pollution Reduction in the Textile Industry. *J. Clean. Prod.* **2015**, *95*, 30–44.

Hashemikiaa, S.; Montazer, M. Sodium Hypophosphite and Nano TiO_2 Inorganic Catalysts along with Citric Acid on Textile Producing Multi-Functional Properties. *Appl. Catal. A Gen.* **2012**, *417–418*, 200–208.

Kamad, K.; Sohb, N. Enhanced Visible-Light-Induced Photocatalytic Activity of α-Fe_2O_3 Adsorbing Redox Enzymes. *J. Asian Ceramic Soc.* **2015**, *3*, 18–21.

Kiwi, J.; Pulgarin C. Innovative Self-Cleaning and Bactericide Textiles. *Catal. Today.* **2010**, *151* (1/2), 2–7.

Mankodi, D. H.; Agarwal, D. B. Studies on Nano-UV Protective Finish on Apparel Fabrics for Health Protection. *Res. J. Text. Apparel.* **2011**, *15* (3), 11–20.

Meilert, K. T.; Laub, D.; Kiwi, J. Photocatalytic Self-Cleaning of Modified Cotton Textiles by TiO_2 Clusters Attached by Chemical Spacers. *J. Mol. Cat. A: Chem.* **2005**, *237*, 101–108.

Montazer, M.; Pakdel, E. Reducing Photoyellowing of Wool Using Nano TiO_2. *Photochem. Photobiol.* **2010**, *86* (2), 255–260.

Montazer, M.; Maali Amiri, M.; Mohammad Ali Malek, R. In Situ Synthesis and Characterization of Nano ZnO on Wool: Influence of Nano Photo Reactor on Wool Properties. *Photochem. Photobiol.* **2013**, *89*, 1057–1063.

Montazer, M.; Morshedi, S. Nano Photo Scouring and Nano Photo Bleaching of Raw Cellulosic Fabric Using Nano TiO_2. *Int. J. Biol. Macromol.* **2012**, *50*, 1018–1025.

Montazer, M.; Morshedi, S. Photo Bleaching of Wool Using Nano TiO_2 under Daylight Irradiation. *J. Ind. Eng. Chem.* **2014**, *20*, 83–90.

Montazer, M.; Seifollahzadeh, S. Enhanced Self-Cleaning, Antibacterial and UV Protection Properties of Nano TiO_2 Treated Textile through Enzymatic Pretreatment. *Photochem. Photobiol.* **2011**, *87*, 877–883.

Qi, K. et al. Photocatalytic Self-Cleaning Textiles Based on Nanocrystalline Titanium Dioxide. *Text. Res. J.* **2011**, *81(1)*, 101-110.

Roy Choudhury, A. K.; Chatterjee, B.; Saha, S.; Shaw, K. Comparison of Performances of Macro, Micro and Nano Silicone Softeners. *J. Text. Inst.* **2012**, *103*, 1012–1023.

Sadeghian Maryan, A.; Montazer, M.; Damerchely, R. Discoloration of Denim Garment with Color Free Effluent Using Montmorillonite Based Nano Clay and Enzymes: Nano Bio-Treatment on Denim Garment. *J. Clean. Prod.* **2015**, *91*, 208–215.

Sadeghian Maryan, A.; Montazer, M.; Harifi, T.; Mahmoudi Rad, M. Aged-Look Vat Dyed Cotton with Anti-Bacterial/Anti-Fungal Properties by Treatment with Nano Clay and Enzymes. *Carbohydr. Polym.* **2013a**, *95*, 338–347.

Sadeghian Maryan, A.; Montazer, M.; Rashidi, A. Introducing Old-Look, Soft Handle, Flame-Retardant, and Anti-Bacterial Properties to Denim Garments Using Nano Clay. *J. Eng. Fibers Fabr.* **2013b**, *8*, 68–77.

Viswanath, C. S.; Ramachandran, T. Comfort Characteristics of Cotton Fabrics Finished with Fluoro-Alkyl Nano Lotus Finish. *Indian J. Fibre Tex. Res.* **2010**, *35* (4), 342–348.

Wong, Y. W. H.; Yuen, C. W. M.; Leung, M. Y. S.; Ku, S. K. A.; Lam, H. L. I. Selected Applications of Nanotechnology in Textiles. *Autex Res. J.* **2006**, *6* (1), 1–8.

Wu, D.; Long, M. Realizing Visible-Light-Induced Self-Cleaning Property of Cotton through Tools toward the Development of Molecular Wires: A Coating N-TiO_2 Film and Loading AgI Particles. *ACS Appl. Mater. Interfaces* **2011**, *3* (12), 4770–4774.

PART II

Nanoparticle Modification Techniques and Applications of Nanotechnology in the Textile Industry

CHAPTER 4

Advanced Nanotechnologies for Finishing of Cellulosic Textiles

FATEN HASSAN HASSAN ABDELLATIF[1*],
MOHAMED MEHAWED ABDELLATIF[2], and HEND M. AHMED[3]

[1]*Textile Research Division, Pre-treatment and Finishing of Cellulosic Fabric Department, National Research Centre, Dokki, Giza, Egypt*

[2]*Chemical Industries Division, Chemistry of Tanning, Materials and Leather Technology, National Research Centre, Dokki, Giza, Egypt*

[3]*Textile Research Division, Dyeing, Printing and intermediates, National Research Centre, Dokki, Giza, Egypt*

Corresponding author. E-mail: tota_nrc@yahoo.com

ABSTRACT

Advanced nanotechnologies have created reasonable progress in different industries. Their applications in textile industry have been widespread in the last few years particularly in producing lightweight and comfortable protective clothing. The innovation of new multifunctional cellulosic textiles is continuously growing with maintaining quality, ecology, and economic concerns. This can be attained by involving various nanoscale inorganic or organic materials in the fabric finishing due to the specific and valuable physiochemical properties of nanoscale materials. The nanomaterials can develop new properties in addition to improving the existing properties. Nanomaterials can induce various potential functionalities, for example, antibacterial, flame-retardant, UV protection, electric conductivity, super-hydrophobicity, and drug delivery. The selection of the nanomaterials is determined by the end use of the textile garments. This chapter will focus on

Fundamentals of Nano-Textile Science. Prashansa Sharma, Devsuni Singh & Vivek Dave (Eds.)
© 2022 Apple Academic Press, Inc. Co-published with CRC Press (Taylor & Francis)

the advancements in finishing of cellulosic textiles by nanotechnology and the available implementation technologies in textile finishing.

4.1 INTRODUCTION

During the twenty-first century, nanotechnology has emerged as worldwide research and is considered as a cutting edge technology involving research on diverse applications. Nanosized materials exhibit valuable behaviors as a result of the better properties of nanoscale materials compared with the bulk materials. Nanosized materials exhibit higher surface areas, better chemical reactivity, and exhibit better magnetic, photoconductive, and photocatalytic abilities. The nanoparticles can be obtained naturally or artificially. The natural nanoparticles can occur from biogenic, atmospheric, or pyrogenic sources, for example, fullerenes, carbon nanotubes, and inorganic nanomaterials. It also includes ultrafine sand grains of mineral origin. Artificial nanomaterials are commercially produced and find usage in diverse industries.

4.2 SYNTHESIS OF NANOMATERIALS

Recently, various synthetic methods have been developed for the production of different nanomaterials. For practical applications, the nanomaterials should be kept individual by preventing the nanostructure agglomerations. Moreover, the nanomaterials should be uniformly distributed and obtained in the desired size to improve the final product quality.

Two main approaches have been utilized for nanomaterial production, top-down and bottom-up. The top-down technique is conducted by reducing the size of the bulk materials while the bottom-up technique starts from the atomic level of the materials. Various methods have been developed for nanoparticle synthesis. Reductions in solutions, chemical and photochemical reactions in reverse micelles are examples for chemical methods for the synthesis of nanoparticles. Furthermore, nanoparticles can be obtained using physical methods, such as thermal decomposition, radiation-assisted, electrochemical, sonochemical and microwave-assisted processes.

These traditional synthetic methods usually need aggressive and hazardous reducing, capping, stabilizing agents, and organic solvent which lead to a negative impact on the environment and also on the economy by excessive consumption of energy and materials.

To overcome the drawbacks of the traditional synthetic methods of nanoparticles, green and eco-friendly approaches have been developed and these have become more attractive alternative to traditional synthetic methods. By continuously developing the green and eco-friendly approaches, it is expected to minimize the wastage of nanoparticle production while maintaining the quality of the resulted nanomaterials by controlling their sizes, shapes, and retaining their monodispersity. Green and eco-friendly approaches could also be achieved by full replacement of the hazardous solvents and chemicals by using clean and nontoxic experimental methods. Synthesis of nanoparticles using inorganic oxides and noble metals by green and eco-friendly approaches could be attained by different routes, such as chemical reduction, electrochemical, bio-based methods or using green alternative of energy sources (Bhattacharya and Gupta, 2005; Hasan, 2015; Ibrahim and Eid, 2015). Moreover, the use of biosynthetic methods produces metal nanoparticles of better sizes and shapes compared with some physicochemical methods utilized in production process (Patil and Kim, 2017). One of the promising approaches of green and eco-friendly synthesis of nanoparticles is the biosynthetic method using microorganisms, enzymes, or plant extracts.

4.2.1 SYNTHESIS OF NANOPARTICLES USING MICROORGANISMS

Most microorganisms, such as bacteria, fungi, algae, and yeast have the ability to synthesize nanoparticles by intracellular or extracellular methods depending on the placement of nanoparticles (Fig. 4.1).

They are considered as biofactories for nanoparticles synthesis. In the intracellular methods, the nucleation of the nanoparticles takes place inside the cells of an organism leading to a reduction in the size of nanoparticles compared with the extracellular method. Nevertheless, the extracellular methods are more applicable than the intracellular methods since they do not need adjoining cellular components from the cell (Narayanan and Sakthivel, 2010). The use of different microorganisms especially fungi in the biosynthesis of nanoparticles has become potential route, since they secrete high amount of enzymes, and further, it is easy to manage them. The catalytic effect of enzymes is responsible for reducing the size of the metal or metal oxide from macro or microscale into nanoscale. Therefore, the microorganisms can attract the cations and act as trigger for the biosynthesis of nanoparticles by their negative electrokinetic potential (Raliya et al., 2015) (Table 4.1).

4.2.2 SYNTHESIS OF NANOPARTICLES USING PLANT EXTRACTS

The attention to use plant extracts in nanoparticle synthesis is continuously growing since such biomaterials offer clean, cheap, nontoxic, and eco-friendly procedure for the production of nanoparticles. Plant-based method can produce nanoparticles with different shapes, sizes, and morphology. Compared with microorganisms-based method, plant-based method is fast and easy to scale up for large-scale production of nanoparticles. Moreover, plant-based method does not require complex procedure or multiple steps while microorganism-based method requires more steps, such as isolation of the microorganism, identification, growth optimization, culture preparation, and maintenance.

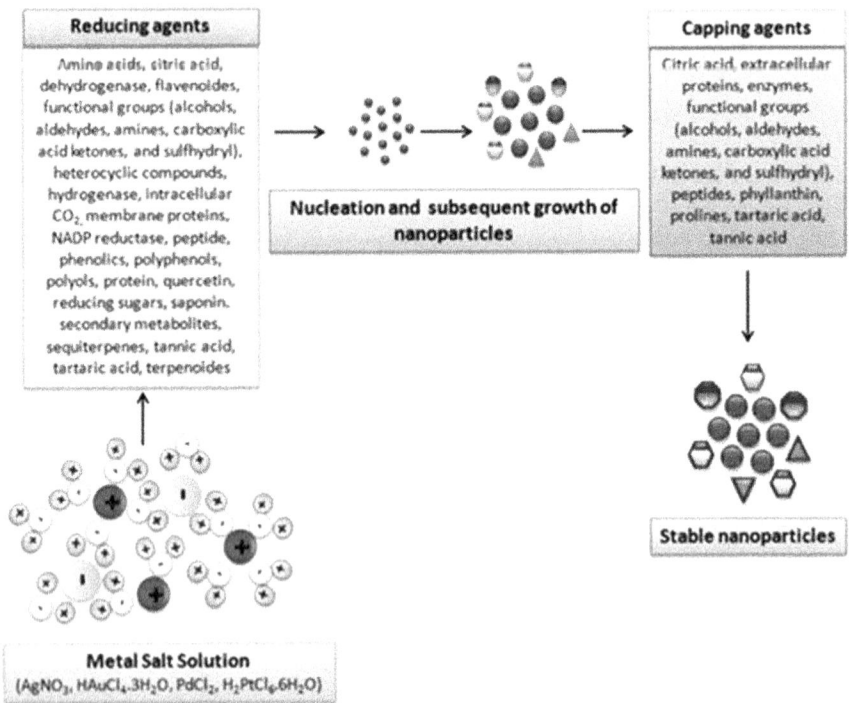

FIGURE 4.1   Mechanism of biosynthesis of nanoparticles. (Reprinted from Akhtar, M. S.; Panwar, J.; Yun, Y. S. Biogenic Synthesis of Metallic Nanoparticles by Plant Extracts. *ACS Sustain. Chem. Eng.* **2013,** *1* (6), 591–602. Copyright (2013) American Chemical Society).

The mechanism of plant-based method has the same principle as the microorganism-based method. The metal salt is reduced to atoms that nucleate in small clusters and grow to particles. Babu and Prabu (2011)have reported the synthesis of silver nanoparticles using the extract of *Calotropis procera* flower as green reducing and stabilizing agent (AgNPs). The phytochemicals present in the flower extract were used as capping agent and

TABLE 4.1 List of Some Microorganisms Used in the Biosynthesis of Nanoparticles.

Nanoparticles	Microorganisms	Size, nm	Ref
	Bacteria		
Silver	*Corynebacterium* sp., *Morganella* sp., *Proteus mirabilis*, *Bacillus* sp., Lactic acid bacteria, *Proteus mirabilis*	<20 nm	(Venkataraman et al., 2011)
	Bacillus cereus, Cuprividus sp.,	≤50 nm	(Husain et al., 2015; Sunkar and Nachiyar, 2012)
	Bacillus subtilis, Escherichia coli, Staphylococcus aureus, Enterobacter cloacae	<100 nm	(Venkataraman et al., 2011)
	Pseudomonas stutzeri AG259, *Lactobacillus* Strains, *Pseudomonas aeruginosa* ATCC 27853	≤500 nm	(Nair and Pradeep, 2002; Peiris et al., 2017; Venkataraman et al., 2011)
Copper	*Pseudomonas stutzeri*	≤150	(Ratnika et al., 2011)
Gold	*Rhodococcus* sp. (Actinomycete), *Rhodopseudomonas capsulate, Nocardia farcinica*	<20 nm	(Abhilash and Pandey, 2011; Oza et al., 2012)
TiO₂	*Escherichia coli, Bacillus subtilis, Pseudomonas aeruginosa, Ureibacillus thermosphaericus, Klebsiella pneumoniae*	≤70 nm	(Deplanche and Macaskie, 2008; Malarkodi et al., 2013; Xiangqian et al., 2011)
ZnONPs	Bacterium *Bacillus subtilis, Lactobacillus* strains	≤70 nm	(Kirthi et al., 2011; Prasad et al., 2007)
SeNPs	*Sphingobacterium thalpophilum*	37 nm	(Rajabairavi et al., 2017)
	Lactobacillus acidophilus	15 nm	(Alam et al., 2020)
	Fungi		
Silver	*Trichoderma longibrachiatum.*	10 nm	(Elamawi et al., 2018)
	Penicillium decumbens (MTCC-2494)	30–60 nm	(Majeed et al., 2016)
	Penicillium janthinellum, Aspergillus flavus	≤30 nm	(Pareek et al., 2020)
Gold	*Penicillium janthinellum Fusarium oxysporum, Colletotrichum* sp., *Yarrowiali polytica, Neurospora crassa*	≤40 nm	(Abhilash and Pandey, 2011; Pareek et al., 2020; Xiangqian et al., 2011)
TiO₂NPs	*A. flavus* TFR 7	12–15 nm	(Raliya et al., 2015)

TABLE 4.1 *(Continued)*

Nanoparticles	Microorganisms	Size, nm	Ref
	Yeast		
Silver	MKY-3	2–5 nm	(Abhilash and Pandey, 2011)
	Candida albicans	50–100 nm	(Xiangqian et al., 2011)
TiO_2	*Sacchharomyces cerevisiae*	12 nm	(Jha et al., 2009)
	Algae		
Sliver	*Plectonema boryanum*	1–200	(Xiangqian et al., 2011)
Gold	*Shewanella algae, Sargassum wightii, Shewanella oneidensis*	<20 nm	(Xiangqian et al., 2011)

the resulted AgNPs were found to have well-defined shape with a size of 35 nm. Fruit extracts of *Eugenia jambolana, Aegle marmelos,* and soursop were used to synthesize gold nanoparticles (AuNPs). The analysis of these extracts exhibited the presence of amino acid, alkaloids, phenol, proteins, tannin, flavonoids, reducing sugars, and total sugars. The flavonoids reduced the gold salt to AuNPs (Fig. 4.2) which was stabilized by the protein carboxylate groups. Furthermore, the electrostatic repulsions that resulted from the presence of like charge around the AuNPs led to keep the particles apart from each other (Vijayakumar, 2019) (Table 4.2).

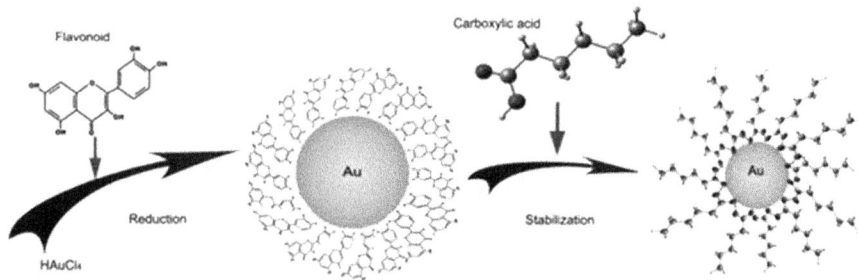

FIGURE 4.2 Using different fruit extracts for reduction and stabilization of AuNPs. (Reprinted from Vijayakumar, S. Eco-Friendly Synthesis of Gold Nanoparticles Using Fruit Extracts and in Vitro Anticancer Studies. *J. Saudi Chem. Soc.* **2019,** *23,* 753–761). Copyright Elsevier Reprinted with Permission.

TABLE 4.2 List of Some Plants Used in the Biosynthesis of Nanoparticles.

Nanoparticles	Plant	Size, nm	Ref
AgNPs	Ginger, garlic, *Indigofera barberi*	<20 nm	(El-Refai et al., 2018; Reddy et al., 2019)
	Calotropis procera flower	35 nm	(Babu and Prabu, 2011)
CuONPs	*Aloe vera, Calotropis gigantea, Phyllanthus amarus*	20 nm.	(Acharyulu et al., 2014; Ibrahim and Eid, 2015; Sharma et al., 2015)
CuNPs	*Azadirachta indica,*	48 nm	(Nagar and Devra, 2018)
	Ginger and garlic	22 nm	(El-Refai et al., 2018)
AuNPs	*Eucalyptus camaldulensis, Pelargonium, Azadirachta indica,*	<10 nm	(Ramezani et al., 2008)
	Camellia sinensis, Aegle marmelos, Eugenia jambolana and soursop	<40 nm	(Sharma et al., 2007; Vijayakumar, 2019)
TiO$_2$NPs	*Terminalia arjuna* bark	5–10 nm	(Gopinath et al., 2016)
ZnNPs	Ginger and garlic	<140 nm	(El-Refai et al., 2018)

4.3   APPLICATION OF NANOPARTICLES IN FINISHING OF CELLULOSIC FABRICS

4.3.1   CELLULOSIC FIBERS CHEMISTRY

Cellulose polymer is a linear organic polysaccharide. Cellulose is the most abundant and cheap polymer found in nature. This polymer is distinguished with its biodegradability, biocompatibility, and renewability. Cellulose polymer has the chemical formula of $(C_6H_{10}O_5)_n$. It consists of β-D-glucopyranose units linked by β-1,4-glycosidic linkage (Fig. 4.3). The degree of polymerization of cellulose varies according to the source of cellulose. It is approximately 10,000 in wood and 15,000 in native cotton fiber. Cellulose polymer chain consists of three pendant –OH groups per anhyroglucose unit. One of the three –OH groups is primary ($-CH_2OH$) and attached to C-6. The other two –OH groups are secondary and are attached at C-3 and C-4. The distributed hydroxyl groups along the polymer chains form hydrogen bonds with oxygen atoms of the same chain or of other chains. These hydrogen bonds hold the chains firmly and impart higher tensile strength to the cellulose polymer. Compared with other polysaccharides, cellulose has high crystallinity and is not soluble in organic or inorganic solvents. Moreover, cellulosic substrates have many drawbacks, for example, dimensional stability, marginal UV-protection, inflammability, and they are easily attacked by microorganisms.

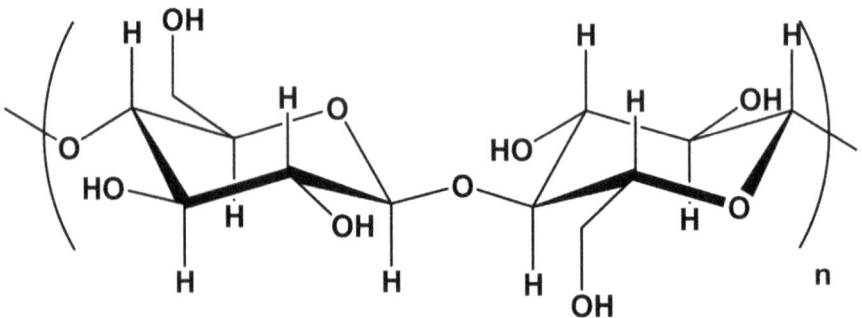

FIGURE 4.3   Chemical structure of cellulose.

To overcome the various drawbacks of native cellulose polymer, cellulose could be modified chemically, for example, esterification, etherification, and copolymer grafting through the reaction on the primary hydroxyl group which is more reactive than the secondary ones.

Cellulosic fibers could be obtained naturally, for example, cotton, flax, jute, or can be regenerated, for example, viscose, bamboo, rayon, Lyocell, acetate, triacetate, modal, and Tenncel. Regenerated fibers are obtained by dissolving cellulose polymer from tree wood, inner pith, and leaves from bamboo plants, and spun again into fibers. An eco-friendly aqueous solvent, that is, N-methyl-morpholine N-oxide could be used in the preparation of regenerated cellulosic fibers.

4.3.2 WET PROCESSING OF CELLULOSIC FABRICS

4.3.2.1 PRETREATMENT OF CELLULOSIC FABRICS

Pretreatment of natural cellulosic fibers is very important to clean the fabrics from the latent non-cellulosic impurities, for example, wax, fats, pectin, lignin, and seed husks. In addition, pretreatment is necessary for removing the extra chemicals which are used during spinning, weaving, or knitting processes for example, spinning oils, sizes, and lubricants. After pretreatment process, the properties of the cellulosic fabrics are improved, for example, the wettability of fabrics, the degree of whiteness enhanced by removing the coloring materials which facilitate the uniform coloration and/or finishing processes while maintaining the physicomechanical properties of the treated fabrics. Pretreatment processes occur in five successive steps (Ibrahim and Eid, 2015):

 ➢ Singeing to eliminate protrusive fiber ends to get softer fabric surface.
 ➢ Desizing to eliminate the sizing materials, for example, starch, CMC, and PVOH.
 ➢ Scouring is necessary to eliminate hydrophobic noncellulosic impurities to improve the swelling and the wettability of the cellulosic fabrics.
 ➢ Bleaching is done to remove the pre-swell seed husks, get rid of any remnant size and to remove oxidized natural coloring materials.
 ➢ Mercerizing to enhance dye ability, dimensional stability, tensile strength, and to increase luster.
 ➢ Optical brightening agents are used in cases where high degree of whiteness is needed.

4.3.2.2 COLORATION OF CELLULOSIC FABRICS

Coloration process of cellulosic fabrics usually requires effective pretreatments of the fabric to obtain high color yield, remarkable fastness properties, along with achieving right-first-time reproducibility. The developments of the existing processes and tools to attain right-first-time production is really important. Moreover, the use of eco-friendly colorant agents and greener coloration auxiliaries is needed to attain eco-friendly production process and to reduce the environmental impact with reasonable costs, and it represents the main prerequisite for sustainable products.

4.3.2.3 *FINISHING OF CELLULOSIC FABRICS*

Great attention has been paid to cellulosic fabrics due to their unique properties, for example, hydrophilicity, comfortability, biodegradability, and their eco-friendly nature. Nevertheless, cellulosic fabrics lack many favorable functional properties, such as antimicrobial activity, self-cleaning, flame-retardant, easy care, UV protection, water, and oil repellent, etc. Therefore, the continuous development of the shape and functionality of the cellulosic fabric with the development of the application techniques are greatly required (Ibrahim et al., 2017).

4.4   FINISHING OF CELLULOSIC FABRICS USING NANOMATERIALS

Currently, nanomaterials are used potentially in textile finishing due to their high surface areas, high surface energy, and their high affinity for the fabrics. Their application in textile industry is continuously growing because the conventional methods of textile treatment do not lead to durable functions, while nanomaterials could enhance the durability of the different functions imparted to the textile fabrics without affecting the comfort properties of the textile substrates (Avila and Hinestroza, 2008). Table 4.3 summarizes the different characteristics of the different nanoparticles that they could confer to the cellulosic substrates (Dastjerdi and Montazer, 2010).

The application of nanoparticles on the surface of cellulosic textiles has great challenges in terms of efficiency and durability. Their incorporation on the fabric surface could be in situ or ex situ. In the case of in situ application of nanoparticles, the chemical treatment of the cellulosic fabric surfaces open up their pores that can be utilized as a template for nanoparticles synthesis. These pores collapse after withdrawing of chemical media and hold the nanoparticles inside them. Ex situ application of nanoparticles on cellulosic fabric surfaces can be attained by pad-dry-cure method, spraying, or foaming technique or using physicochemical method, for example, coating, laser, plasma or surface treatment (Ibrahim et al., 2018; Vigneshwaran et al., 2006).

TABLE 4.3 Summary of the Characteristics of the Common Nanomaterials Used in Textile Finishing (Dastjerdi and Montazer, 2010).

Nanoparticles	Characteristics
Titanium dioxide	Antibacterial, photocatalyst, self-cleaning, UV protecting, superhydrophobic, cocatalyst for cotton
Zinc oxide	Antibacterial, UV-blocking, superhydrophobic, and photocatalyst
Silver	Antimicrobial, electrical conductive, and UV protection
Copper	Antibacterial, UV protection, and electrical conductive
Gold	Antibacterial, antifungal, and electrical conductive
Carbon nanotubes	Antimicrobial, electrical conductive, fire-retardant, antistatic, and chemical absorber
Nanoclay	Fire-retardant

4.4.1 ANTIMICROBIAL APPLICATION OF NANOPARTICLES IN CELLULOSIC FABRICS FINISHING

Cellulosic fabrics commonly provide a suitable environment for microorganisms' growth. The higher absorbency of cellulosic fabrics creates a good trapping of bacteria inside the fiber structure (Midha et al., 2014). Microorganisms can exist in air, water, or on the contact skin. Contamination of cellulosic fabrics with microorganisms leads to detrimental effects on both fabric and its user. Therefore, the developing of cellulosic fabrics capable of inhibiting or killing microorganisms, preserving the fabric from microbial degradation and eliminating body odor release is continuously growing (Ibrahim et al., 2017). For this purpose, a lot of antimicrobial agents have been developed for finishing cellulosic fabrics to impart them antimicrobial functionality. The antimicrobial agents can be natural, for example, chitosan and medicinal plants and their extracts or synthetic, for example, quaternary ammonium compounds, polybiguanides, N-halamines, and triclosan (Ibrahim et al., 2018).

Recently, nanomaterials have been used for antimicrobial finishing of cellulosic fabrics. The antimicrobial activity of nanoparticles has three different mechanisms of action (Fig. 4.3). These mechanisms include (1) formation of reactive oxidative species, (2) releasing process of ions, and (3) interaction of NPs with the cell membrane. In the first mechanism (Dutta et al., 2012), the metal nanoparticle ions are attached to the bacterial cell via trans-membrane protein leading to structural changes in the cell membrane and blocking the transport channels. This process depends on the size of the nanoparticles (Guzman et al., 2012; Ivask et al., 2014). The smaller nanoparticles can penetrate the cell wall easily while the bigger nanoparticles attach to the cell wall by Vander Waals forces. In this mechanism (Prabhu and Poulose, 2012), the nanoparticles ionize within the cell, leading to cell death by damaging the intracellular structures (Dutta et al., 2012). In the second mechanism, the generation of reactive oxidative species by nanoparticles is responsible for the effective antibacterial activity of nanoparticles. The reactive oxidative species, for example, superoxide radicals (O^{2-}), hydrogen peroxide (H_2O_2), and hydroxyl radicals (OH^-) (Pelgrift and Friedman, 2013) have high reactivity that can damage the cell membranes, peptidoglycan, ribosomes, DNA, mRNA, and proteins (Baek and An, 2011). In addition, reactive oxidative species have the ability to inhibit enzymatic activity, transcription, translation, and the electron transport chain (Dutta et al., 2012).

In the third mechanism, the antibacterial activity of nanoparticles emerges from the tendency of nanoparticles to bind with thiol group present in the enzymes causing the enzymes to be deactivated. Moreover, the nanoparticles have the tendency to attack DNA molecules causing their destruction (Jung et al., 2008). Figure 4.4 depicts various mechanisms of antimicrobial activity (Dizaj, 2014).

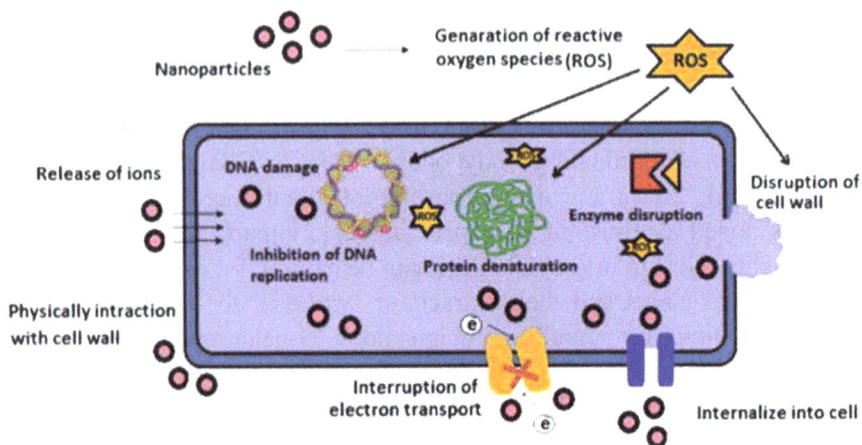

FIGURE 4.4   Different mechanisms of antimicrobial activity of the metal nanoparticles. (Reprinted from Dizaj, S. M.; Lotfipour, F.; Jalali, M. B.; Zarrintan, M. H.; Adibkia, K. Antimicrobial Activity of the Metals and Metal Oxide Nanoparticles. *Mater. Sci. Eng. C* **2014,** *44,* 278–284. Copyright Elsevier Reprinted with permission).

AgNPs are potentially used as antimicrobial agent in a wide range of biomedical applications. Comparing with their ionic form, AgNPs show a remarkable antimicrobial activity. Many studies have been reported about the utilization AgNPs in cellulosic textile finishing. Rehan et al. (2017) reported a green method for the deposition of AgNPs onto cotton fabrics. In this method, UV irradiation was used for the reduction of $AgNO_3$ solution. In this system, cotton fabrics acted as seed for AgNPs nucleation. The particle size distribution of AgNPs was 50–100 nm. The finished cotton exhibited excellent protection against the growth of *Escherichia coli* (99%). In situ synthesis approach has been used by El-Naggar to load cotton fabric with AgNPs. In this technique, the cotton fabrics were preactivated using ethanolamine. The aminated fabrics were immersed in different concentration of $AgNO_3$ The deposited AgNPs on the surface of the fabrics were fixed at 150°C for 2 min. The treated fabrics showed good sustainable bactericidal properties (El-Naggar et al., 2018). Pivec et al. described a green in situ procedure for loading AgNPs on the surface of regenerated cellulosic fibers. This procedure consisted of three steps. In the first step, the fibers were activated by alkaline treatment to increase the swelling of the fibers. This step is important to open up the surface and internal pores of cellulose fibers and loaded them from inside with $AgNO_3$ which was reduced to AgNPs in the presence of NaOH in the second step. Finally, the fibers were washed and neutralized. The resulted fibers displayed superior durability against washing and maintained excellent antimicrobial properties (>99) even after 20 washing cycles against different pathogens, for example, *Staphylococcus aureus*, *Klebsiella pneumoniae*, and *Candida albicans* (Piveca et al., 2017).

In situ approach was also used for loading jute fabrics with AgNPs. Jute fabrics were immersed in a bath containing $AgNO_3$ with 6.5 pH. The AgNPs were formed at 90°C. The size of AgNPs was in the range of 40–100 nm. The zones of inhibition of treated jute fabrics were investigated against *B. subtilis* and *E. coli*. The fabrics exhibited resistance against the mentioned pathogens and resistance to washing up to 15 cycles (Lakshmanan and Chakraborty, 2017).

Recently, AuNPs have been used in cellulosic textile finishing. Ibrahim et al. (2016) reported an eco-friendly green method for the fabrication of AuNPs using chlolroauric acid as precursor and marine bacterial isolates as reducing/capping/stabilizing bioagent. Cotton- and viscose-knitted fabrics were activated with O_2-Plasma. The activated fabrics were subsequently treated with biosynthesized AuNPs alone and in combination with TiO_2NPs or ZnONPs. The treated fabrics displayed remarkable antibacterial resistance against G +ve (*S. aureus*) and G −ve (*E. coli*) bacteria (Fig. 4.5).

a   b   c   d

FIGURE 4.5   Photos of flasks containing: (a) HAuCl$_4$·3H$_2$O (3 mM), (b) *Streptomyces* sp. Culture filtrate, (c) Biomass/HAuCl$_4$·3H$_2$O mixture, and (d) AuNPs formation. (Reprinted from Ibrahim, N. A.; Eid, B. M.; Abdel-Aziz, M. S. Green Synthesis of AuNPs for Eco-Friendly Functionalization of Cellulosic Substrates. *Appl. Surf. Sci.* **2016,** *389,* 118–125. Copyright Elsevier Reprinted with permission).

TiO$_2$NPs displayed extraordinary photocatalytic properties, high activity, strong oxidizing agent, and unique optical and electronic characteristics. Many researchers have functionalized cellulosic fabrics with TiO$_2$NPs to impart them new functionality. Karimi et al. created a new nanocomposite of graphene oxide/titanium dioxide. Cotton fabric was coated with this nanocomposite to obtain multifunctional cellulose substrate. The nanocomposite enhanced the properties of cotton fabric more than using only TiO$_2$NPs. The antimicrobial activity of the coated fabrics was evaluated against both G +ve (*S. aureus*) and G -ve (*E. coli*) bacteria. The treated cotton fabrics exhibited strong antimicrobial activity against the latter pathogens (Karimi et al., 2016).

El-Naggar et al. (2016) has loaded TiO$_2$NPs on cotton fabrics using in situ synthesis of TiO$_2$NPs from titanium isopropoxide as precursor for TiO$_2$NPs. Urea nitrate was used as a peptizing agent to release nitric acid which converts titanium hydroxide to TiO$_2$NPs. The resulted samples revealed excellent antibacterial activity against G +ve (*S. aureus*) and G -ve (*E. coli*) bacteria (>95%). Moreover, these samples exhibited washing resistance up to 20 cycles.

ZnONPs are widely used in different fields, for example, cosmetics, depollution, and protective medical textiles application due to their attractive properties. ZnONPs have good photocatalytic activity, high stability, and nontoxicity. ZnONPs are excellent candidate for cellulosic textile finishing because of their low cost and white appearance (Dastjerdi and Montazer, 2010).

Shaheen et al. created a single-stage procedure for the synthesis (ZnONPs). In this procedure, hexamethyltriethylene tetramine (HMTETA) provided alkaline medium which facilitated the formation of uniform ZnONPs. The nanoparticles were formed by in situ method within cotton fabrics without stabilizing or capping agents. Cotton fabrics acted as a template for the nanoparticles which were confined in the fibril and microfibrils. The antibacterial activity of ZnONPs-loaded cotton fabric was evaluated against *E. coli* and *S. aureus* bacteria. The bacterial reduction of the treated cotton fabrics ranged from 94.9% to 97.5% depending on the ZnONPs concentration. After repeated washing of treated cotton fabrics, it was found that the antibacterial reduction decreased to 90% after 20 washing cycles (El-Naggar et al., 2016).

Petkova et al. proposed sonochemical/enzymatic process for cotton fabrics with ZnONPs. In this process, the fabric was enzymatically preactivated to provide better adhesion to the nanoparticles. The sonochemical nanocoating was followed by the activation process to produce "ready-to-use" antibacterial medical textiles in a single step. The nanocoated cotton fabrics could reduce the bacterial growth by 67% and 100% for *S. aureus* and *E. coli*, respectively. After multiple washing of the treated fabrics, only 33% of the initially deposited ZnONPs remained on the surface of the fabrics. Nevertheless, the remaining amount of ZnONPs was effective against bacteria and exhibited growth reductions of 100% and 50% toward *E. coli*, and *S. aureus*, respectively (Petkova et al., 2016).

Ghayempour and Montazer (2017) used the natural polymer, Tragacanth gum, as green and eco-friendly reducing, stabilizing, and binding agent for green in situ synthesis of ZnONPs on cotton fabrics through ultrasound waves. The resulted antimicrobial and photocatalytic cotton fabrics exhibit 100% bacterial reduction against *E. coli*, *S. aureus* and *C. albicans* (Fig. 4.6).

The attention toward copper nanoparticles CuNPs is continuously increased due to unique catalytic and antimicrobial properties of CuNPs. Copper metal is abundant and relatively inexpensive compared with Ag. Copper can be a cofactor for metalloproteins and enzymes. Its toxicity is low compared with other metals. Various techniques have been utilized for the immobilization of Cu-based NPs onto cellulosic fabrics. Moazami and Montazer (2016) could synthesize Cu_2ONPs from $CuSO_4$ using cotton fabrics as template and glucose as a reducing and capping agent in alkaline medium. The treated fabrics displayed excellent antibacterial efficiency against *S. aureus* and *E. coli* and resistance to washing up to 10 cycles. This can be attributed to the capping effect of glucose surrounding the synthesized nanoparticles on the fabric surface.

Marković et al. proposed in situ process for the immobilization of CuNPs on cotton fabrics which was preactivated with different carboxylic acids. The carboxylic acids were succinic, citric, and 1, 2, 3, 4-butanetetracarboxylic.

It was found that increasing the number of free carboxylic groups led to increase in the number of the Cu^{2+}-ions attached to the fabrics and subsequently increasing the amount of CuNPs on the cotton substrates. The resulted nanocomposites exhibited maximum antibacterial activity against G –ve *E. coli* and G +ve *S. aureus* bacteria. The fabricated nanocomposites displayed controlled release of Cu^{2+}-ions in physiological saline solution that are essential for preventing fiber infection (Marković et al., 2018). The same group proposed another procedure for in situ synthesis of Cu/Cu_2ONPs on cotton fabrics. In this procedure, cotton fabrics were activated by TEMPO-mediated oxidation which provided carboxylic groups on the fabric. Carboxylic groups were needed for adsorption of Cu^{2+}-ions which reduced to Cu/Cu_2ONPs by sodium borohydride. The fabricated nanocomposites exhibited 99.9% growth reduction against *E. coli* and *S. aureus*. Moreover, they displayed acceptable growth reduction for *C. albicans* (Marković et al., 2018).

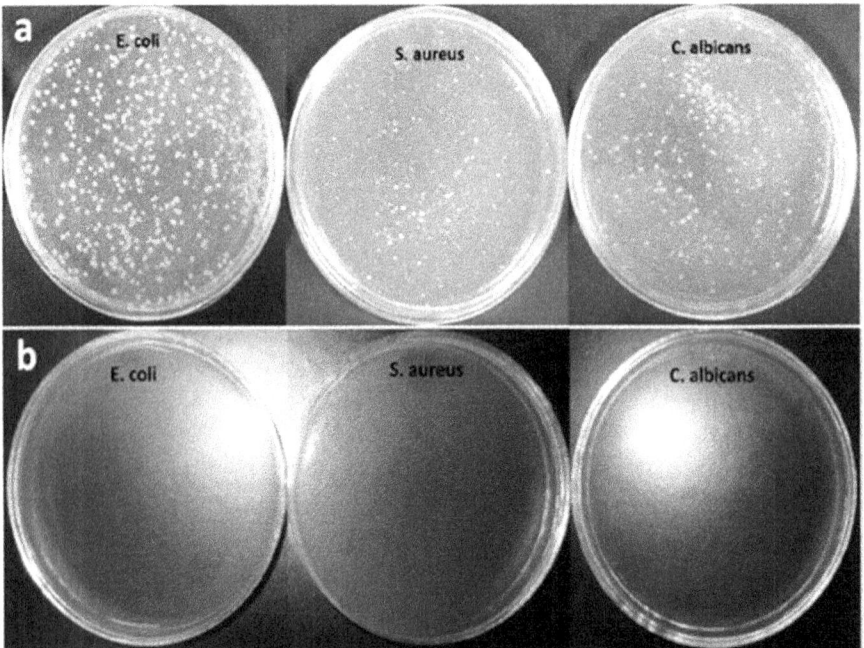

FIGURE 4.6 Colony count method for antimicrobial activity estimation of (a) untreated cotton and (b) treated cotton fabrics against *E. coli, S. aureus*, and *C. albicans*. (Reprinted from Ghayempour, S.; Montazer, M. Ultrasound Irradiation Based In Situ Synthesis of Star-Like Tragacanth Gum/Zinc Oxide Nanoparticles on Cotton Fabric. *Ultrason. Sonochem.* **2017,** *34,* 458–465. Copyright Elsevier Reprinted with permission).

4.4.2 UV PROTECTION APPLICATION OF NANOPARTICLES IN CELLULOSIC FABRICS

UV protection application is one of the most essential requirements for the protection of cellulosic substrates against extended exposure to sunlight. UV radiation consists of three radiation categories: $\lambda = 320\text{--}400$, $\lambda = 280\text{--}320$, and $\lambda = 100\text{--}290$ nm. These categories are known as UV-A, UV-B, and UV-C, respectively. Ozone layer usually absorbs the UV-C radiation while UV-A and UV-B reach the earth. Overdose of UV-A and UV-B radiations has dangerous effects on human skin, such as skin reddening, sunburns, carcinogenesis, skin damage, aging, and DNA damage (Ahmed et al., 2019). Therefore, great efforts have been made for developing suitable functional finishes to have cellulosic fabrics with UV protection property. UV protection factor (UPF) reflects the amount of UV protection provided by the fabric. The UPF for protective textile substrates should be between 40 and 50 to have textile substrates with excellent UV protection functionality. The UPF value is affected by construction, type, composition of the fabric, and used dyes in addition to using UV protection additives (Ibrahim and Eid, 2015).

UV protection of fabrics can be classified according to ASTMD6603 and AS/NZS standards (Dubrovski and Golob, 2009) into four groups. UPF of 40–50 is excellent, UPF of 25–39 is very good, UPF of 15–24 is good and UPF <10 is nonratable.

Inorganic UV blockers are usually better than organic UV blockers since they are nontoxic and stable against prolonged exposure to UV radiation or to high temperature. Different nanoparticles, especially TiO_2 and ZnO exhibited efficient absorbing and scattering of UV radiations. Recently, numerous studies have proved the UV protection functionalization of cellulosic fabrics using nanoparticles. Among the different nanoparticles, TiO_2NPs and ZnONPs exhibited outstanding durable UV protection.

4.4.3 FLAME-RETARDANT APPLICATION OF NANOPARTICLES IN CELLULOSIC FABRICS

The need for the development of protective cellulosic textile with flame-retardant functionality is continuously required since textiles are the main ignition sources in fire accidents. The best flame-retardant agent should retard the textile ignition and should have the ability to decrease the flame spreading. Nonintumescent and intumescent systems are the two main flame-retardant systems. Phosphorus, halogenated, silicone, nitrogen, or inorganic metal compounds-based flame- retardants are utilized to have nonintumescent systems. Recently, nanoparticles have been used as a new alternative for conventional nonintumescent flame-retardant agents. Nanoparticles can enhance mechanical, physical, and thermal stability properties.

Yazhini and Prabu (2015) developed cotton fabrics coated with polypyrrole–zinc oxide–carbon nanotube (ppy–ZnO–CNT) nanocomposites. The cotton fabrics were loaded with the nanocomposite by pad-dry-cure method. The flame-retardant functionality of the treated cotton fabric was evaluated in comparison with the untreated cotton fabric by vertical flammability test. The cross-linking of cotton fabrics with (ppy–ZnO–CNT) nanocomposite led to decrease the flammability of the cotton fabrics while untreated cotton fabric catches fire, and with the continuing of burning, the untreated cotton burns completely to ashes.

Moazami and Montazer (2016) produced multifunctional cotton fabrics using ZrO_2NPs along with cetyltrimethyl ammonium bromide and urea. Cotton fabrics were treated with the finishing formula at 40°C for 30 min in an ultrasonic bath using maleic acid (MA) as a cross-linking agent and sodium hypophosphite (SHP) as catalyst. Urea was used as a source of nitrogen to improve the flame-retardant property and photoactivity of the nano-ZrO_2-treated fabric. Furthermore, addition of MA and SHP decreased the fabric flammability. Therefore, use of this finishing formula led to a significant decrease in char length during the flammability test.

Ortelli et al. created a novel and durable intumescent flame-retardant coatings for cotton fabrics. These coatings were based on TiO_2NPs and two different proteins, for example, whey proteins and caseins. TiO_2NPs were used as specific cross-linking agents to fix whey proteins and caseins to cotton fabrics. The nanocomposites were deposited on cotton fabrics by dip-pad-dry-cure technique. Flammability and cone calorimetry tests were used for assessing the flammability behavior of the treated fabrics. The different coatings enhance the flame-retardant functionality of cotton fabrics by increasing the burning time and a decrease of the burning rate. Moreover, casein-based coating displayed better characteristics compared with the whey proteins (Ortelli et al., 2018) (Fig. 4.7).

4.4.4 SELF-CLEANING APPLICATION OF NANOPARTICLES IN CELLULOSIC FABRICS

Self-cleaning has been introduced by nanotechnology as a new concept in textile industry. Self-cleaning protective property can be attained by

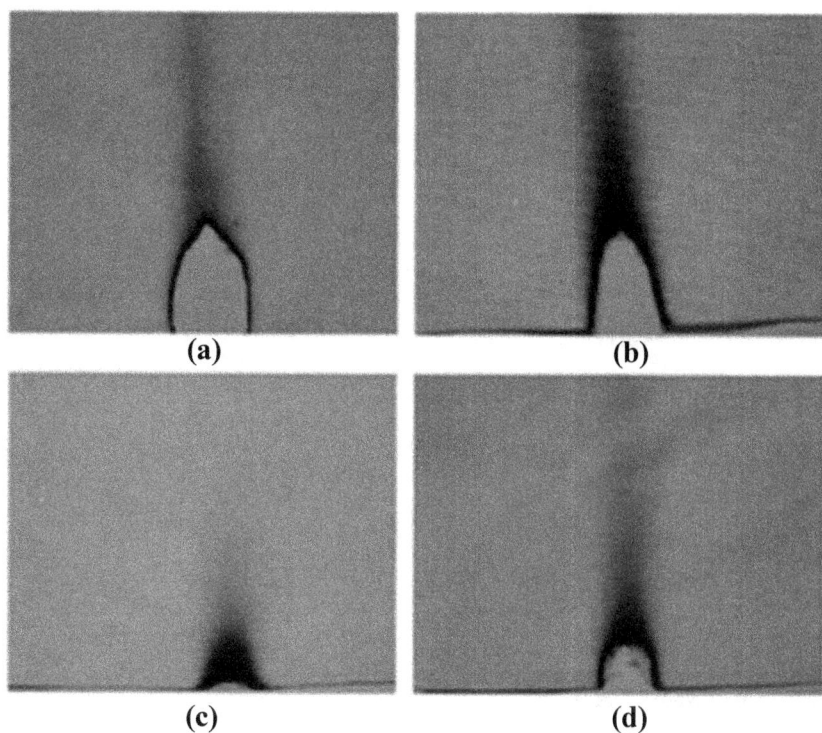

FIGURE 4.7 Vertical flame test of (a) untreated cotton fabric, (b) cotton fabric/ MA, SHP, (c) cotton fabric/ZrO_2, Urea, MA, SHP in one bath, and (d) cotton fabric/MA, SHP followed by ZrO_2, Urea in two baths. (Reprinted from Moazami, A.; Montazer, M. A Novel Multifunctional Cotton Fabric Using ZrO_2 NPs/Urea/CTAB/MA/SHP: Introducing Flame-Retardant, Photoactive, and Antibacterial Properties. *J. Textile Instit.* **2016,** *107* (10), 1253–1263. Copyright Taylor & Francis Reprinted with permission).

superhydrophobicity or superhydrophilicity. Both techniques lead to self-cleaning of clothing by the action of water. In superhydrophobicity, the water droplets roll over the fabric surface carrying the dirt away. Surface with water contact angle higher than 150° is an excellent water-repellent surface. This property is called lotus effect where superhydrophobic textile mimics lotus leaves which are the best examples for the self-cleaning (Fig. 4.8) (Zhang, 2016).

Superhydrophilic surfaces can clean themselves by sheets of water that can remove dirt. Moreover, use of a photocatalyst can provide superhydrophilic surface an additional characteristic. They can chemically breakdown the attached organic dirt in sunlight which is known as hydrophilic photocatalytic coating (Cheung and Li, 2018). Self-cleaning textiles can provide

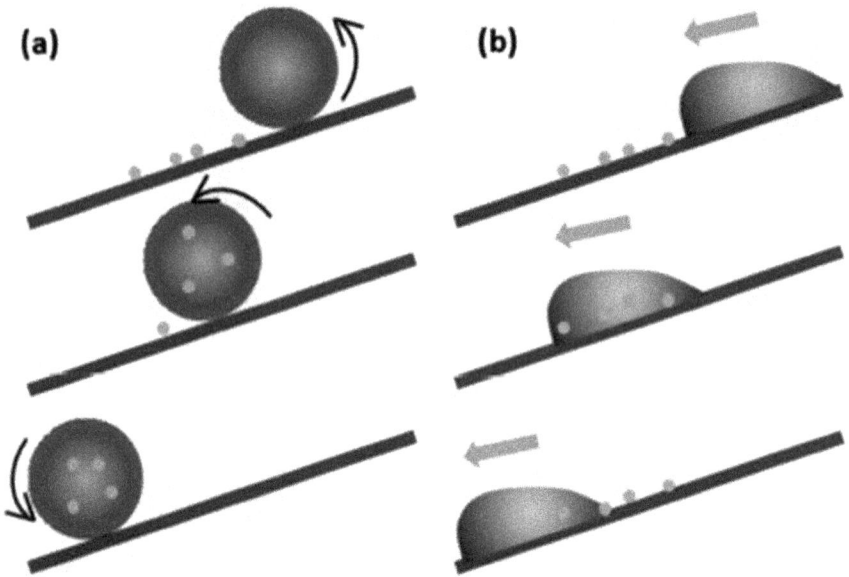

FIGURE 4.8   Self-cleaning of a low energy superhydrophobic surface. (Reprinted from Zhang, M.; Feng, S.; Wang, L.; Zheng, Y. Lotus Effect in Wetting and Self-Cleaning. *Biotribology* **2016**, *5*, 31–43. Copyright Elsevier Reprinted with permission).

hygiene and stop the pathogenic infection spreading. The self-cleaning garment can stay bright, fresh, and they are durable than the usual garments. Self-cleaning textiles find applications in medical textiles, sport wear, and defense textiles. Figure 4.9 shows the mechanism of self-cleaning textiles by photocatalytic coating (Banerjee, 2015).

Photocatalytic self-cleaning fabrics can be obtained using TiO_2NPs, or ZnONPs. Therefore, the fabrics possess several functionalities, for example, antibacterial, UV protection. So self-cleaning fabrics can be evaluated by several test methods. Determining the color strength of stain or pollutant dye before and after exposing to UV light reflects the photocatalytic activity of the functionalized fabric (Table 4.4).

4.5   FUTURE PROSPECTS

- Creating green and cost-effective synthetic methods for nanoparticles production and to reduce the negative impact of the widely used

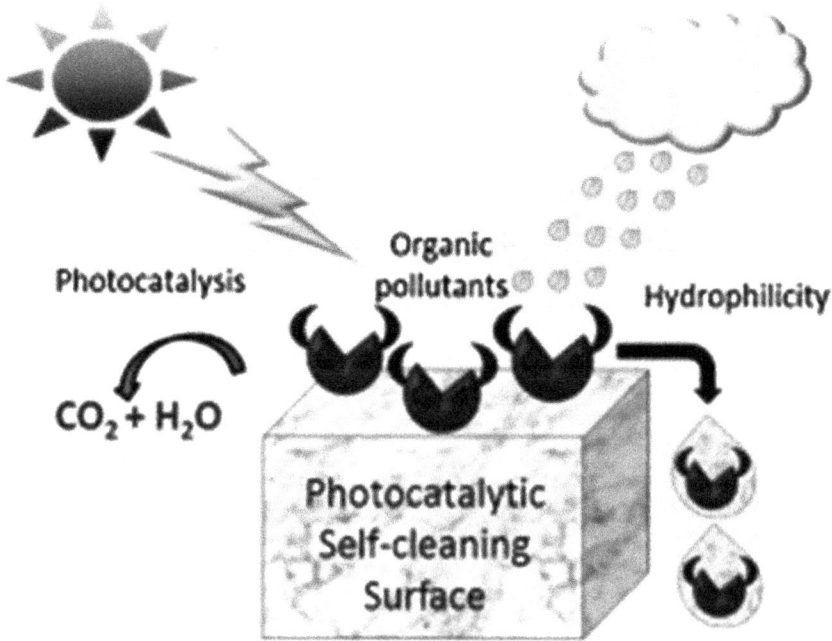

FIGURE 4.9 Mechanism of self-cleaning textiles by photocatalytic coating. (Reprinted from Banerjee, S.; Dionysiou, D. D.; Pillai, S. C. Self-Cleaning Applications of TiO_2 by Photo-Induced Hydrophilicity and Photocatalysis. *Appl. Catalys. B Environ.* **2015,** *176–177,* 396–428). Copyright Elsevier Reprinted with permission.

chemical methods as well as new generation of green cellulosic textile finished with nanoparticles to create multifunctional textile products for household, apparel, and smart textile using.

- Innovation of large-scale production methods of nanomaterials to use the nanomaterials in industrial scale.
- Merging nano and biotechnologies to create new materials used for the development of sustainable multifunctional textiles to meet the consumer needs.
- Using an eco-friendly method for surface modification of cellulosic substrates and treating them with proper functionalizing agents to impart them one or more functional characteristics, such as antimicrobial, UV protection, flame-retardant, self-cleaning, insect repellent, etc.

TABLE 4.4 Using the Different Nanoparticles in Imparting Cellulosic Textile New Functionalities.

Nanoparticles	Substrate	Method of application	UPF	Self-cleaning	Ref
TiO$_2$NPs	Cotton	In situ	Very good		(El-Naggar et al., 2016)
TiO$_2$NPs	Cotton	Dip-drying	good		(Karimi et al., 2016)
TiO$_2$NPs	Cotton and Viscose	Padding		Degradation of Orange II dye obtained by photocatalytic activity	(Kale et al., 2016)
ZnONPs	Cotton	In situ	Very good		(El-Naggar et al., 2016)
ZnONPs	Cotton	Pad-dry-cure	good		(Raza et al., 2016)
ZnONPs	Cotton	Pad-dry-cure	Very good		(Pandimurugan and Thambidurai, 2017)
ZnONPs	Cotton	Coating by microwave	Excellent	Degradation of coffee stain obtained by photocatalytic coating	(Ibrahim et al., 2017)
ppy–ZnO–CNT	Cotton	Pad-dry-cure	Excellent		(Yazhini and Prabu, 2015)
AuNPs	Cotton	Loading in sonicator bath	Very good		(Ibrahim et al., 2016)
AuNPs	viscose	Loading in sonicator bath	good		(Ibrahim et al., 2016)
AuNPs+ TiO$_2$NPs	Cotton or viscose	Loading in sonicator bath	Excellent for cotton Very good for viscose		(Ibrahim et al., 2016)
AuNPs+ ZnONPs	Cotton or viscose	Loading in sonicator bath	Excellent		(Ibrahim et al., 2016)
CuNPs+ stearic acid	Cotton	Exhaustion		Stain is removed by Superhydrophobocity characteristic	(Suryaprabha and Sethuraman, 2017)
CuO$_2$NPs	Cotton	In situ		Degradation of methylene blue dye obtained excellent by photocatalytic activity	(Moazami and Montazer, 2016)

KEYWORDS

- **nanomaterials**
- **cellulosic fabric**
- **finishing**
- **antimicrobial**
- **UV protection**
- **flame-retardant**
- **self-cleaning**

REFERENCES

Abhilash, R. K.; Pandey, B. D. Microbial Synthesis of Iron-Based Nanomaterial's—A Review. *Bull. Mater. Sci.* **2011,** *34* (2), 191–198.

Acharyulu, N. P. S.; Dubey, R. S.; Kollu, P.; Swaminadham, V.; Kalyani, R. L.; Pammi, S. V. N. Green Synthesis of CuO Nanoparticles using Phyllanthus Amarus Leaf Extract and their Antibacterial Activity Against Multidrug Resistance Bacteria. *Int. J. Eng. Technol.* **2014,** *3* (4), 639–641.

Ahmed, H. M.; Abdellatif, M. M.; Ibrahim, S.; Abdellatif, F. H. H. Mini-Emulsified Copolymer/Silica Nanocomposite as Effective Binder and Self-Cleaning for Textiles Coating. *Prog. Org. Coat.* **2019,** *129*, 52–58.

Alam, H.; Khatoon, N.; Khan, M. A.; Husain, S. A.; Saravanan, M.; Sardar, M. Synthesis of Selenium Nanoparticles Using Probiotic Bacteria Lactobacillus Acidophilus and Their Enhanced Antimicrobial Activity against Resistant Bacteria. *J. Clust. Sci.* **2020,** *31* (5), 1003–1011.

Avila, A. G.; Hinestroza, J. P. Tough cotton. *Nat. Nanotechnol.* **2008,** *3* (8), 458–459.

Babu, S. A.; Prabu, H. G. Synthesis of AgNPs Using the Extract of *Calotropis procera* Flower at Room Temperature. *Mater. Lett.* **2011,** *65* (11), 1675–1677.

Baek, Y.-W.; An, Y.-J. Microbial Toxicity of Metal Oxide Nanoparticles (CuO, NiO, ZnO, and Sb2O3) to *Escherichia coli*, *Bacillus subtilis*, and *Streptococcus aureus*. *Sci. Total Environ.* **2011,** *409* (8), 1603–1608.

Bharathi Yazhini, K.; Gurumallesh Prabu, H. Study on Flame-Retardant and UV-Protection Properties of Cotton Fabric Functionalized with ppy-ZnO-CNT Nanocomposite. *RSC Adv.* **2015,** *5* (61), 49062–49069.

Bhattacharya, D.; Gupta, R. K. Nanotechnology and Potential of Microorganisms. *Crit. Rev. Biotechnol.* **2005,** *25*, 199–204.

Cheung, T. W.; Li, L. Sustainable Development of Smart Textiles: A Review of 'Self-Functioning' Abilities Which Makes Textiles Alive. *J. Text. Eng. Fash. Technol.* **2018,** *4* (2), 151–156.

Dastjerdi, R.; Montazer, M. A Review on the Application of Inorganic Nano-Structured Materials in the Modification of Textiles: Focus on Anti-Microbial Properties. *Colloids Surf. B,* **2010,** *79* (1), 5–18.

Deplanche, K.; Macaskie, L. Biorecovery of Gold by Escherichia coli and Desulfovibrio Desulf Uricans. *Biotechnol. Bioeng.* **2008,** *99* (5), 1055−1064.

Dubrovski, P. D.; Golob, D. Effects of Woven Fabric Construction and Color on Ultraviolet Protection. *Text. Res. J.* **2009,** *79,* 351–359.

Dutta, R. K.; Nenavathu, B. P.; Gangishetty, M. K.; Reddy, A. V. R. Studies on Antibacterial Activity of ZnO Nanoparticles by ROS Induced Lipid Peroxidation. *Colloids Surf. B* **2012,** *94,* 143–150.

El-Naggar, M. E.; Shaarawy, S.; Hebeish, A. A. Bactericidal Finishing of Loomstate, Scoured and Bleached Cotton Fibres via Sustainable in-Situ Synthesis of Silver Nanoparticles. *Int. J. Biol. Macromol.* **2018,** *106,* 1192–1202.

El-Naggar, M. E.; Shaheen, T. I.; Zaghloul, S.; El-Rafie, M. H.; Hebeish, A. Antibacterial Activities and UV Protection of the in Situ Synthesized Titanium Oxide Nanoparticles on Cotton Fabrics. *Ind. Eng. Chem. Res.* **2016,** *55* (10), 2661–2668.

El-Refai, A. A.; Ghoniem, G. A.; El-Khateeb, A. Y.; Hassaan, M. M. Eco-Friendly Synthesis of Metal Nanoparticles Using Ginger and Garlic Extracts as Biocompatible Novel Antioxidant and Antimicrobial Agents. *J. Nanostruct. Chem.* **2018,** *8* (1), 71–81.

Elamawi, R. M.; Al-Harbi, R. E.; Hendi, A. A. Biosynthesis and Characterization of Silver Nanoparticles Using Trichoderma Longibrachiatum and Their Effect on Phytopathogenic Fungi. *Egypt. J. Biol. Pest Control* **2018,** *28* (1), 28.

Ghayempour, S.; Montazer, M. Ultrasound Irradiation Based in-Situ Synthesis of Star-Like Tragacanth Gum/Zinc Oxide Nanoparticles on Cotton Fabric. *Ultrason. Sonochem.* **2017,** *34,* 458–465.

Gopinath, K.; Kumaraguru, S.; Bhakyaraj, K.; Thirumal, S.; Arumugam, A. Eco-Friendly Synthesis of TiO_2, Au and Pt doped TiO_2 Nanoparticles for Dye Sensitized Solar Cell Applications and Evaluation of Toxicity. *Superlattice Microst.* **2016,** *92,* 100–110.

Guzman, M.; Dille, J.; Godet, S. Synthesis and Antibacterial Activity of Silver Nanoparticles against Gram-Positive and Gram-Negative Bacteria. *Nanomedicine: N. B. M.* **2012,** *8* (1), 37–45.

Hasan, S. A Review on Nanoparticles: Their Synthesis and Types. *Res. J. Recent Sci.* **2015,** *4,* 1–3.

Husain, S.; Sardar, M.; Fatma, T. Screening of Cyanobacterial Extracts for Synthesis of Silver Nanoparticles. *World J. Microbiol. Biotechnol.* **2015,** *31* (8), 1279–1283.

Ibrahim, N. A.; Eid, B. M. Chapter 6 -Potential Applications of Sustainable Polymers in Functionalization of Cellulosic Textile Materials. In *Polymer Polymers Handbook of Processing and Applications*; Thakur, V. K., Thakur, M. K., Eds.; CRC Press, 2015; pp 215–252.

Ibrahim, N. A.; Eid, B. M.; Abdel-Aziz, M. S. Green Synthesis of AuNPs for Eco-Friendly Functionalization of Cellulosic Substrates. *Appl. Surf. Sci.* **2016,** *389,* 118–125.

Ibrahim, N. A.; Eid, B. M.; Abdel-Aziz, M. S. Effect of Plasma Superficial Treatments on Antibacterial Functionalization and Coloration of Cellulosic Fabrics. *Appl. Surf. Sci.* **2017,** *392,* 1126–1133.

Ibrahim, N. A.; Eid, B. M.; Abdellatif, F. H. H. Advanced Materials and Technologies for Antimicrobial Finishing of Cellulosic Textiles. In *Handbook of Renewable Materials for Coloration and Finishing*; **2018**; pp 303–356.

Ibrahim, N. A.; El-Zairy, E. M.; Eid, B. M.; Emam, E.; Barkat, S. R. A New Approach for Imparting Durable Multifunctional Properties to Linen-Containing Fabrics. *Carbohydr. Polym.* **2017**, *157*, 1085–1093.

Ivask, A.; Kurvet, I.; Kasemets, K.; Blinova, I.; Aruoja, V.; Suppi, S.; et al. Size-Dependent Toxicity of Silver Nanoparticles to Bacteria, Yeast, Algae, Crustaceans and Mammalian Cells In Vitro. *PLoS One* **2014**, *9* (7), 102–108.

Jha, A. K.; Prasadb, K.; Kulkarnic, A. R. Synthesis of TiO_2 Nanoparticles Using Microorganisms. *Colloids Surf. B* **2009**, *71*, 226–229.

Jung, W. K.; Koo, H. C.; Kim, K. W.; Shin, S.; Kim, S. H.; Park, Y. H. Antibacterial Activity and Mechanism of Action of the Silver Ion in *Staphylococcus aureus* and *Escherichia coli*. *Appl. Environ. Microbiol.* **2008**, *74* (7), 2171.

Kale, B. M.; Wiener, J.; Militky, J.; Rwawiire, S.; Mishra, R.; Jacob, K. I. et al. Coating of Cellulose-TiO_2 Nanoparticles on Cotton Fabric for Durable Photocatalytic Self-Cleaning and Stiffness. *Carbohydr. Polym.* **2016**, *150*, 107–113.

Karimi, L.; Yazdanshenas, M. E.; Khajavi, R.; Rashidi, A.; Mirjalili, M. Functional Finishing of Cotton Fabrics Using Graphene Oxide Nanosheets Decorated with Titanium Dioxide Nanoparticles. *J. Text. Inst.* **2016**, *107* (9), 1122–1134.

Kirthi, A. V.; Abdul Rahuman, A.; Rajakumar, G.; Marimuthu, S.; Santhoshkumar, T.; Jayaseelan, C.; et al. Biosynthesis of Titanium Dioxide Nanoparticles Using Bacterium *Bacillus subtilis*. *Mate. Lett.* **2001**, *65*, 2745–2747.

Lakshmanan, A.; Chakraborty, S. Coating of Silver Nanoparticles on Jute Fibre by in Situ Synthesis. *Cellulose* **2017**, *24*, 1563–1577.

Majeed, S.; Abdullah, M. S. b.; Dash, G. K.; Ansari, M. T.; Nanda, A. Biochemical Synthesis of Silver Nanoparticles Using Filamentous Fungi *Penicillium decumbens* (MTCC-2494) and its Efficacy against A-549 Lung Cancer Cell Line. *Chin. J. Natur. Med.* **2016**, *14* (8), 615–620.

Malarkodi, C.; Rajeshkumar, S.; Vanaja, M.; Paulkumar, K.; Gnanajobitha, G.; Annadurai, G. Eco-Friendly Synthesis and Characterization of Gold Nanoparticles Using *Klebsiella pneumoniae*. *J. Nanostruct. Chem.* **2013**, *3* (1), 30–37.

Marković, D.; Deeks, C.; Nunney, T.; Radovanović, Z.; Radoičić, M.; Šaponjić, Z.; et al. Antibacterial Activity of Cu-Based Nanoparticles Synthesized on the Cotton Fabrics Modified with Polycarboxylic Acids *Carbohydr. Polym.* **2018**, *200*, 173–182.

Marković, D.; Korica, M.; Kostić, M.; Radovanović, Z.; Saponjić, Z.; Mitrić, M.; et al. In Situ Synthesis of Cu/Cu_2O Nanoparticles on the TEMPO Oxidized Cotton Fabrics. *Cellulose* **2018**, *25*, 829–841.

Midha, V. K.; Vashisht, R.; Midha, V. Durability of Fluoropolymer and Antibacterial Finishes on Woven Surgical Gown Fabrics. *Fash. Text.* **2014**, *1* (1), 12.

Moazami, A.; Montazer, M. A Novel Multifunctional Cotton Fabric Using ZrO_2 NPs/Urea/ CTAB/MA/SHP: Introducing Flame Retardant, Photoactive and Antibacterial Properties. *J. Text. Inst.* **2016**, *107* (10), 1253–1263.

Nagar, N.; Devra, V. Green Synthesis and Characterization of Copper Nanoparticles Using *Azadirachta indica* Leaves. *Mater. Chem. Phys.* **2018**, *213*, 44–51.

Nair, B.; Pradeep, T. Coalescence of Nanoclusters and Formation of Submicron Crystallites Assisted by *Lactobacillus strains*. *Cryst. Growth Design* **2002**, *2* (4), 293–298.

Narayanan, K. B.; Sakthivel, N. Biological Synthesis of Metal Nanoparticles by Microbes. *Adv. Colloid Interface Sci.* **2010**, *156*, 1–13.

Ortelli, S.; Malucelli, G.; Cuttica, F.; Blosi, M.; Zanoni, I.; Costa, A. L. Coatings Made of Proteins Adsorbed on TiO₂ Nanoparticles: A New Flame Retardant Approach for Cotton Fabrics. *Cellulose* **2018**, *25* (4), 2755–2765.

Oza, G.; Pandey, S.; Gupta, A.; Kesarkar, R.; Sharon, M. Biosynthetic Reduction of Gold Ions to Gold Nanoparticles by *Nocardia farcinica*. *J. Microbiol. Biotechnol. Res.* **2012**, *2* (4), 511–515.

Pandimurugan, R.; Thambidurai, S. UV Protection and Antibacterial Properties of Seaweed Capped ZnO Nanoparticles Coated Cotton Fabrics. *Int. J. Biol. Macromol.* **2017**, *105*, 788–795.

Pareek, V.; Bhargava, A.; Panwar, J. Biomimetic Approach for Multifarious Synthesis of Nanoparticles Using Metal Tolerant Fungi: A Mechanistic Perspective. *Mater. Sci. Eng.: B* **2020**, *262*, 114771.

Patil, M. P.; Kim, G. Eco-Friendly Approach for Nanoparticles Synthesis and Mechanism Behind Antibacterial Activity of Silver and Anticancer Activity of Gold Nanoparticles. *Appl. Microbiol. Biotechnol.* **2017**, *101*, 79–92.

Peiris, M. K.; Gunasekara, C. P.; Jayaweera, P. M.; Arachchi, N. D. H.; Fernando, N. Biosynthesized Silver Nanoparticles: Are They Effective Antimicrobials? *Memórias do Instituto Oswaldo Cruz* **2017**, *112*, 537−543.

Pelgrift, R. Y.; Friedman, A. J. Nanotechnology as a Therapeutic Tool to Combat Microbial Resistance. *Adv. Drug Deliv. Rev.* **2013**, *65* (13), 1803–1815.

Petkova, P.; Francesko, A.; Perelshtein, I.; Gedanken, A.; Tzanov, T. Simultaneous Sono-chemical-Enzymatic Coating of Medical Textiles with Antibacterial ZnO Nanoparticles. *Ultrason. Sonochem.* **2016**, *29*, 244–250.

Piveca, T.; Hribernika, S.; Kolarb, M.; Kleinschek, K. S. Environmentally Friendly Procedure for In-Situ Coating of Regenerated Cellulose Fibres with Silver Nanoparticles. *Carbohydr. Polym.* **2017**, *163*, 92–100.

Prabhu, S.; Poulose, E. K. Silver Nanoparticles: Mechanism of Antimicrobial Action, Synthesis, Medical Applications, and Toxicity Effects. *Int. Nano Lett.* **2012**, *2* (1), 32.

Prasad, K.; Jha, A. K.; Kulkarni, A. R. Lactobacillus Assisted Synthesis of Titanium Nanoparticles. *Nanoscale Res. Lett.* **2007**, *2* (5), 248.

Rajabairavi, N.; Raju, C. S.; Karthikeyan, C.; Varutharaju, K.; Nethaji, S.; Hameed, A. S. H.; et al. *Biosynthesis of Novel Zinc Oxide Nanoparticles (ZnO NPs) Using Endophytic Bacteria Sphingobacterium thalpophilum*; Cham, **2017**.

Raliya, R.; Biswas, P.; Tarafdar, J. C. TiO₂ Nanoparticle Biosynthesis and Its Physiological Effect on Mung Bean (*Vigna radiata* L.). *Biotechnol. Rep.* **2015**, *2*, 22–26.

Ramezani, N.; Ehsanfar, Z.; Shamsa, F.; Amin, G.; Shahverdi Hamid, R.; Esfahani Hamid, R. M.; et al. Screening of Medicinal Plant Methanol Extracts for the Synthesis of Gold Nanoparticles by Their Reducing Potential, *Zeitschrift für Naturforschung B* **2008**, *63*, 903–908.

Ratnika, V.; Bhadauria, S.; Gaur, M. S.; Renu, P. Copper Nanoparticles Synthesis from Electroplating Industry Effluent. *Nano Biomed. Eng.* **2011**, *3* (2), 115–119.

Raza, Z. A.; Anwar, F.; Ahmad, S.; Aslam, M. Fabrication of ZnO Incorporated Chitosan Nanocomposites for Enhanced Functional Properties of Cellulosic Fabric. *Mate. Res. Express* **2016**, *3*, 1–12.

Reddy, G. S.; Saritha, K. V.; Reddy, Y. M.; Reddy, N. V. Eco-Friendly Synthesis and Evaluation of Biological Activity of Silver Nanoparticles from Leaf Extract of *Indigofera barberi* Gamble: An Endemic Plant of Seshachalam Biosphere Reserve. *SN Appl. Sci.* **2019**, *1* (9), 968.

Rehan, M.; Barhoumc, A.; Asschec, A. G.; Dufresnef, A.; Gätjenb, L.; Wilken, R. Towards Multifunctional Cellulosic Fabric: UV Photo-Reduction Andin-Situ Synthesis of Silver Nanoparticles Into Cellulose Fabrics. *Int. J. Biol. Macromol.* **2017,** *98*, 877–886.

Sharma, J. K.; Akhtar, M. S.; Ameen, S.; Srivastava, P.; Singh, G. Green Synthesis of CuO Nanoparticles with Leaf Extract of *Calotropis gigantea* and Its Dye-Sensitized Solar Cells Applications. *J. Alloys Compd.* **2015,** *632*, 321–325.

Sharma, N. C.; Sahi, S. V.; Nath, S.; Parsons, J. G.; Gardea- Torresde, J. L.; Pal, T. Synthesis of Plant-Mediated Gold Nanoparticles and Catalytic Role of Biomatrix-Embedded Nano-materials. *Environ. Sci. Technol.* **2007,** *41* (14), 5137–5142.

Sunkar, S.; Nachiyar, C. V. Biogenesis of Antibacterial Silver Nanoparticles Using the Endophytic Bacterium *Bacillus cereus* Isolated from *Garcinia xanthochymus*. *Asian Pacif. J. Trop. Biomed.* **2012,** *2* (12), 953–959.

Suryaprabha, T.; Sethuraman, M. G. Fabrication of Copper-Based Superhydrophobic Self-Cleaning Antibacterial Coating Over Cotton Fabric. *Cellulose* **2017,** *24*, 395–407.

Venkataraman, D.; Kalimuthu, K.; Sureshbabu, R. K. P.; Sangiliyandi, G. *Metal Nanoparticles in Microbiology*; Vol. XI; Springer, 2011.

Vigneshwaran, N.; Kumar, S.; Kathe, A. A.; Varadarajan, P. V.; Prasad, V. Functional Finishing of Cotton Fabrics Using Zinc Oxide-Soluble Starch Nanocomposites. *Nanotechnology* **2006,** *17* (20), 5087–5095.

Vijayakumar, S. Eco-Friendly Synthesis of Gold Nanoparticles Using Fruit Extracts and in Vitro Anticancer Studies. *J. Saudi Chem. Soc.* **2019,** *23*, 753–761.

Xiangqian, L.; Huizhong, X.; Chen, Z.; Chen, G. Biosynthesis of Nanoparticles by Micro-organisms and Their Applications. *J. Nanomater.* **2011,** *2011*, 1–16.

CHAPTER 5

Nano-Finishing on Woolens

AJAY KUMAR*, D. B. SHAKYAWAR, SEIKO JOSE, VINOD KADAM, and
N. SHANMUGAM

*Textile Manufacturing and Textile Chemistry Division,
ICAR Central Sheep and Wool Research Institute, Avikanagar, India*

Corresponding author. Email: kumarajay8171@gmail.com

ABSTRACT

Nanoparticles (NP) are gaining importance day by day in the textile research. Textile industries started working on specialty nano finishes in bulk production. Metal NPs, viz., Ag, ZnO, TiO_2, are being used for various functional finishes. The finishing of textile goods ultimately depends on the end uses. The apparel quality fabric needs softer finish, whereas an industrial fabric may need fire-retardant finish. Apart from the conventional textile finishes, another important area is the treatment of textile fabric with NPs for various industrial applications. Water-repellent and conductive polymer-coated industrial textiles are widely used for specific applications. Many NPs could find application on composites. NP-embedded spinning of synthetic fibers is another emerging area of research interest.

Wool accounts for only 1.6% of world's total fiber production; even then it has held the status of luxury fiber due its uniqueness,—such as high elasticity, high moisture content—aesthetics, and self-extinguishing properties that enables its use in premium apparels, carpets, upholstery, etc. Out of the total 2 million tonnes of global wool production, fine wool (18–22 µ) shares 60% and goes into apparel industries. The medium-coarse and coarse wool (24–40 µ) account for 35% and used for furnishing fabrics, carpets, upholstery, blankets and industrial felts, etc. The remaining wool is high-coarse quality (above 40 µ) and underutilized. Reports are available on

Fundamentals of Nano-Textile Science. Prashansa Sharma, Devsuni Singh & Vivek Dave (Eds.)
© 2022 Apple Academic Press, Inc. Co-published with CRC Press (Taylor & Francis)

the application of NPs on wool for fire retardancy, self-cleaning, ultraviolet protection, moth repellence, and antimicrobial finishes. Keeping the above points in mind, this chapter deals with different nano finishes on wool.

5.1 INTRODUCTION

Woolen fabrics are popular among society due to their unique functionality of thermal insulation. They also have aesthetic properties of lightweight, wrinkle resistance, hydrophobicity, etc. Usually, wool fabrics are not radially washed, rather they are dry cleaned after numbers end-use. Being a keratin protein, woolen are not infected by most of the insect and microorganism species but prone to attack by certain insects (wool moth) and fungi, and the microorganisms produce pungent odor during prolong storage. This may cause deterioration in fabric functionality, hygiene, and aesthetic appearance (Lam et al., 2012). Microorganisms can grow under favorable conditions, viz., moisture and temperature, availability of dust, dirt, and sufficient food through perspiration and other body secretions, fats, and skin particles (Yu et al., 2015). Most of them are not safe for both fabrics and human. These pathogens may cause diseases or decay the fabric quality. Gram-positive bacteria on undergarments produce odor and gram-negative bacteria produce odor and cause ulcer on skin and diaper rash, etc. The potential threat to cotton and wool fabrics is to their mechanical and functional properties (Zang et al., 2018). Similar to other fabrics, woolens are also in direct contact with consumer skin and infested fabric can affect the skin health. A fabric in contact with infected skin serves as a carrier for fungi and bacteria, thereby facilitating their propagation. McQueen et al. (1997) investigated the bacterial population in cotton and polyester fabrics. They found bacterial presence in all fabrics up to 28 days once infested. The bacterial population is relatively stable in the case of wool fabric while it declined in polyester fabrics. The consumers' growing awareness toward hygiene and aesthetics in their active lifestyle increased the demand for antimicrobial fabrics; for better hygiene in the products antimicrobial fabrics have different end uses like medical, pharmaceutical, packed food industries, and so on.

To produce an antimicrobial/antifungal fabric is a challenging task for researchers, and to achieve this highly effective synthetic anti-moth agents are envisaged. The microorganisms have cell walls made of polysaccharides to protect it from external factors and intercellular components, and it consist of several enzymes and nucleic acids. Gao and Cranston (2008) found that all the biochemical reaction responsibility for microbial growth and functionality (odor, infection, etc.) of the microorganism is facilitated by available enzymes within the polysaccharides cell. Antimicrobial agents attack and damage the cell walls or disrupt the normal functionalities of microorganisms/bacterial cells. According to chemical classifications, there are good agreements for biocides or biostat to be used as an antimicrobial agent as their mechanism of action is well correlated to impact active substance to antimicrobial fabrics (Simoncic and Tomsic, 2010). The commercially applicable synthetic antimicrobial agents are quaternary ammonium compounds, polyhexamethylene biguanide, triclosan, and N-halamines. These chemicals have very high antimicrobial efficacy of 90%–100% (Zain et al., 2018). The use of synthetic antimicrobial/antifungal/anti-moth agents is restricted due to environmental issues and ecotoxicity. Sometimes, they are also harmful to human health causing skin irritation (itching), and continuous use of these antimicrobial-treated fabrics results in induction of resistance to bacteria. Keeping in view these adverse effects, the synthetic antibacterial agent's research are looking into ancient processes and practices to impart antimicrobial/antifungal properties. The well-proven example is spices- and herbs-treated cotton fabric used in wrappings to preserve mummies in the Egyptian dynasty. Several metals are also reported to have antimicrobial/antibacterial activity. Among them, silver is extensively reported to the highest level of antimicrobial/antibacterial activity. The mechanism of silver that causes highest antimicrobial activity is due to the binding of silver ions with disulfide (S-S) and sulfhydryl (SH) groups in the microbial cell wall proteins which disrupts the metabolic processes and ultimately causes the death of microbes (Silvestry-Rodriguez et al., 2007). Some of the natural dyes having high tannin and naphthoquinones content, viz., pomegranate rind, curcumin, henna, and walnut, have antimicrobial properties. Curcumin dye extract is commonly used to impart antimicrobial ability to wool fabric and food items. Han and Yang (2005) found durability of the curcumin dye extract treatment against *Staphylococcus aureus* and *Escherichia coli* to 45% and 30% respectively even after 30 domestic washing cycles. The natural dye sources with high fashion content also imparted anti-moth efficacy to woolens (Shakyawar et al., 2015). Zhao (2015) found very good antimicrobial activity against *S. aureus* and *E. coli* when treated with chitosan extracted from outer shells of crustaceans like crabs and shrimp, the cell walls of fungi, etc.

Like silver other metal salts like titanium, zinc, copper, magnesium, and gold are also having antibacterial efficacy. But most of these metal salts are nonreactive and need binding or cross-linking agents for the desired functionality. However, when metallic silver is exposed to aqueous conditions, some ionic silver (Ag^+) is released as in the case of silver nitrate. It is water soluble and used as an antiseptic agent over the years. The use of metal salt is associated with environmental and health issues. Silver ions have low toxicity to the human cell, high thermal stability, and low volatility (Marambio-Jones et al., 2010).

Nanotechnology is a promising way to textiles due to low chemical usage, low energy costs, and remarkable charge in physical and mechanical properties of the fabrics without affecting its breathability and handle (Sundarrajan et al., 2010). Nanoparticles (NP) can be prepared by synthesis and stabilization of various metal ions, like Ag, Au, Fe, Co, Se, Pt, Pd, and metal oxides like CuO, ZnO, Fe_3O_4, and so on. The most important characteristic of NPs is its high surface area, and thus, for the treatment of fabric for any of the required functionality very little quantity of NPs is required with the advantage of more uniform distribution in the fabric system. The green synthesis of these metal ions and metal oxides NPs will further improve the hygiene and customer satisfaction level. Plant extracts, biopolymers, glucose, etc. are the naturally occurring materials that provide suitable reducing agents for NPs' synthesis and stabilization (Virkute and Verma, 2011; Iravani et al., 2011; Hebbalalu et.al, 2013; Raveendran et al., 2003).

5.2   APPLICATION OF NANO FINISHES ON WOOLEN FABRICS AND THEIR ANTIMICROBIAL PROPERTIES

Barani et al. (2014) applied AgNPs/Lecithin on wool fabric for imparting antibacterial properties. The scoured fabric is treated with both pad-dry-cure and exhaust method. In pad-dry-cure wool fabric were impregnated in AgNPs/Lecithin with M:L Ratio of 1:40 for 15 min. In colloidal AgNPs, dispersion was found to have a silver content of 300 ppm and lecithin to silver ratio varied to 0, 0.2, 1, and 2. The fabrics were squeezed to 80% wet pick-up and dried at 80°C for 20 min, followed by curing at 130°C. In exhaust method MLR was kept the same and the scoured fabric was introduced in the bath at 40°C. The treatment was performed at 90°C for another 30 min. The treated wool fabrics were then washed and dried at 80°C. The bactericidal properties of AgNPs-loaded wool fabrics were evaluated against gram-positive and gram-negative bacteria, *S. aureus* and *E. coli*, respectively. Fabrics with high silver content showed the larger inhabitation zone. The antibacterial efficacy of AgNPs is found more with exhaust treated fabrics. The bacterial reduction efficiency for exhaust treated wool fabric against *S. aureus* increased from 91.15% to 98.8% compared to pad-dry cure method treated fabrics which showed bacterial reduction from 40.55% to 97.3%.

The antibacterial efficiency against gram-negative bacteria (*E. coli*) for the same lecithin to silver ratio was observed to increase from 18.1% to 87.3% for exhaust treated fabrics, and for a lecithin to silver ratio of 2 with pad-dry-cure method only to 61.2%.

Hebeish et al. (2011) investigated the antibacterial activity of green-synthesized AgNPs on cotton fabrics. The 500-ppm AgNP was synthesized by mixing $AgNO_3$ and hydroxypropyl starch under optimum conditions. Before administrating this mixture to cotton fabrics, the solution was vigorously stirred for 15 min. The colloidal dispersion of AgNPs was diluted in two concentrations of 100 and 50 ppm with distilled water. The fabric was immersed in colloidal dispersions for 30 s. The fabric was padded to 100% wet pick-up and dried at 70°C for 3 min followed by curing at 150°C for 2 min. Both the treatments were found to be highly efficient with 96% and 98% against gram-positive bacteria (*S. aureus*) and gram-negative bacteria (*E. coli*), respectively. They also tested the durability of treatment against washing and found quite high antibacterial efficacy of 80% after five washing cycles, which remains to 60% even after 20 washing cycles. In their experiment, Saran and Demir (2016) first prepared 0.2% chitosan solution by dissolving chitosan in 1% acetic acid solution. Second, wool fabric was impregnated by the pad-dry-cure method by keeping a wet pick-up of 50%, followed by drying at 80°C for 5 min and subsequent curing at 100°C for 3 min. The wool fabrics were pretreated by three different methods, viz., untreated, plasma, and enzyme. The pretreated fabrics were applied to chitosan, chitosan NPs, and chitosan-AgNPs. The bare chitosan-treated wool fabric observed bacterial reduction of 73.15% and 81.83% for *S. aureus* and *E. Coli* respectively. The plasma and enzyme pretreatment to wool fabric showed higher antibacterial efficacy of 77.02% and 80.64% respectively for *S. aureus* and 83.16% and 89.21% respectively for *E. coli*. The application of chitosan in nano form showed higher antibacterial efficiencies for gram-positive (*S. aureus*) to 84.86%, 89.99%, and 90.35% respectively and for gram-negative (*E. coli*) to 89.56%, 91.55%, and 93.98% respectively. The combination of chitosan-AgNPs was found to contain highly effective antibacterial properties. The antibacterial efficiencies for wool fabric finished with chitosan-AgNPs without pretreatment was observed to have bacterial reduction of 96.38% for *S. aureus* and 97.66% for *E. coli*. Whereas, the pretreatment of enzyme and plasma to wool fabric followed by chitosan-AgNPs finishing observed enhancement of antibacterial efficacy to more than 99% for both *S. aureus* and *E. coli*. Similar observations were also reported (Samant et al., 2020).

To impart antibacterial finishing to another protein fiber material, viz., silk fabric, Chakarborty et al. (2015) applied chitosan-neem nano emulsion (CNNE) to silk fabric. The fresh and clear CNNE was prepared by adding 30 g chitosan-neem nano complex in phosphate buffer solution (150 mL, pH 7.4), followed by continuous stirring for 10–15 min. The silk fabric has been pad-dry-cure with 70% wet pickup, dried at 95°C for 5 min and cured at 150°C for 3 min sequentially. The antibacterial activity of CNNE to silk fabric was optimized to 1.5% concentration of CNNE. The antibacterial efficacy improved with increasing concentration of CNNE for both gram-positive bacteria (*S. aureus*) and gram-negative bacteria (*E. coli*). The observed bacterial reduction at 0.5% CNNE is 82.3% and 80.2% for *S. aureus* and *E. coli* respectively. Whereas, bacterial reduction improved to 99.98% and 99.88% for *S. aureus* and *E. coli* respectively for silk fabric treatment with 1.5% concentration of CNNE.

5.3   ANTI-MOTH PROPERTIES TO WOOLENS USING NANOPARTICLES

Wool is one of the most important natural fibers. Chemically, wool is a protein fiber containing a complex structure of 18 amino acids. Moth attack is one of the inherent issues associated with woolens. The larvae of carpet beetle (*Anthrenus verbasci*) attack the wool and cause severe damages to woolens. This pest takes the wool as its preferred food because of the cystine linkage. That is why moth attack is only visible in wool, but not in silk, though both are protein fibers. Entomologists, chemists, and textile scientists are working for a long time to impart moth repellency to woolens for the last few decades. Many of the chemical formulations were tried and found successful. The use of NPs for pest repellency is still an under exploited area. Very few works have been reported in the literature on the uses of NPs for attaining moth repellency.

Stadler et al. (2010) studied the insecticidal activity of nano alumina against two insect pests, viz., *S. oryzae L. and Rhyzopertha dominica (F.)*. These two pests are generally found in food supplies. The study revealed that there is a significant hike in the mortality rate of these pests after continuous exposure of NPs for 3 days. One of the main drawbacks of the use of NPs in the crops is the lack of proper knowledge of the toxicity of the NPs, which restrict its uses. Jose et al. (2018) studied the efficiency of nano kaolinite, naturally occurring clay, for achieving moth repellency on *A. verbasci*. The nano kaolinite was applied to lightweight wool fabric with two different methods. In the first method, the fabric was treated with nano kaolinite at room temperature by pad-dry method, keeping 100% expression. In the second method, the treatment was performed at 80°C for 30 min. In both cases, the MLR was kept at 1:30. After treatment, the fabric was dried at ambient temperature. The anti-moth test was performed as mentioned elsewhere. In addition to that, the visual assessment of the moth-attacked fabric was also performed. The bioassay results showed that application of the 0.1% solution of nano kaolinite caused no death to the moth. The mortality rate of larvae was observed to reach a maximum of 30% with 1.0% application of nano kaolinite dispersion on moth after 60 min of observation. The anti-moth test revealed that the untreated fabric had a weight loss of 12.85%. A decrease in weight loss was observed with the increase in the concentration of nano kaolinite. The fabric weight loss after the incubation period with the treated samples was observed in a range of 1.52–4.28 mg with an increasing mortality rate according to the concentration of nano kaolinite. The least fabric damage was observed for a 1.5% nano kaolinite application by the pad-dry method. While comparing the two methods of applications, the room temperature application was found to deliver better results than exhaust method of application. Concerning mortality rate, after the incubation period, it was found that 86% of the larvae were alive with control fabric, however; in the nano-kaolinite-treated fabrics, only 30% were alive. While compiling the bioassay results and the anti-moth test, it was concluded that since the nano kaolinite is not toxic to the larvae, the least fabric damage and the higher mortality rate may be because of the antifeedant property of nano clay. In another work, Nazari et al. (2014) reported the efficiency of nano TiO_2 against *A. verbasci*. The scoured fabric was treated with 75 g/L citric acid and 45 g/L sodium hypophosphite at 75°C for 60 min. After treatment, the fabric samples were dried at 95°C, followed by curing at 120°C for 3 min. The cured fabrics were then impregnated with nano TiO_2 dispersion with various concentrations (0.1%–1.5 %). The treatment was performed at 75°C for 55 min, followed by drying and curing at 110°C for 3 min. After treatment, the unfixed NPs were removed using 1 g/L Na_2CO_3 and 1 g/L nonionic detergent at 70°C and dried at room temperature. To check the anti-moth efficacy, 2.0 g of treated wool fabrics were placed in the plastic breeding plate and 10 larvae of *A. verbasci* were introduced into it. The experiment was conducted at 25 ± 2°C, 22%–26% RH, and a photo-period (L:D) of 16:8 h for 6 months. After the completion of the experiment, the samples were visually assessed for its visual damage and weight loss according to ISO

3998-1977. The results showed that the control sample registered a weight loss of 17.33%. The nano-TiO$_2$-treated samples showed less fabric damage, and the least damage was observed with 1.15% of nano TiO$_2$ application. The presence of the NPs was further confirmed by SEM analysis.

5.4   WATER-REPELLANT NANO FINISHES FOR WOOLENS

Water repellency is the phenomenon by which water does not penetrate or wet the fabric surface. Water-repellent finishing chemicals reduce the free energy at fiber surface. The surface properties of textiles can be altered according to the requirement by implementing appropriate techniques using nanotechnology (Jose et al., 2018). The introduction of nanotechnology and nano chemicals explored new ways to impart water repellency to textiles. Conventionally, hydrophobizing agents, such as paraffin waxes, silicones, silanes, and fluorinated polymers, are used to impart water repellency to textiles. However, recent researches show the possibility of using nano compounds like SiO$_2$, ZnO, and TiO$_2$. The benchmark for water-repellent finish to fabric is expressing a contact angle of a drop above 150°, the phenomenon also known as "Lotus-Effect". Hydrophobic/superhydrophobic surface on textile fabrics can be imparted either by creating rough structures on a surface or by modifying a rough surface using materials with low surface free energy. While applying water-repellent nano finishes on the textile, the surface of the fabric becomes hydrophobic and drops of water easily drips off. The water-repellent nano finishes significantly maintain air permeability or the breathability of the fabric.

The wool fabric surface is having an inherent scale structure, which prevents the immediate water penetration inside the fiber. Thus, in comparison with the other natural fibers, the wool fabric is having better water-repellent properties. The literature on imparting the water-repellent finishes on wool fabric is scanty. Wang et al. (2010) synthesized silica NPs from Tetraethyl orthosilicate (TEOS) by sol-gel method. The dispersion was then padded on the fabric and dried. Subsequently, curing was performed at 110°C for 1 h. The presence of NPs is confirmed by SEM. As a result of nano silica treatment, the water contact angle was raised up to 172.6 degree. The study concluded that the nano silica can be effectively utilized for imparting superhydrophobic finishes to woolens. Chattopadhyay et al. (2016) imparted water-repellent finishing using synthesized colloids of nano paraffin and nano copper on jute fabric. The nano copper was applied through the exhaust method followed by the application of nano paraffin dispersion. The efficiency of the treatment was assessed in terms of contact angle. The result inferred that the application of nano copper enhanced the water repellency from 78 to 130 degree. Xue et al. (2008) reported superhydrophobicity of the cotton fabric using nano TiO_2. The nano TiO_2 was prepared from tetrabutyl titanate by sol-gel method. The coated fabrics were analyzed with SEM. The water contact angle of the cotton fabric was found to be elevated from 95 to 160 degree, with the coating of nano TiO_2 along with stearic acid.

Borah et al. (2019) extracted nano silica from rice husk and applied on eri silk fabric to achieve superhydrophobicity. To enhance the water-repellent properties, a silicone-based polymer was also used. The effect of nano silica application on the water-repellent properties, as well as on the physio-mechanical properties, was analyzed. The study revealed that the application of nano silica showed hydrophobicity by elevating the contact angle from 87.5 to 112.9 degree. The application of 1.0% silicon polymer further enhanced it to superhydrophobic level (up to 154 degree). The nano silica coating slightly enhanced the whiteness and brightness of eri silk fabric. The physio-mechanical properties of the treated fabric were found to be intact.

5.5 ULTRAVIOLET PROTECTION USING NANOPARTICLES

The solar ultraviolet (UV) radiation is composed of UV-A (400–315 nm), UV-B (315–290 nm) and UV-C (290–200 nm) region (Altun and Becenen, 2017). Among these, UV-A and UV-B are hazardous to human health. Prolonged UV exposure may cause sunburn, skin cancer, photo keratitis, and cataract (Yang et al., 2004; Rezaie et al., 2017; Sankaran et al., 2020). Textiles are good means of UV protection which is measured in terms of solar protection factor or ultraviolet protection factor (UPF). Fabrics with a UPF value in the range of 15–24 are referred as good UV protection while 25–39 as very good, and 40 or greater as excellent UV protection (Mong-kholrattanasit, 2011). Another measurement is transmittance (%) of dyed fabrics. Shabbir et al. (2018) deduced that if it is less than 5%, it means good UV protection. The level of UV protection depends on fiber type and fabric construction parameters (Pandey et al., 2018; Jose et al., 2019). The wool fabric has poor photo stability and exposure to light causes yellowing of the fabric (Montazer and Pakdel, 2010). UV protection can be enhanced by a finishing process in which fabric is coated with UV-blocking material such as TiO_2. TiO_2 has a higher refractive index than most of the other semiconductors (Yang et al., 2004). However; conventional finishing may alter the inherent fabric properties like bending and air permeability.

NPs can improve photo stability and absorb UV rays more than submicron size particles without affecting other fabric properties. Inorganic nano materials like metal and metal oxides have received considerable attention recently. Zinc NPs improved the photo stability of wool while the stability between zinc NPs and fabric surface has been increased by creating the linkages between them (Becheri et al., 2008). Ibrahim et al. (2018) observed improvement in UPF values of zinc NPs when attached to wool; this reflected their ability to block the harmful UV-B radiation due to their increased surface area and intense absorption in the UV range rather than reflecting and/or scattering. The UPF further improved (up to 74) with cross-linking with citric acid.

TiO_2 NP also presents good affinity with fabrics because of its very large specific surface area and high surface energy and activity (Yang et al., 2004; Lin et al., 2006). Montazer and Pakdel (2010) applied TiO_2 NPs to wool fabric with the aid of cross-linking with polycarboxylic acid and a catalyst. TiO_2 NPs have shown higher UV absorption in the range of 300–350 nm. TiO_2 NPs encapsulated in SiO_2 shell reduced its photocatalytic activity which retained the photo stability for longer duration (Zhang et al., 2017).

Recently, NPs from cupric oxide have been studied for UV protection. The high surface area to volume ratio of cupric oxide semiconductor NPs adsorbs equal or higher energy than the band gap energy under exposure to UV light. Thus, in situ synthesized cupric oxide NPs showed better UV protection (Rezaie et al., 2017; Rezaie et al., 2018). Copper-oxide-coated wool fabric showed better UV protection than the untreated fabric (Altun and Becenen, 2017).

5.6 FIRE-RETARDANT NANO FINISH TO WOOLENS

Wool has an inherent flame resistance characteristic having a high LOI value of 25 compared to other fibers (Scott, 2005). Keeping this in view, most of the research work is associated to impart the flame-resistant property to fibers other than wool. Although wool fibers are the most preferred for making furnishing products like carpets, curtains, upholstery, it need to be more fire-retardant. Fire-retardant chemicals are mostly added to the textile product to suppress, reduce, and delay the combustion of materials (Samanta et al., 2017). Different chemicals and methods were used in developing fire-retardant wool, viz., titanium complexes, zirconium complexes, methylated phosphnamides, and so on. The major issues with these chemicals are a higher add-on, noticeable strength loss, stiffness, in addition to toxicity.

They also need to be replaced with halogen-free fire-retardant formulations. The environment-friendly additives, viz., organic and inorganic nano composites based on silicon dioxide and clay materials, are more appropriate materials which are not only ecofriendly but also on applying with NPs resolve the issue of high add-on for desired functionality. Azlina et al. (2016) synthesized silicone dioxide (SiO_2) nano structure with sol-gel method using tetra orthosilicate, tetra exthoxyislone as the precursor. The calcination temperature and time of ageing were found to be the factors deciding the size of NPs. The parameters optimized to 600°C temperature for 2 h. The size of NP was analyzed to be in the range of 79.68 to 87.35 nm respectively.

Wei et al. (2014) prepared silica sol to improve fire retardant properties to wool fabric. The preparation of dispersion can use TEOS as precursor, ethanol as co-solvent, and hydrochloric acid as a catalyst. The mixture is added to distilled water dropwise to get pure sol. To enhance the efficacy of silica sol, doped silica sol was also prepared by adding ammonium biphosphate. The wool fabrics were treated with prepared sols for the duration of 1 h under vibrating conditions afterward cured at 160°C for 6 min. The dosage of water added to the reaction solution has significant effect on team retardant properties of the treated fabric. For $n(TEOS):n(H_2O)$ less than 1:4, the LOI value of treated fabric showed a mixed trend; a higher value of 26.6% was observed for a ratio of 1:1, it then decreased to 25.8% at a ratio of 1:2, and again increased to 26.1% for 1:4. It was deduced that at lower ratios, the hydrolysis was not completed and particles are fairly small in size and better penetrated and adhered tightly in the fabric porosities, viz., yarns interstices in fabric and fibers interstices in the yarns. With increasing water content, $n(TEOS):n(H_2O)$ ratio higher than 1:4, the dilution of sol and completion of hydrolysis result in larger size of silica NPs causing to nonsignificant increment to the LOI value to the treated fabric. Dosages of ethanol also showed the dilution effect. For pure silica sol treatment to wool fabric, the dosage of $n(EtOH):n(TEOS)$ was optimized to 3, which observed the highest LOI value of 26.7%. Further dilution with co-solvent drastically reduced the LOI value. The reaction condition was optimized to 70°C temperature for 3 h with an observed LOI value of 26.7%. However, high LOI value of 28.1% was found in the reaction condition of 80°C for 3 h with yellowing and strength loss in fabric.

The obtained higher seam retardant properties the fabric is also treated with ammonium bisulfate doped silica sol. The stability of ammonium bisulfate render to molar ratio n(H2O):n(TEOS) as 20:1 was selected with different concentrations of LOI value between 2% to 12% of ammonium bisulfate was doped in pure silica sol. The wool fabric was doped in silica sol keeping other conditions constant as in the pure silica sol treatment. The doping with ammonium bisulfate showed significant enhancement of LOI value even with low concentration (2% of ammonium bisulfate), to LOI value 28.7% compared to pure silica sol treatment (LOI26.7%). The maximum LOI value of 29.7% has been observed at 8% concentration of ammonium bisulfate. The doping of ammonium bisulfate synergized the fire-retardant properties to the wool fabric. The overall enhancement of flame-retardant characteristics of wool fabric was achieved by 18% to LOI 29.7% against 25.2% for raw wool.

Jose et al. (2018) studied the fire-retardant performance of nano kaolinite (nano-clay) on lightweight wool fabric. Nano kaolinite dispersion of different concentration, viz., 0.5%, 1.0%, 1.5%, 2.0%, and 2.5% were prepared and applied by exhaust and pad-batch method. In the exhaust method, the fabric to dispersion ratio (MLR) was kept to 1:20. The treatment was carried out in a closed steel beaker attached with infrared heating at 80°C for 30 min, followed by drying in room temperature. Whereas, in the pad-batch method, wool fabrics were immersed in the kaolinite dispersion for 30 min, followed by padding with 100% expression. The fabrics were rolled with aluminum foil and kept for 24 h to avoid any moisture-evaporation losses. The conditioning would help in adhering/anchoring of nano kaolinite particles in the fiber interstices of fabric system. After the required time, the fabric was unwrapped and dried at room temperature. The results revealed that high LOI value of 33 was achieved for 2.0% nano kaolinite treatment on wool fabric through exhaust method, against LOI value of 25 of the control sample. The improvement in LOI value with pad-batch method achieved maximum to 31 with 2.5% nano kaolinite treatment. The flammability characteristic in terms of burning time(s) and char length (mm) was tested on inclined flammability tester as per ASTM D 1230 (1994). The result showed that the untreated fabric sample was burned completely, once ignited. At lower concentrations of nano kaolinite (i.e., 0.5% and 1%), the treatment was not found effective, however, a higher concentration of nano kaolinite imparted effective flame-retardant properties to fine wool fabric. The treatment resulted in self-extinguishing of flame with no afterglow after burning. The lowest value of burning time and char length was observed for 2.5% concentration of nano kaolinite for both exhaust and pad-batch method. The results of burning time and char length are 10 s and 2.5 cm and 8 s and 3.3 cm respectively. The add-on of nano kaolinite was confirmed by higher ash content of 1.92% and 2.01% for 2.5% concentration for exhaust and pad-batch methods respectively against ash content of 0.73% for the control sample. To investigate the effect of treatment on physio-mechanical properties, the treated fabrics were tested for their strength value on a universal tensile testing machine (Instron-4695). No loss in fabric strength value (cN/tex) was observed for any of the treated samples with both the methods of treatment. Rather the fabric strength was observed to increase slightly by 8.3% and 8.9%, for 1.5% nano kaolinite (exhaust) and 0.5% nano kaolinite (pad-batch) treated fabric to 4.02 cN/tex and 4.04 cN/tex, respectively, against 3.71 cN/tex of the control fabric. Thus, the flame-resistance treatment with nano kaolinite is an easy and more effective way to achieve higher flame-retardant property to fine wool fabric without any strength loss.

5.7 MULTIFUNCTIONAL PROPERTIES TO WOOLENS USING NANOPARTICLES

The nano metal ions and metal oxides impart not only layered targeted functionality but also multifunctional properties to the woolens. Ki et al. (2007) finishes Marino wool fiber with sulfur nano silver-based collide (SNSE). Nano silver collides include sulfur compounds on nano-sized silver particles. Before treatment, wool fibers were cleaned with dichloromethane at 40°C for 30 min, followed by twice rinsing at 25°C for 10 min and equilibration in conditioning chamber at 20°C in 65% RH. The finish (SNSE) was applied by the dip-pad-dry-cure method at four different levels of 5, 10, 20, and 30 ppm. The wool fiber samples were dipped in a fresh colloidal solution for 10 min and squeezed in laboratory padding mangle under constant pressure. The treated wool fibers were dried at room temperature for 30 min and cured at 120°C for 5 min. The loading of sulfur nano silver particles is dependent on pick-up ratio of different concentration of nano silver. The treated wool fibers were used to prepare worsted yarn and fabric and wool/acrylic (50/50) blend to knitted fabric. The treatment of SNSE to wool fiber at a level of even 5 ppm is found to cause a significant bacterial reduction of *S. aureus* and *Klebsiella Pneumoniae* to 99.70%. For the wool and blended knit fabric, treatment concentration of 20 ppm is considered and tested the efficiency of SNSE for anti-moth efficacy. The specimen of a known weight is placed in a covered container for 14 with alive larvae of *Tinea pellionellea*. After incubation period, the container of control specimen observed all alive moths without any pupation. The fabric weight loss due to insect feeding waste was 40.54 mg, with visible surface damage and formation of holes in the fabrics. Whereas the treated wool fabric container has only one alive moth larvae in the container, the fabric weight loss observed reduction to 6.32 mg only (1/6 to control) with relative good difference in visible appearance compared to control sample. Being a conducting material, Ag treated fabric was also evaluated for anti-static propensity; the anti-static effect was observed with high concentration of SNSE on the wool surfaces. The anti-static effect is decreased for a higher concentration of 50 ppm of silver. Ibrahim et al. (2018) envisaged the ecofriendly multifunctional finishing of wool-blended fabrics using AgNPs and/or zinc oxide NPs. Scoured and bleached wool/ cotton and wool/viscose blend (50/50) were used for the experiment. An aqueous finishing dispersion was prepared by mixing carboxylic acid (citric or succinic acid) with known amount of AgNPs and/or zinc oxide NPs and distilled water using sodium hyphophosphates (SHP) as a catalyst in the ultrasonic bath for 20 . The fabrics were padded twice with 80% wet pick-up and microwave dried at 1300 W for 4 min. To remove the excess and unfixed reactant, treated fabrics were washed, rinsed, and dried at 50°C for 10 min. The SEM and EDX analysis confirmed the existence of Ag/Zn on the fabric substrate. The treatment extended multifunctional properties, viz., anti-bacterial, UV protection, and wrinkle resistance, to the wool blended fabrics. The nano finishing formulation of AgNPs/ZnONPs/CA/SHP showed a high level of durability. The multifunctional properties sustain even after 10 washing cycles.

Wool and silk are having considerable use in the textile industry both as independent material and blended with other fibers for specific end-use (Okeila et al., 2008). Titanium dioxide has a very good functionality as TiO_2 has a great affinity for hydroxyl and carboxyl groups (Meilert et al., 2005; Dhananjayan et al., 2001; Yu et al., 2002). Being keratinous structure, wool and silk fibers have scanty availability of these functional groups compared to other fibers, the free functional groups are about to 50% only, and for improving the affinity of NPs special pre-treatment is desirable to increase the number of functional groups. The acylation of wool by succinic acid anhydride increases the number of carboxyl groups on the substrate. This shows enhanced binding of TiO_2 NPs and durable finishing treatment.

The wool fiber also has lower photo stability compare to other textile fibers. The photo-oxidation of wool (yellowing process) largely depends on irradiation exposure time (Altun and Becenen, 2017). The rate of photoyellowing of wool can be controlled by the application of UV blocker. Inorganic metal oxides, like TiO_2 ZnO, are some of the universal UV blockers. Nano TiO_2 is found most suitable for self-cleaning and as UV-blocker agent.

To enhance the functionality, Montazer and Seifollah Zadeh (2011) envisaged the enzymatic pretreated wool-blended fabric to increase the adsorption of TiO_2 NPs for multifunctional properties to the fabric. To increase the reactive sites/negative groups on fabric, the samples were treated with 5% (w/w) protease enzyme at 60°C for 60 min to modify the wool surface. The blended fabric was also treated with Lipex 1% (w/w) for surface modification of polyester component at 30°C for 30 min with continuous stirring, separately. The enzymatic pretreatment reduces fabric weight by 5%. The dispersions were prepared with the combination of nano TiO_2 at different levels (0.25, 0.5, 0.75, and 1.0), butane tetra carboxylic acid (BTCA) at different level (0, 5, 10, and 15), SHP (0, 2.5, 5, 7.5, and 10) in distilled water and then sonicated for 10 min. The pretreated fabric was immersed in nano TiO_2 dispersion for five durations 0, 15, 30, 45 and 60 min followed by drying at 70°C and curing at 180°C for 1.5 min. Least photoyellowing $\Delta Y1$ (UV) value 0.13 was observed with enzymatic pretreated fabric finished to 30 min duration with dispersion solution by nano TiO_2 (1.0%), BTCA (10%), and SHP (5%) against 13.40 for untreated fabric sample. Thus, nano TiO_2 treatment not only imparts UV protection but also makes the fabric UV absorbent. The antibacterial efficacy of nano-TiO_2-treated fabric was tested against gram-negative bacteria *E. coli*. The bacterial reduction is found to be 100% with 0.75% nano TiO_2 and 99% with 0.25% nano TiO_2. This proved the highly effective antibacterial properties of nano TiO_2 to the wool fabric.

The self-cleaning efficiency of treated fabric was evaluated in terms of photocatalytic decomposition of stained acid blue 113(dye) under UV and day light irradiation. The semiconductor photocatalysis through TiO_2 or UV irradiation is very effective for dye degradation. It absorbs the light energy for the catalysis and makes atmospheric oxygen highly active and the degradable hydroxide free radicals are followed to dye (acid blue 113) decomposition. The nano-TiO_2-treated sample (0.75%) observed color difference value 22.33 after 12 h of UV irradiation and 28.60 after 24 h of daylight irradiation against untreated samples having lower values of color difference, viz., 6.28 and 14.52, respectively. Thus, bio-catalysis or enzymatic modification helps to increase the NPs absorption on the fabric surface. The use of BTCA along with nano TiO_2 improves self-cleaning properties by allowing partial elimination of chromophores of stained acid blue 113 under UV and day light irradiation; this treatment not only protects the fabric from UV degradation but also absorbs UV irradiation. The nano TiO_2 has a lethal effect on microbes, bacteria or fungi.

The antibacterial feature of nano TiO_2 on oxidized wool fabric with the use of cross-linking agent for higher absorption of TiO_2 NPs is studied by Montazer et al. (2011) by using potassium permanganate as an oxidizing agent which would diffuse in wool and oxidizes all the cystine available in the outer layer of wool fabric, that is, epi-cuticle. The oxidized wool sample was treated by the dip-dry-cure method with dispersion solution prepared by using nano TiO_2, cross-linking agent (BTCA or citric acid), and SHP as a catalyst through sonication. The major cross-linking reactions between wool fabric and cross-linking agent are carried during the curing process at 120°C for 2 min. Oxidization and finishing allowed higher cross-linking wool surfaces. The BTCA treatment has higher deposition of cross-linking agents. The higher affinity of nano TiO_2 to the carboxyl group results in higher deposition of nano TiO_2 on the wool surface. The higher amount of available nano TiO_2 results in better antimicrobial properties to the samples and those treated with BTCA as cross-linking agent observed microbial reduction of most gram-negative bacteria *E. coli*. The mechanism of higher antibacterial efficacy of nano TiO_2 is a positive change (Ti^{4+}), especially under UV irradiation which initiated its photocatalytic reaction. The nano TiO_2 reacts with the negatively charged cell surface of bacteria and leads to leakage of intercellular substance, resulting in loss of habitat, metabolism of microorganism, and finally death. The antibacterial value in terms of % reduction of habitat zone for 0.75% nano TiO_2, 10.5% BTCA as cross-linking agent, and 6% SHP have been observed to be 60%. This is attributed to the higher amount of TiO_2 NPs on the substrate in comparison to nonoxidized wool sample treated with nano TiO_2, 10.5% citric acid, and 6% SHP that have observed 40% reduction of habitat zone.

The multifunctional finishing properties is also imparted to wool fabric through natural polymer zirconium dioxide (ZrO_2) as mordant for dyeing with natural dye. Taheri et al. (2015) investigated thyme for synthesizing ZrO_2 name particles with concentrations of 1%, 3%, 6%, and 9% as mordant for pre-, post-, and meta-mordanting for wool fabric during dyeing with thyme. The MLR for mordanting, that is, pre- and post-mordanting was kept to 1:50 and treatment conditions are temperature of 85°C for 1 hr. For natural dyeing with thyme, the pre-mordanted sample was immersed in dye liquor at temperature 40°C and the dyeing was performed at 85°C. The dyeing was carried out in acidic pH and followed by rinsing to have a neutral pH. The water droplet absorption time which represent water-repellent properties of fabric was observed higher for post-mordanted fabric samples. This is attributed to the formation of linkage between thyme and nano zirconium oxide and protein chain of wool. The hydrophobicity of wool fabric for all the mordant treatment was found increasing with increasing concentration of nano zirconium oxide (ZrO_2). The dye affinity of thyme was optimized at 6% concentration of ZrO_2 NPs.

The mordanting of wool fabric can cause chemical, physical, and morphological changes in wool fibers and may change the flammability properties of the substrate. The higher ignition time of more than double, that is, 5.71, 7.8, and 9.4 s for pre-, meta-, and post-mordanted fabric, respectively, was observed against 2.71 s for control sample. The burning rate after ignition (mm/s) and burn length (mm) for post-mordanted (ZrO_2) wool fabric was observed to be 1.91 mm/s and 18 mm against 29.52 mm/s and 30 mm of the control sample, respectively. The flame-resistance properties are attributed to the high heat insulation characteristics of nano zirconium oxide (ZrO_2 NPs). The pre- and meta-mordanting have comparatively lower deposition/ embedding of nano zirconium and accordingly observed less flame-resistance properties. However, the mordanting of either of the methods with nano zirconium dioxide (ZrO_2 NPs) can significantly improve the flame resistance properties of the wool fabrics.

The treatment of nano zirconium as mordant for dying with thyme showed good antibacterial properties on wool fabric. The presence of nano zirconium has synergic effect on antibacterial efficacy to the thyme dyed fabric. The bacterial reduction (%) for pre-mordanted fabric samples was observed to be 43.75% and 32.10% for gram-positive and gram-negative bacteria respectively. The meta- and post-mordanting method were more efficient method as the bacterial reduction observed was 98.90% and 99.53% for *S. aureus* and 96.44% and 97.11% for *E. coli* respectively. The high concentration of nano zirconium oxide made it uninhabitable for the growth of the microbes in the treated wool fabric. The mordanting of wool fabric with a higher concentration of nano zirconium oxide (ZrO_2 NPs), that is, above 6% is observed to decrease fabric bending length by order of 40% compared to untreated fabric. This may infer to the improvement of handle properties of treated wool fibers.

5.8 INDUSTRIAL APPLICATIONS OF NANOTECHNOLOGY IN TEXTILES

Nanotechnology is converting the material into nano scale of the size (10^{-9} m) and using that material for improving the functions of a product. It is well established that materials behave differently when its size is reduced to nano scale from bulk due to structure and size-dependent relationships.

Research on nanotechnology in textiles has resulted in products with various functionalities, viz., antimicrobial, flame resistance, wrinkle resistance, stain resistance, UV protection and electrical properties (Vigneshwaran et al., 2018; Hosne and Hasan, 2018; Patra and Gouda, 2013; Das et al., 2013; Mahmud and Nabi, 2017; Hassan et al., 2019; Sawhney et al., 2008; Srinivasan et.al., 2018; Joshi and Bhattacharyya, 2011).

Commercializing nanotechnology research outputs involves scaling up of production forms with nano-enabled functionalities. Textile industries are using nanotechnology in their products through three major routes, namely, applying nano coating in the base textiles, integrating NPs during the manufacture of fiber, and making fibers of nano size and incorporating in product matrix. Adhesion of the fabric surface depends on the contact area of fabric. By coating fabric with NPs, the contact area can be reduced and fabric surface can be converted into fine structure with subdued adhesion. Schoeller Textil AG, textile manufacturer in Switzerland, has patented a "NanoSphere" finishing technology that, on application, the fabric becomes water and dust repellent (NanoSphere—Technologies, 2020). Sportswear, business wear, and work wear are the application areas. Schoeller applies "NanoSphere" on their fabrics, and this technology is also available for licensing to other fabric manufacturers. The claim made by the technology provider is that the nano finish coating will withstand up to 50 wash cycles. "NanoSphere" uses C6 fluoro chemicals, and it is free from perfluorooctane sulfonate and perfluorooctanoic acid.

Nano Textile, an Israeli company has developed single-step coating process for the application of zinc oxide NPs and has built two machines on this technology and supplied the machines to Italy and Romania (Israeli Co. Nano Textile, 2016). ZnO NPs impart antibacterial properties to fabrics and protect against bacteria such as *S. aureus* which are even resistant to antibiotics like methicillin. The sonochemical process is used in this technology for the application of NPs. The application of ZnO NPs does not in any way affect the color of the initial fabric as the solution is colorless. The coated fabric meets the medical norms and withstands up to 100 wash cycles at 75°C. Dust- and oil-repellent water-based spray is developed and marketed under the trade name "Nanoflex Tex 2" by Nano-Care Deutschland AG, a company located in Germany ("Neverwet" fabric, nano-care.com, 2020). The liquid coating is based on modified silicon dioxide that can be applied on the textile surface using a pump spray. The spray thickness is between 100 and 150 nm and has a water resistance of 130 to 150 degree and oil resistance of 120 to 130 degree. Nano-Care also markets other products namely Nanoflex F-Bond (C6 fluorocarbon), Nanoflex Repel Eco (nano-modified resin), Nanoflex Repel Eco (hydrophobic-arranged methyl groups in a hydrocarbon matrix), Nanoflex L-Care (combination of modified silicon dioxide and wax), Nanoflex Wash-on (fluorinated polymer) and Nanoflex Tex barrier (nano-modified paraffin) for application in textiles and leather for water-, oil-, and dirt-repellent finishes. Nanotex, a U.S. company, is offering various nanotechnologies to fabric finishers: water repellent, spill resistance, stain release, moisture wicking, odor control, and wrinkle free (Nanotex, nanotex.com, 2020). Nanotex claims that it has operations in United States, Asia, and Europe, and 80 textile mills worldwide are using their technologies. Nanophase technologies corporation, USA markets NanoArc Zinc Oxide for enhancing the UV absorption of textiles (Textiles, nanophase.com, 2020). ZnO powder is made available at three particle sizes as 20, 40, and 60 nm. It is also made available in aqueous dispersions, solvent dispersions, and monomer dispersions. NanoArc Zinc Oxide coated fabric's sun protection effect remains intact even after wetting due to perspiration. Elmarco, Czech Republic, has developed unique nanospider electrospinning technology for the production of industrial scale nanofibers and the machines are supplied to nanofiber-based product development companies in air filtration, liquid filtration, biomedical, and traditional product development with enhanced performance (Nanospider™, Elmarco, 2020). Electrospinning machine meant for air filtration end-use nanofiber substrate manufacture has a width of 1.6 m with a production capacity of 20 m/min. Nanovia, Czech Republic, adopted this technology and made nanofiber-based functional fabrics and markets its products under its brand name as Nanovia anti-allergy fabric, Nanovia antibacterial fabric, Nanovia waterproof fabric, and Nanovia filter material (NANOVIA s.r.o.—nanofibers plant, nanovia. cz, 2020). RTI International, United States, AquaCure Ltd., South Africa, and Tricol Biomedical (formerly HemCon) are producing nanofibers using nanospider technology and developed products for filtration and wound care (Kala et.al., 2020). Two types of nanofibers namely organic (polymer) fibers and inorganic (ceramic) are produced at industrial scale by Kertak Nanotechnology s.r.o., Czech Republic, and marketed by Thorson Chemical (Kertak Nanotechnology s.r.o., 2020). Ceramic nanofibers find applications in Li-ion batteries, fuel cell separator, catalyst, gas sensors, thermal insulators, ceramics nano-composites, dehumidifiers, abrasives, thermal barrier coatings, and filtration. Organic nanofibers after conversion into membranes find applications in water and air purification due to their high filtration efficiency.

Poor breathability is one of the issues faced in traditional protective clothing. Use of nanofiber felt materials improves breathability due to its tightly packed structure accompanied by large number of nano pores. Polartec, United States, produces Power Shield Pro jackets that are made from nanofiber felt material which excels in protective performance compared to traditional products (Power Shield Pro, Polartec, 2020). Filtration is another area in which nanofibers are used. Cummins Filtration, United States, distributes NanoForce Air Filters made from nanofibers to automobiles (NanoForce Air Filters, Cummins Filtration, 2020). NanoForce Air Filter claims to remove submicron level particles, which if not filtered out will cause serious damage to engine piston, ring, liners, and other engine parts and has a filtration efficiency of 99.99%. Ultra-Web technology (Donaldson Company, Inc., United States) is another example of commercial use of nanofiber (Ultra-Web, Donaldson Company, 2020). Electrospinning process is used to produce fine web with nanofibers having diameter of 0.2-0.3 μ in the diameter with web capable of trapping fine dust on the surface due to fine inter-fiber spaces. Compared to other cartridge filter media, ultra-web technology based filter is efficient due to high capturing of submicron dust particles, lower energy requirement, less compressed air use, and lower operating pressure drop.

5.9 CONCLUSIONS

For the last decades, a tremendous amount of research works has been reported on nano technology. This is in the headings because of low chemical usage, lower energy costs, high surface area, and remarkable changes in physical and mechanical properties. The properties of the elements will extremely change when it reaches to nano scale. Textile is a highly explored field, where nano-technological interventions are highly exploited. Various NPs are highly established in the field of antimicrobial, anti-moth, UV resistant, water repellent, fire retardant, and dyeing. Many of the formulations are commercially established. Apart from the aesthetic appeal, the textile industries and apparel buying houses are very much interested in the functional finishing of textiles and huge research works are progressing in this area. One of the major constraints of the metal NPs in the textiles is the poor adhesion, due to lack of chemical interaction. This causes a decrease in the desired functional properties during washing. However, binders and adhesive are extremely successful in the fixation of these NPs. The debate is still going among researchers about the toxicity of NPs to the human and the environment. However, we can expect more nano formulation and technological intervention in textiles in the upcoming future.

KEYWORDS

- **nanoparticles**
- **wool**
- **anti-moth**
- **textile finishes**
- **antimicrobial**
- **ultraviolet protection factor**

REFERENCES

Altun, Ö.; Becenen, N. Antioxidant, Antibacterial and UV-Resistant Activities of Undyed and Dyed Wool Fabrics Treated with CuO Nanoparticles. *J. Nanosci. Nanotechnol.* **2017**, *17* (6), 4204–4209.

Azlina, H. N.; Hasnidawani, J. N.; Norita, H.; Surip, S. N. Synthesis of SiO_2 Nano Structures Using Sol-Gel Method. *Acta Phys. Polon. A.* **2016**, *129* (4), 842–844.

Barani, H.; Montazer, M.; Samadi, N.; Toliyat, T.; Zadeh, M. K.; Smeth, B. Application of Nano Silver/Lecithin on Wool through Various Methods: Antibacterial Properties and Cell Toxicity. *J. Eng. Fibers Fabrics* **2014**, *9* (4), 126–133.

Becheri, A.; M. Durr, M.; Nostro, P.; Baglioni, P. Synthesis and Characterization of Zinc Oxide Nanoparticles: Application to Textiles as UV-Absorbers. *J. Nanoparticle Res.* **2008**, *10* (4), 679–689.

Borah, M. P.; Jose, S.; Kalita, B. B.; Shakyawar, D. B.; Pandit, P. Water Repellent Finishing on Eri Silk Fabric using Nano Silica. *J. Textile Instit.* **2019**, *111* (5), 701–708.

Chakarborty, A.; Chakarborty, R.; Sen, B.; Kumari, P.; Sarkar, P. Antibacterial Finishing of Silk Fabric Using Chitosan-Neem Nano Emulsion. *Int. J. Curr. Biotechnol.* **2015**, *3* (6), 10–20.

Chattopadhyay, D. P.; Patel, B. H. Imparting Water Repellency to Jute Fabric by Nano Paraffin and Nano Copper Colloid Treatment. *Univ. J. Chem.* **2016**, *1* (1), 24.

Das, S. C.; Paul, D.; Hassan, S. M.; Chowdhury, N.; Sudipta, B. Application of Nanotechnology in Textiles: A Review. In *Proceedings of the International Conference on Mechanical Engineering and Renewable Energy*; Chittagong, Bangladesh, 2013, 1–3 May 2014; ICMERE2013-PI-120.

Dhananjayan, M. R.; Mielczarski, E.; Thampi, K. R.; Buffat, P. H.; Bensimon, M.; Kulik, A.; Mielczarski, J.; Kiwi, J. Photodynamics and Surface Characterization of TiO_2 and Fe_2O_3 Photocatalyst Immobilized on Modified Polyethylene Films. *J. Phys. Chem. B.* **2001**, *105*, 12046–12055.

Gao, Y; Cranston, R. Recent Advances in Antimicrobial Treatments of Textiles. *Textile Res. J.* **2008**, *78* (1), 60–72.

Han, S.; Yang, Y. Antimicrobial Activity of Wool Fabric Treated with Curcumin. *Dyes Pigment.* **2005**, *64* (2), 157–161.

Hassan, B.S; Islam, G. M. N; Haque, A. N. M. A. Applications of Nanotechnology in Textiles: A Review. *Adv. Res. Textile Eng.* **2019**, *4* (2), 1038, 1–9.

Hebbalalu, D.; Lalley, J.; Nadagouda, M. N.; Varma, R. S. Greener Techniques for the Synthesis of Silver Nanoparticles Using Plant Extracts, Enzymes, Bacteria, Biodegradable Polymers, and Microwaves. *Am. Chem. Soc.-J. Sustain. Chem. Eng.* **2013**, *1* (7) 703–712.

Hebeish, A.; Elnaggar, M. E.; Fouda, M. M. G.; Ramadan, M. A.; Al-Deyab, S. S.; El-Rafie, M. H. Highly Effective Antibacterial Textiles Containing Green Synthesized Silver Nanoparticles. *Carbohydr. Polym.* **2011**, *86* (2), 936–940.

Hosne Asif, A. K. M. A.; Hasan, Md. Z. Application of Nanotechnology in Modern Textiles: A Review. *Int. J. Curr. Eng. Technol.* **2018**, *8* (02), 227–231.

Ibrahim, N. A.; Emam, E. A. M.; Eid, B. M.; Tawfik, T. M. An Eco-Friendly Multifunctional Nano-Finishing of Cellulose/Wool Blends. *Fibers Polym.* **2018**, *19* (4), 797–804.

Iravani, S.; Klefenz, H.; Chan, W. C. W.; Nie, S.; Tian, Z. Ren, B. Green Synthesis of Metal Nanoparticles Using Plants. *Green Chem. J. R. Soc. Chem.* **2011**, *13* (10), 2638–2650.

Jose, S.; Nachimuthu, S.; Das, S.; Kumar, A. Moth Proofing of Wool Fabric using Nano Kaolinite. *J. Textile Instit.* **2018**, *109* (2), 225–231.

Jose, S.; Shanmugam, N.; Das, S.; Kumar, A.; Pandit, P. Coating of Light Weight Wool Fabric with Nano Clay for Fire Retardancy. *J. Textile Instit.* **2018**, *110* (5), 764–770.

Jose, S.; Pandey, R.; Pandit, P. Chickpea Husk–A Potential Agro Waste for Coloration and Functional Finishing of Textiles. *Indust. Crops Prod.* **2019**, *142*, 111833.

Joshi, M.; Bhattacharyya, A. Nanotechnology- a New Route to High Performance Functional Textiles. *Textile Progress* **2011**, *43* (30), 155–233.

Kala, D.; Athira, M. S.; Mohan, M. Nanospider Technology and Its Applications. *Int. J. Pharm. Sci. Res.* **2020**, *1* (4), 1557–1561.

Kertak Nanotechnology, S. R. O. (Ltd.), Nanofibers—Gate to the Beginning. http://www.thorson.cz (accessed Jan 8, 2020).

Ki, H. Y.; Kim, J. H.; Kwon, S. C.; Jeong, S. H. A Study on Multi Fractional Wool Textiles Treated with Name-Sized Silver. *J. Mater. Sci.* **2007**, *42* (19), 8020–8024.

Lam, Y. L.; Kan, C. W.; Chun, W.; Yuen, M. Effect of Metal Oxide on Anti-Microbial Finishing of Cotton Fabric. *Bioresources* **2012**, *7* (3), 3960–3983.

Lin, H.; Huang, C. P.; Li, W.; Ni, C.; Shah, S. I.; Tseng, Y. H. Size Dependency of Nanocrystalline TiO_2 on Its Optical Property and Photocatalytic Reactivity Exemplified by 2-Chlorophenol. *Appl. Catalys. B Environ.* **2006**, *68* (1–2), 1–11.

Mahmud, R.; Nabi, F. Application of Nanotechnology in the Field of Textile. *IOSR J. Polym. Textile Eng.* **2017**, *4* (01), 1–6.

Marambio-Jones, C.; Hoek, E. M. V. A Review of the Antibacterial Effects of the Silver Nanomaterials and Potential Implications for Human Health and the Environment. *J. Nanoparticle Res.* **2010**, *12* (5), 1531–1551.

McQueen, R. H.; Laing, R. M.; Brooks, H. J. L.; Niven, B. E. Odor Intensity in Apparel Fabrics and the Link with Bacterial Populations. *Textile Res. J.* **2007**, *77* (7), 449–456.

Meilert, K. T.; Laub, D.; Kiwi, J. Photocatalytic Self-Cleaning of Modified Cotton by TiO_2 Clusters Attached by Chemical Spacers. *J. Mol. Catalys.* **2005**, *237* (1–2), 101–108.

Mongkholrattanasit, R.; Kryštůfek, J.; Wiener, J.; Viková, M. Dyeing, Fastness, and UV Protection Properties of Silk and Wool Fabrics Dyed with Eucalyptus Leaf Extract by the Exhaustion Process. *Fibres Text East Eur.* **2011**, *19* (3), 94–99.

Montazer, M.; Pakdel, E. Reducing Photoyellowing of Wool using Nano TiO_2. *Photochem. Photobiol.* **2010**, *86* (2), 255–260.

Montazer, M.; Seifollah Zadeh, S. Enhanced self-cleaning, Antibacterial and UV Protection Properties of Nano TiO_2 Treated Textile through Enzymatic Pretreatment. *Photochem. Photobiol.* **2011**, *87* (4), 877–883.

Montazer, M.; Pakdel, E.; Behzadnia, A. Novel Feature of Nano Titanium Dioxide on Textiles: Anti Felting and Anti-Microbial. *J. Appl. Polym. Sci.* **2011**, *121* (6), 3407–3413.

NanoForce Air Filtration|Cummins Filtration. https://www.cumminsfiltration.com (accessed Jan 8, 2020).

NanoSphere–Technologies, Schoeller Textiles AG, © Schoeller Switzerland 2020. https://www.schoeller-textiles.com/en/technologies/nanosphere (accessed Jan 3, 2020).

Nanospider™ needle-free technology, Elmarco. https://www.elmarco.com (accessed Jan 7, 2020).

Nanotex–Stain, Moisture, Odor & Wrinkle Resistant Apparel Fabrics © 2020 Nanotex LLC. https://www.nanotex.com/ (accessed Jan 8, 2020).

Nano Textile Unveils Antibacterial Fabric. Globes. June 14, 2016. Published by Globes, Israel business news. www.globes-online.com

NANOVIA, s. r. o.—výroba nanovlákna, nanofibers plant. http://www.nanovia.cz/en/ (accessed Jan 8, 2020).

Nazari, A.; Montazer, M.; Jafari, F.; Dehghani-Zahedani, M. Optimization of Wool Mothproofing with Nano TiO_2 Using Statistical Analysis. *J. Textile Instit.* **2014**, *105* (1), 74–83.

"Neverwet" Fabric with Nano Coating © NanoCare Deutschland AG, 2019. https://nano-care.com/products/neverwet-fabric/ (accessed Jan 3, 2020).

Okeila, A. A. Citric Acid Crosslinking of Cellulose Using TiO_2 Catalyst by Pad-Dry-Cure Method. *Polym.-Plast. Technol. Eng.* **2008**, *47* (2), 174–179.

Pandey, R.; Patel, S.; Pandit, P.; Nachimuthu, S.; Jose, S. Coloration of Textiles Using Roasted Peanut Skin-an Agro Processing Residue. *J. Clean. Prod.* **2018**, *172*, 1319–1326.

Patra, J. K.; Gouda, S. Application of Nanotechnology in Textile Engineering: An Overview. *J. Eng. Technol. Res.* **2013**, *5* (5), 104–111.

Power Shield Pro. https://polartec.com/fabrics/weather-protection/power-shield-pro (accessed Jan 8, 2020).

Raveendran, P.; Fu, J.; Wallen, S. L. Completely "Green" Synthesis and Stabilization of Metal Nanoparticles. *J. Am. Chem. Soc.* **2003**, *125* (46), 13940–13941.

Rezaie, A. B.; Montazer, M.; Rad, M. M. A Cleaner Route for Nano Colouration of Wool Fabric via Green Assembling of Cupric Oxide Nanoparticles along with Antibacterial and UV Protection Properties. *J. Clean. Prod.* **2017**, *166*, 221–231.

Rezaie, A. B.; Montazer, M.; Rad, M. M. Scalable, Eco-friendly and Simple Strategy for Nano-Functionalization of Textiles Using Immobilized Copper-based Nanoparticles. *Clean Technol. Environ. Policy* **2018**, *20* (9), 2119–2133.

Sankaran, A.; Kamboj, A.; Samant, L.; Jose, S. *Synthetic and Natural UV Protective Agents for Textile Finishing: In Innovative and Emerging Technologies for Textile Dyeing and Finishing*; Scrivener-Wiley Publishers, 2020.

Samant, L.; Jose, S.; Rose, N. M.; Shakyawar, D. B. Antimicrobial and UV Protection Properties of Cotton Fabric Using Enzymatic Pretreatment and Dyeing with *Acacia catechu*. *J. Natur. Fibres* **2020**.

Samanta, A. K.; Bhattacharyya, R.; Jose, S.; Basu, G.; Chowdhury, R. Fire Retardant Finish of Jute Fabric with Nano Zinc oxide. *Cellulose* **2017**, *24*, 1143–1157.

Saran, G.; Demir, A. A Green Application of Nano Sized Chitosan in Textile Finishing. *Tekstil ve Konfeksiyon* **2016**, *26* (4), 414–420.

Sawhney, A. P. S.; Condon, B.; Singh, K. V.; Pang, S. S.; Li, G.; Hui, D. Modern Applications of Nanotechnology in Textiles. *Textile Res. J.* **2008**, *78* (8), 731–739.

Scott, R. A. Textile and Protection, a Book Published by Wood head Publishing Company, 1st ed., 2005; **ISBN:** 9781855739215(cross-referred).

Shabbir, M.; Rather, L. J.; Mohammad, F. Economically Viable UV-Protective and Antioxidant Finishing of Wool Fabric Dyed with Tagetes Erecta Flower Extract: Valorization of Marigold. *Indust. Crops Prod.* **2018**, *119* (4), 277–282.

Shakyawar, D. B.; Raja, A. S. M.; Kumar, A.; Pareek, P. K. Antimoth Finishing Treatment for Woolens Using Tannin Containing Natural Dyes. *Ind. J. Fibre Textile Res.* **2015**, *40* (2), 200–202.

Silvestry-Rodriguez, N.; Sicairos-Ruelas, E. E.; Gerba, C. P.; R. Bright, K. Title NA. *Rev. Environ. Contam. Toxicol*; Ware, G., Ed.; Springer: New York, **2007**, *191*; pp 23–45.

Simoncic, B.; Tomsic, B. Structures of Novel Antimicrobial Agents for Textiles—A Review. *Textile Res. J.* **2010**, *80* (16), 1721–1737.

Srinivasan, K.; Kumar, K. R.; Sheetal, K. B.; Kumari, B. L.; Murthy, C. Nanotechnology Trends in Fashion and Textile Engineering. *Curr. Trends Fashion Technol. Textile Eng.* **2018**, *2* (3), 56–59.

Stadler, T.; Buteler, M.; Weaver, D. K. Novel Use of Nanostructured Alumina as an Insecticide. *Pest Manage. Sci.* **2010**, *66* (6), 577–579.

Sundarrajan, S.; Chandrasekharan, A. R.; Ramakrishna, S. An Update on Nanomaterials- Based Textiles for Protection and Decontamination. *J. Am. Ceram. Soc.* **2010**, *93* (12), 3955–3975.

Taheri, M.; Malekina, N.; Ghamsari, A.; Almasian, A.; Chizarifard, G. Effect of Zirconium Dioxide Nanoparticles as a Mordant on Properties of Wool with Thyme: Dyeing, Flammability and Antibacterial. *Orient. J. Chem.* **2015**, *31* (1), 85–96.

Textiles|Nano Zinc Oxide for UV Protection in Textiles © 2020 Nanophase Technologies Corporation. http://nanophase.com/markets/textiles/ (accessed Jan 7, 2020).

Ultra-Web—High Efficiency Fine Fibre Filters ©2016–2020 Donaldson Company, Inc. https://www.donaldson.com (August 2020).

Vigneshwaran, N.; Bharimalla, A.; Arputharaj, A. Application of Functional Nanoparticle Finishes on Cotton Textiles. *Trends Textile Eng. Fashion Technol.* **2018**, *3* (4), 358–362.

Virkute, J.; Varma, R. S. Green Synthesis of Metal Nanoparticles: Biodegradable Polymers and Enzymes for Stabilization and Surface Functionalization. *Chem. Sci. J. R. Soc. Chem.* **2011**, *2*, 837–846.

Wang, H.; Ding, J.; Lin, T.; Wang, X. Super Water Repellent Fabrics Produced by Silica Nanoparticle-Containing Coating. *Res. J. Textile Apparel*. **2010**, *14* (2), 30–37.

Wei, Z.; Qiang-hua, Z.; Guo-qiang, C.; Tie-ling, X. Flame Retardancy Finishing of Wool Fabric via Sol-Gel Method. *Adv. Mater. Res.* **2014**, *989–994*, 607–610.

Xue, C. H.; Jia, S. T.; Chen, H. Z.; Wang, M. Superhydrophobic Cotton Fabrics Prepared by Sol–gel Coating of TiO_2 and Surface Hydrophobization. *Sci. Technol. Adv. Mater.* **2008**, *9* (3), 035001.

Yang, H.; Zhu, S.; Pan, N. Studying the Mechanisms of Titanium Dioxide as Ultraviolet-Blocking Additive for Films and Fabrics by an Improved Scheme. *J. Appl. Polym. Sci.* **2004**, *92* (5), 3201–3210.

Yu, J. C.; Zhang, L.; YU, J. Rapid Synthesis of Mesoporous TiO_2 with High Photocatalytic Activity by Ultrasound Induced Agglomeration. *N. J. Chem.* **2002**, *26* (4), 416–420.

Yu, Q.; Wu, Z.; Chen, H. Dual-function Antibacterial Surfaces for Biomedical Applications. *Acta Biomater.* **2015**, *16* (1), 1–13.

Zain, N.; Akindoyo, J.; Beg, M. Synthetic Antimicrobial Agent and Antimicrobial Fabrics: Progress and Challenges. *IIUM Eng. J.* **2018**, *19* (2), 10–29.

Zang, S.; Yang, X.; Tang, B.; Yuan, L.; Wang, K.; Liu, X.; Zhu, X., et al. New Insights into Synergistic Antimicrobial and Antifouling Cotton Fabrics via Dually Finished with Quaternary Ammonium Salt and Zwitterionic Sulfobetaine. *Chem. Eng. J.* **2018**, *336*, 123–132.

Zhang, M.; Xie, W.; Tang, B.; Sun, L.; Wang, X. Synthesis of TiO_2 & SiO_2 Nanoparticles as Efficient UV Absorbers and Their Application on Wool. *Textile Res. J.* **2017**, *87* (14), 1784–1792.

Zhao, X. Microwave-Assisted Antimicrobial Finishing of Wool Fabric with Chitosan Derivative. *Ind. J. Fibre Textile Res.* **2015**, *40* (1), 51–56.

CHAPTER 6

Application of Nanotechnology-Based UV Finishes on Textiles and Their Evaluation

M. S. PARMAR

Director (LABS), Northern India Textile Research Association (NITRA), Rajnagar, Ghaziabad, Uttar Pradesh, India

E-mail: drmsparmar@nitratextile.org

ABSTRACT

Finishing of textiles material using nanotechnology has given a new dimension into overall finishing processes. One of the significant differences among traditional and nanotechnology finishing processes is that in the traditional- or conventional-finished textile materials (fabric or garment), it often do not lead to permanent effects. The effect of such functional finishes deteriorated after laundering or wearing. However, in the case of fabric finished with nanotechnology, it acquires better durability. This enhance in durability of particular functional effect is due to the large surface area of nanoparticles as compared to the same volume of material manufactured with bigger particles. Beside this, coating with nanoparticles does not alter the porosity or softness of the finished fabric. This chapter comprises various nanotechnology-based finishes such as oil and water impermeability, soil and wrinkles resistance, antimicrobial, electrical resistivity, ultraviolet protection, fire and flame protection. Various methods to evaluate finished fabrics using these finishes are also discussed in this chapter.

6.1 INTRODUCTION

Nanotechnology can be applied in textile industry to create performance and functional-based textiles. It also helps in developing intelligent textiles

Fundamentals of Nano-Textile Science. Prashansa Sharma, Devsuni Singh & Vivek Dave (Eds.)

with unique characteristics and properties. The traditional finishing processes generally do not provide permanent effect on textile materials. These treated materials start losing their properties after repeated washing and uses. The physical and chemical properties of conventional (bulk form) as well as nanomaterials are dependent on their surface. When material is at the nanoscale, there is an enhancement in the surface-to-volume ratio, which ultimately provides large area on surface for chemical reaction. If a textile is treated with a normal finishing chemical and a nanomaterial-based finishing chemical, the surface of the textile will have a higher number of nanoparticles compared to the normal finishing chemical. This implies that the reactivity of nanomaterial is much higher than its bulk form. Nanotechnology can give excellent durability to textile materials. The high surface area-to-volume ratio of nanoparticles contributes to this extra ordinary quality. Due to this, they possess high surface energy, therefore, providing better bonding of nanoparticles to the fabric surface, which ultimately leads to improved durability of particular functional property (Becher et al., 2008). Besides, the application of nanoparticles on textiles does affect the breathability and hand feel. The word "nano" has a Greek origin meaning "dwarf" and it symbolize a factor of 10^{-9}. Nanomaterials consist at least one dimension in between 1 and 100 nm. Depending upon the number of dimensions limited to the nanoscale (up to 100 nm), there are four classifications of the nano-based material (Pokropivny and Skorokhod, 2007) as shown in Table 6.1.

TABLE 6.1   Classification of Nano-Based Material.

Class	Shape of nanomaterial	Number of dimensions confined to the nanoscale
Zero D	Spherical	Nanomaterials with all dimension up to 100 nm (e.g., fullerene)
One D	Tubular	1D nanomaterial with one dimension more than 100 nm (e.g., carbon nanotube)
Two D	Nano plate	Nanomaterial with two dimensions more than 100 nm (e.g., graphene nanosheet)
Three D	0D/1D/2D nanomaterial reinforced bulk material	Nanomaterials with 0D/1D/2D nanoparticle reinforced material (e.g., carbon nanotube reinforced polymer nanocomposite).

There are two ways to manufacture nanomaterials. These are: top down and bottom up (Poole and Owens, 2006). In the case of top-down approach, nanomaterials are formed by splitting or cutting down large piece using grinding, bombardment with high energy, etc., into nano-sized material. While in the case of bottom-up approach, the nanoparticles are used as manufactured material, and are organized into more complex way until the required dimensions of the material are obtained.

The use of nanomaterials in textiles is very vast. Its application on textiles may convert textiles into functional textiles by making them water repellant, antistatic, wrinkle resistance, UV resistant, antibacterial, etc. (Yetisen et al., 2016). It was found that self-assembly coating of CsxWO3 nanostructured material on cotton fabric enhanced its UV blocking, heat insulation, and self-cleaning properties. When wool is treated with some particular nanostructured material, its shrinkage resistance property improved significantly (Bakker et al., 2018). To impart antimicrobial property in textiles, copper nanoparticles, zinc oxide, and silver nanoparticles are used (Ditaranto et al., 2016; Emam et al., 2014; Emam et al., 2015; Emam et al., 2016; Yalcinkaya et al., 2016). It was also found that silver nanoparticle imparts antimicrobial property in textile materials (Martin et al., 2011). There are some nanomaterials which show self-cleaning property. Titanium dioxide nanomaterial is one of them (Saif et al., 2013).

6.2 ULTRAVIOLET RADIATION PROTECTION TEXTILES

It is well-known fact that sunlight is one of the main sources to synthesize vitamin D naturally by human skin. About 90% of the vitamin D needed by the human body is generated by the action of UV radiation (UVR) (Passeron et al., 2019). If someone undergoes high exposure of UVR, there are chances of adverse effect on health. One of the earlier studies revealed that the high exposure of UVR may bring skin tanning, ageing and if exposure is too high then lead to skin cancer (Reichrath, 2009; Shi et al., 2018; Sayre et al., 2010). The UVR can be classified into three classes (Fig. 6.1) on the basis of their wavelength. These are:

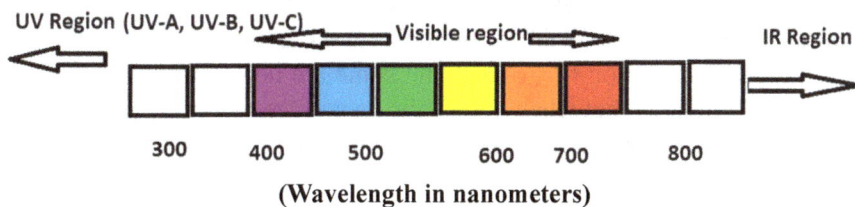

(Wavelength in nanometers)

FIGURE 6.1 Ultraviolet spectra.

It is clear from the Table 6.2; only UV-A and small portion of UV-B reaches to the earth surface, while UV-C doesn't reach to the earth surface as it gets absorbed by the ozone layer. However, it should be noted that if the ozone layers' thickness is reduced, the chances of reaching UV-C to the earth will be high. The effect of UV-C on human health is more severe than other radiations. The effect of UVR s may vary from one person to other person skin. Beside this, the effect of UVR on individual also depends on location, genetic factor, duration and day time etc. Some time it was also seen that the repeated UVR may also effect DNA of individual adversely (Sample and He, 2018; Williams et al., 2014). The cumulative exposure of UV-A can cause the skin to tan. On the other hand, UV-B can cause sunburn, skin cancer and melanoma (Moshammer et al., 2017; Ruegemer et al., 2002; Situm et al., 2008). It has been studied that the high exposure of UV-B exposure in childhood may cause Basal cell carcinoma (BCC). Beside this, the family history of skin cancer, skin type, radiotherapy history, exposure to other carcinogenic substances, etc., may also be the reasons for developing BCC (Situm et al., 2008; Dore and Chignol, 2007). Squamous cell carcinoma (SCC) is another category of skin cancer, which can also be caused by chronic exposure to UVR (Dore and Chignol, 2007).

TABLE 6.2   Classification of UV Radiation.

UV radiation (UVR)	Wavelength (nm)	Effect on human skin	UVR reaching on the earth (%)
UV-A	315–400	Tanning	Approx. 90
UV-B	280–315	Sun burn, cancer, melanoma	Approx. 10
UV-C	200–280	Nil	Nil (absorbed in the atmosphere)

Therefore, UV protective clothing or textile material is needed in order to be saved from harmful UVR. It is a well-known fact that clothing covers most of the body parts and thus helps in reduction of the hazardous effects that UVR has on human skin. The protection level of clothing depends on various characteristics of fabric used for making clothing. Some fabrics provide good protection from UVR and some are not very little protection level. It was noticed that when UVR falls on surface of cloths, it can divide into three parts. These are transmission, absorption, and reflection as shown in the Figure 6.2.

For making UV protective clothing, the transmission of UVR through the fabric should be stopped. To get this type of clothing, the fabric should either reflect or absorb UVR. The casual clothing does not provide safety against UVR due to its porous structure. It is not only porous structure of clothing but there are also many properties that influence protection from UVR. These are type of weave, type of yarn, fiber type, color, finishing application, etc. However, finishing treatment with UV blocking nanomaterial is one of the most preferred ways to render fabric UV protective. There are various types of nanomaterials which can be used for making UV protective textiles. Clay nanoflakes, which are composed of hydrous alumnio oscillates, zinc oxide, Nanostructure TiO_2, multi-walled carbon nanotube (MWCNT), boron and nitrogen co-doped carbon dot (BN-CD), ZnO-Ag, ZnO-Sb, silver nanoparticles, etc., are used to develop UV protective clothing.

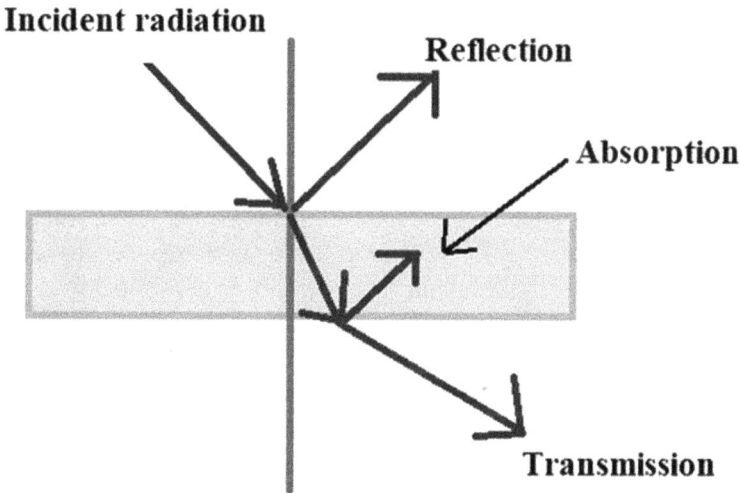

FIGURE 6.2 Division of UV rays on fabric surface.

6.2.1 SOME IMPORTANT FACTORS THAT INFLUENCE ULTRAVIOLET PROTECTION FACTOR RATINGS

There are many factors that influence ultraviolet protection factor (UPF) ratings. Some of them are given below:

i) Fabric porosity is directly linked with fabric cover factor (CF). If the construction of fabric is dense or CF is high, it will work as a barrier for UV light and increase UPF rating. The percentage of UVR transmission and UPF can be determined (Crews et al., 2008) using following formulas:

a) Percentage UVR transmission = 100 Cover factor

$$UPF = \frac{100}{100 - \text{cover factor}}$$

If we want to have minimum UPF rating 15, then the CF of fabric should be around 93% (Pailthrope, 1998). If the CF is 95%, the UPF rating would be 20. And the further increase in CF, e.g., 98%, there is a drastic increase in UPF rating up to 50. The chemical or mechanical finished fabric having same CF may give different UPF rating. Knitted fabric is having more open structure compare to woven. So the CF of knits will be lower than woven fabric. In knitted fabric, the stitch density is one of the crucial characteristics which affects UPF. It is very well known that the stitch density is depended on count of yarn used during manufacturing of fabric, gauge of the knitting machine, and the structure of the fabric taken. An inverse relationship exists between stitch density and pore size volume of the knitted fabric (Ogulata and Mavruz, 2010). Since UPF is dependent on porosity where UPF is $\alpha 100$/porosity (Kan, 2014), stitch density also affects UPF. The multiplication of course density and wales density is the resultant stitch density as shown below:

Stitch density = Courses per inch × wales per inche

Porosity is usually expressed in percentage (%); it can be determined as per the following equation (Wong et al., 2013):

$$\text{Porosity } (\varphi) = 100 \left[1 - (\rho_t / \rho_m) \right]$$

Where ρ_t is the bulk density (g/cm^3) and ρ_m is the fiber density (g/cm^3). For calculation purpose, the cotton fiber density (1.54 g/cm^3) is substituted in the above equation.

Following equation can be used to calculate the bulk density (ρ_t) of knitted fabric: $\rho t = M/V$

where M is the mass per unit area of fabric and V is the volume of the unit area of fabric (Postel, 1971, 1974).

The thickness (t) of geometrical fabric is equivalent to the volume of the unit area of fabric. Following equation is used to determine the bulk density of knitted fabric:

$$\text{Bulk density (g/cm}^3) = M/t$$

On the other hand, if we compare various weaves of woven fabric, the plain weave fabric is having higher CF than other weaves. A formula was given (Booth, 1968) for plain weave fabric to get some idea of fabric CF. In this formula, it was presumed that if all yarns are in close proximity and touching each other than the fabric CF would be 28. In the commercially available fabrics, it was seen that the CF ranges from 8 to 28. The formula given by Booth for fabric CF is given below:

$$\text{Cover factor (CF)} = \text{CFwarp} + \text{CFweft} - [\frac{\text{CFwarp x CFweft}}{28}]$$

$$\text{CFwarp} = \frac{\text{Ends per inch}}{\text{warp count in Ne}}$$

$$\text{CFweft} = \frac{\text{Picks per inch}}{\text{weft count in Ne}}$$

Researchers have found that CF was not accurately representing porosity of fabric, therefore, not reliable to predict UVR transmission through fabric.

Other methods (Hilfiker et al., 1996; Wong et al., 1997; Sliney et al., 1987; Crewa et al., 1999) like UV-visible spectrophotometer, image analysis, etc., were also used to determine porosity of fabric. Similarly, thick fabric reduces UV light penetration more than thin fabric.

ii) It is a popular notion that the color of any dye is dependent on the absorbing properties of dyes in the visible band of electromagnetic spectrum between 380 and 770 nm. The UVR band lies in between 290 and 400 nm. So visible band of spectrum also contain little bit UVR spectrum. This indicates that all dyes are having little or more property to increase UV protective property of fabric after dyeing with particular dye. Dark colors are having properties to absorb more UV rays compare to light colors. There are some dyes and chemicals available when they are applied (Nigar et al., 2012) on the fabric, absorb UV light, and thus improve UPF rating. Wong et al. (2015) carried out study on dyeing of loose and compact structure-knitted fabric samples using reactive dyes. It was found that if the structure of fabric was compact and having light in color, the UV protection was better or similar to the dark colored fabric with open structure. It is revealed from the study that only color of the fabric is not the factor of providing better UV protection, fabric structure also plays a key role.

iii) Types of fiber used in the construction of fabric also play an important role to influence UPF rating. Polyester and nylon fibers improve UPF rating while wool and silk are having moderate effect. Whereas, cellulosic fibers such as cotton, hemp, rayon, and flax provide low-UPF rating.

UPF is directly linked with chemical structure of the fibers (https://www.technicaltextile.net/articles/a-review-uv-radiation-and-textiles-5136, 28.04.2020). The UVR absorption of fibers such as cotton, jute, silk, and wool found to be lower than synthetic fibers like polyester, nylon. It is also found that in the grey stage, cotton gives better UPF value as it contains various natural additives like pigments, waxes, pectin, and other impurities. These additives act as UV absorber. In the case of bleached cotton as it has high-UV transparency, show low UPF. Natural fibers like linen and hemp in raw state, are not good UV protector, contain UPF of 20 and 10–15, respectively, despite containing lignin. On the other hand, the existence of lignin in jute gives it a good UV protection property and makes it act as a natural UV absorber. Protein fibers such as wool and silk are also having mixed effects in permitting UVR. It has been seen that PET fibers absorb more in the UV-A and UV-B regions compared to aliphatic polyamide fibers.

iv) Presence of extra moisture or wetness in the fabric causes reduction in UPF rating (Gambichler et al., 2020). The hygroscopicity of individual fibers directly influences moisture content in the fiber. These types of fibers swell with as the time of conditioning at high moisture increases. It was seen that these fibers swell in the presence of high-relative humidity or moisture and thus decreases interstices, which facilitates UV transmission. Besides, wet fiber decreases scattering effect as the reflective index of water is nearby to that of the polymeric material, and hence increases UV transmission. The overall impact of moisture on fiber is to have a reduction in UV protection factor. It is revealed that fabric made out of cotton at normal stage (without wet) can transmit 15–20% UVR. On the other hand, in the wet condition it can transmit more than 50% UVR of the normal stage.

If the fabric shrinks, it enhances UPF rating as fabric become denser.

v) Old clothing or clothing used many times or fabric become permanent stretched is also become less effective to protect from UV light. The used clothing mostly losses color and also become thin due to abrasion; due to this reason their UPF value is reduced. In one of the study by Kan et al., 2013, it was found that if the fabric is stretched during wearing, the UPF rating is decreased.

vi) The presence of optical brightening agent (OBA) in the detergent or in laundering enhances UPF rating. It is well known that OBA is a common additive to home laundry detergent formulations. The whitening of textiles increases in the presence of OBA; this occurs due to the conversion of UV excitation into the visible blue range of the electromagnetic spectrum. The occurrence of excitation and emission is brought about by the transition of electrons including p-orbitals from either conjugated or aromatic compounds. It is known that majority of OBA have excitation maxima between the range of 340 and 400 nm. These optical brightening agents may result an increase in the UPF of cotton and its blends. However, these optical brightening agents are not effective on fabrics made out of 100% polyester or nylon. OBA treatment is not a permanent treatment. Its effect decreases with repeated washing. After 10 washing cycles, fabric loses almost all UPF effect and become just like an untreated one, this indicates the semi-permanent UPF property of OBA finish. Beside its semi-permanent nature for providing UV protection, it has another shortcoming as OBAs usually absorb in the UVA part of the day light spectrum (93%) but having a very low absorption in UV absorption around 308 nm (92%), which results in skin-related ailments (Ying and Patricia, 1998).

6.3 EVALUATION OF UV PROTECTIVE PROPERTIES OF TEXTILES

The UV protection property is based on the protection of human skin from sunburn. There are two types of measuring factors that are being used for this purpose. These are sun protection factor (SPF) and UPF.

SPF determines the effectiveness of sunscreens on humans. It helps in gauging the total time duration for which a person can be exposed to the sunlight before having a sunburn effect. For instance, if one is exposed to sunlight for 10 min without using sunscreen and then he/she applies a recommended dose of sunscreen having an SPF value of 15, he/she should be protected from sunburn for 150 min.

Spectrophotometer is one of the reliable tools to determine UPF rating of textile material in fabric form. With the help of UPF rating, one can understand that how much UVR of the sun is absorbed by the textile material. For an example, a textile material having a rating of UPF 50 will only allow 1/50th of the UVR to cross through it. This shows that these textile materials will reduce the skin's exposure to UVR by 50 times (98% UV block). UV protective factor is the ratio of the average effective UVR irradiance determined for unprocessed skin to the mean effective UVR irradiance calculated for skin protected by the textile material. Effective UVR irradiance is calculated by multiplying the relative erythemal spectral effectiveness and the relative energy value of solar irradiance reaching skin. CIE (International Commission on Illumination) has provided information of relative spectral erythemal effectiveness between 290 and 400 nm with 5-mm interval. The solar spectral distribution generally symbolizes a "noonday" solar spectra (Labshere, 2000; Richard, 2005).

Following equation is used to determine UPF of textiles:

$$UPF = \frac{\sum_{290}^{400} E_\lambda . S_\lambda . \Delta_\lambda}{\sum_{290}^{400} E_\lambda . S_\lambda . T_\lambda . \Delta_\lambda}$$

E_λ = Erythemal action spectrum
S_λ = Solar spectrum [Wm^{-2}/nm]
T_λ = Transmittance [%]
$\Delta\lambda$ = Range of wavelength (5 nm)

UV protective properties of textiles in terms of UPF can be measured using various methods. Some of them are discussed over here:

6.3.1   AUSTRALIAN/NEW ZEALAND STANDARD (AS/NZS 4399, 2017)

It is very well-known fact that the overexposure to ultraviolet radiation may cause skin cancer and premature skin aging. The effect of UVR is decreased by wearing suitable wear. The suitability of UV protection clothing can be tested using AS/NZS 4399 standard. According to this, the UPF defined as the ratio of UV transmittance under the wavelength range of 290–400 nm.

This standard is very useful to evaluate the performance of textile materials that are worn by next to the skin to protect from solar ultraviolet radiation (UVR). The sun protective ability of textiles is described as UPF. The UPF value is given in rating. As per AS/NZS 4399 method, UPF of textiles can be classified or rated as good, very good, or excellent with UPF of 12–24, 25–39, or above 40, respectively. The test is carried out using spectrophotometer and specially designs software, which automatically provides UPF rating of fabric. In case of Australia and New Zealand, the role of UV protective clothing is very important as these countries are having highest rates of skin cancer all over the world. It is found that if an UPF rating of fabric/clothing is 50 or more, the skin of people can be protected.

Following (Table 6.3) is the classification as per AS/NZS 4399 which measurement is carried out on fresh, dry, and un-stretched state of fabric:

TABLE 6.3 Classification of UPF Rating.

UPF range	UV protection category	Effective UV transmission (T%)	UPF rating
15–24	Good protection	6.7–4.2	15, 20
25–39	Very good protection	4.1–2.6	25, 30, 35
40–50, 50, and over	Excellent protection	≤2.5	40, 45, 50, 50+

6.3.2   ISO/EN 13758-1 AND EN 13758-2

EN 13758-1 provides a test method for UPF and EN 13758-2 is a specification that gives the required values of UPF to meet the requirements of textile material. According to this standard, UPF as the ratio of UV transmittance under the wavelength range of 280–400 nm. UPF rating as per EN 13758-2 is good, very good, or excellent with UPF of 20–29, 30–40, or above 40, respectively.

6.3.3   AATCC 183

American Association of Textile Chemists and Colorists test method (AATCC 183-2004) defined UPR as the ratio of UV transmittance under the wavelength range of 280–400 nm. According to AATCC 183-2004 method, UPF of 15–24, 25–39, or over 40 is referred to as good, very good, or excellent. UPF of 40 means, 1/40 or 2.5% of UVR penetrates the textile.

6.4   APPLICATION OF NANOPARTICLES ON TEXTILES

During the finishing of textile material with nanomaterial, the inorganic colloidal nanosuspension (nanosols) converts fabric into such type of material which acquires properties of inorganic nanosuspension. To retain nanoparticles on the surface of textiles, it is required to bind these particles tightly on the surface. The tightly binding of nanoparticles on the surface of textiles ensures better durability of desired property. To achieve better durability, it is also required to ensure better compatibility between nanoparticles and the surface of textile material. Pad-dry-cure method is more often used for this purpose to form bonds between fabric surface and nanoparticles. Following (Table 6.4) are some processes used to treat textile material:

TABLE 6.4   Various Nano-Based Treatment Processes for Textile Material.

Substrate	Treatment method	References
Cotton fabric	Nano TiO_2 colloid of different concentrations was prepared. Cotton fabric was immersed in this liquor for 10 min. This wet fabric is then allowed to pass through the padding mangle. This is then dried for 30 min and cured at 140°C for 3 min	Chaudhari et al. (2012)
62% cotton and 38% polyester	Fabric treated with TiO_2 nanosol for 3 min and then passed through padding mangle followed by oven drying and curing at 130°C for 10 min	Ortelli et al. (2015)
Wool fabric	TiO_2/SiO_2 nanocomposite was applied on wool fabric using dip-pad-dry-cure method	Pakdel et al. (2013)
Cotton fabric	Cotton fabric is dipped in a 1:1 solution of colloidal TiO_2 and SiO_2 Ludox SM-30. Followed by drying at 100°C for 60 min. SiO_2 works as a binder and improves durability	Yuranova et al. (2005)
Cotton fabric	Cotton fabric was immersed for 1 min in aqueous solution of 0.5% acrylic binder. After this treatment, the treated fabric is squeezed using padding mangle. This fabric was again dip in TiO_2 nanosol and pass through laboratory padding mangle at speed of 15 m/min by maintaining 15 kg/cm^2 of pressure followed by drying and curing at 120°C for 3 min	Sivakumar et al. (2013)
Modified cotton fabric	In this process, sodium monochloroacetate (MCAA) and NaOH were used to incorporate carboxylic acid on the cotton surface. This process is called carbomethylation. Then, grafting of cotton fabric is carried out by the treatment with TiO_2 nanoparticles. After the grafting process, the unattached particles were removed using sonication process followed by drying at 60°C. The outcome indicated that TiO_2 nanoparticles loading were enhanced with increase of MCAA concentration taken to treat the cotton fabric	Wijesena et al. (2014)

TABLE 6.4 *(Continued)*

Substrate	Treatment method	References
Cotton fabric	Treatment of plasma-treated cotton fabric with nanostructured ZnO improves UPF value more than 250	Wang et al. (2019), Jazbec et al. (2015), and Gorjanc et al. (2014)
Cotton fabric	Coating of cotton fabric with nano ZnO (ZnO concentrations varies from 0.3 to 3 g/100 mL) in polyvinylsilsesquioxane polymer solution increases UPF from 58 to 158. In one of the work, UV protective effect of cotton fabric treated with ZnO and TiO_2 nanoparticle were compared. Nanoparticles were applied on textiles substrate by pad-dry-cure technique using binder.	Mai et al. (2018) and Khan et al. (2018)
Film	Nanostructured thin ZnO film on textiles substrate demonstrated excellent UV blocking property	Ghamsari et al. (2017)
Cotton fabric	Bleached cotton fabrics treated with 2% ZnO nanoparticle (particle size 40 nm) in acrylic binder showed 75% of UV blocking. Nano ZnO with increased surface area and uniformly distributed particles on the fabric surface can enhance the UV blocking of coated fabrics.	Yadav et al. (2006)
Cotton fabric	When cotton fabric is treated with nanostructure hybrid ZnO and polystyrene (PS), the UPF increases to more than 900 as compared with 4.9 for untreated fabric	Lu et al. (2006)
Cotton fabric	The cotton fabric coated with ZnO-SiO_2 nanorods improved UP protection factor up to 101.51	Wang et al. (2011)
Textile material	Nanostructure TiO_2 has been widely used as UV blocking agent for the preparation of UV protective textiles	Radetic (2013)
Cotton fabric	Cotton substrate was treated with TiO_2 nanoparticle using sol-gel method in order to prepare UV protective textiles. Undyed control sample showed UPF of 8.85 whereas nanoparticle-treated fabric showed excellent UV protective (UPF +50) property	Paul et al., (2009)
Cotton fabric	The coating of cotton textile with polymer solution containing multi-walled carbon nanotube (MWCNT) increases UPF to 46 as compared with untreated fabric for which UPF is 5.6. The coated textile having 1% MWCNT shows 174 UPF rating. With the increase of MWCNT percentage up to 2.5, the UPF rating enhanced to 421. The UV-shielding properties of MWCNT leads to the high UPF rate for CNT having the PU coat. The result revealed that UV absorption in the wavelength range of 200–700 nm increases with the rise of MWCNT in the coating solution	Mondal and Hu (2007)

TABLE 6.4 *(Continued)*

Substrate	Treatment method	References
Silk fabric	Silk fabrics were coated with nano TiO_2 coating by atomic layer deposition (ALD) which is a low-temperature vacuum deposition procedure. In a typical ALD process, at first titanium (IV) isopropoxide (TIP) was purged onto the reaction chamber and chemisorb onto the surface of silk fabrics. Afterward, nitrogen was purged to liberate the residual reagent. Then, water was passed into the chamber for the reaction of TIP with silk fabrics. Finally, nitrogen was used as purged gas again. UPF of resultant fabric was found to be around 50.	Yang et al. (2019)

6.5 CONCLUSION

Nano-based materials are classified into zero D, one D, two D, and three D. There are three types of UVR s. These are UV-A, UV-B, and UV-C. SPF and UPF are two types of measuring factors used to see the effect on human skin. SPF is used to measure the effectiveness of sunscreens. While on the other hand, UPF is used to measure UV protection ability of textile materials. Various factors that influence UPF ratings are also discussed. UPF rating classification as per AS/NZS 4399 are also given. Evaluation of UV protective properties of textiles using various test standards and specifications are also discussed. The application of nanoparticles to get various effects on textile material is tabulated in this chapter.

KEYWORDS

- **antimicrobial**
- **durability**
- **functional finishes**
- **nanoparticles**
- **UV protection**

REFERENCES

AATCC 183: Transmittance or Blocking of Erythemally Weighted Ultraviolet Radiation through Fabrics.

AS/NZS 4399: 2017 Sunprotective Clothing—Evaluation and Classification.

Bakker, C.; Ghosh, A.; Tandon, S.; Ranford, S. Surface Modification of Wool Fabric with POSS® Nanomaterial. *Fibers Polym.* **2018**, *19* (10), 2127–2133.

Becheri, A.; Durr, M.; Lo, Nostro P.; Baglioni, P. Synthesis and Characterization of Zinc Oxide Nanoparticles: Application to Textiles as UV-Absorbers. *J. Nanopart. Res.* **2008**, *10* (4), 679–689.

Booth, J. E. *An Introduction to Physical Methods of Testing Textile Fibres, Yarns and Fabrics*, 3rd ed.; Newnes-Butterworths: Boston, 1968.

Chaudhari, S. B.; Mandot, A. A.; Patel, B. H. Effect of nano TiO_2 Pretreatment on Functional Properties of Cotton Fabric. *Int. J. Eng. Res. Dev.* **2012**, *1* (9), 24–29.

Crewa, P. C.; Kachman, S.; Beyer, A. G. Influence on UVR Transmission of Undyed Woven Fabrics. *Text. Chem. Colorist* **1999**, *31* (6), 17–26.

Ditaranto, N.; Picca, R. A.; Sportelli, M. C.; Sabbatini, L.; Cioffi, N. Surface Characterization of Textiles Modified by Copper and Zinc Oxide Nano-Antimicrobials. *Surf. Interf. Anal.* **2016**, *48* (7), 505–508.

Dore, J. F.; Chignol, M. C. Cancer, Sun and UV: What's the Best Protection? *Oncologie* **2007**, *9* (5), 348–351.

Emam, H. E.; Manian, A.; Siroka, B.; Duelli, H.; Redl, B.; Merschak, P.; Bechtold, T. Copper(I) Oxide Surface Modified Cellulose Fibers-Synthesis, Characterization and Antimicrobial Properties. *Surf. Coat. Technol.* **2014**, *254*, 344–351.

Emam, H. E.; Rehan, M.; Mashaly, H. M.; Ahmed, H. B. Large Scaled Strategy for Natural/ Synthetic Fabrics Functionalization via Immediate Assembly of AgNPs. *Dyes Pigm.* **2016**, *133*, 173–183.

Emam, H. E.; Saleh, N. H.; Nagy, K. S.; Zahran, M. K. Functionalization of Medical Cotton by Direct Incorporation of Silver Nanoparticles. *Int. J. Biol. Macromol.* **2015**, *78*, 249–256.

Gambichler, T.; Hatch, K. L.; Avermaete, A.; Altmeyer, P.; Hoffmann, K. Influence of Wetness on the Ultraviolet Protection Factor (UPF) of Textiles: in Vitro and in Vivo Measurements. *Photodermatol. Photoimmunol. Photomed.* **2020**, *18* (1), 29–35.

Ghamsari, M. S.; Alamdari, S.; Han, W.; Park, H. H. Impact of Nanostructured Thin ZnO Film in Ultraviolet Protection. *Int. J. Nanomed.* **2017**, *12*, 207–216.

Gorjanc, M.; Jazbec, K.; Salam, M.; Zaplotnik, R.; Vesel, A.; Mozetic, M. Creating Cellulose Fibres with Excellent UV Protective Properties Using Moist CF4 Plasma and ZnO Nanoparticles. *Cellulose* **2014**, *21* (4), 3007–3021.

Hilfiker, R.; Kaufman, W.; Reinert, G.; Schmidt, E. Improving Sun Protection Factors of Fabrics by Applying UV-Absorbers. *Text. Res. J.* **1996**, *66* (2), 62–69.

ISO 13758-1: Textiles—Solar UV Protective Properties—Part 1: Method of Test for Apparel Fabrics.

ISO 13758-2: Textiles—Solar UV Protective Properties. Classification and Marking of Apparel.

Jazbec, K.; Sala, M.; Mozetic, M.; Vesel, A.; Gorjanc, M. Functionalization of Cellulose Fibres with Oxygen Plasma and ZnO Nanoparticles for Achieving UV Protective Properties. *J. Nanomater.* **2015**, *6*, 1–9.

Kan, C.; Yam, L.; Ng, S. The Effect of Stretching on Ultraviolet Protection of Cotton and Cotton/Coolmax-Blended Weft Knitted Fabric in a Dry State. *Materials.* **2013**, *6* (11), 4985–4999.

Kan, C. W. A Study on Ultraviolet Protection of 100% Cotton Knitted Fabric: Effect of Fabric Parameters. *Sci. World J.* **2014**, 1–10.

Khan, M. Z.; Baheti, V.; Ashraf, M.; Hussain, T.; Ali, A.; Javid, A.; Rehman, A. Development of UV Protective, Superhydrophobic and Antibacterial Textiles Using ZnO and TiO$_2$ Nanoparticles. ***Fibers Polym.*** **2018,** *19* (8), 1647–1654.

Labshere. *Technical Notes. SPF Analysis of Textiles*; Labshere: North Sutton, NH 03260, 2000.

Lu, H. F.; Fei, B.; Xin, J. H.; Wang, R. H.; Li, L. Fabrication of UV-Blocking Nanohybrid Coating via Miniemulsion Polymerization. *J. Colloid Interf. Sci.* **2006,** *300* (1), 111–116.

Mai, Z. H., Xiong, Z.; Shu, X.; Liu, X.; Zhang, H.; Yin, X.; Zhou, Y.; Liu, M.; Zhand, M.; Xu, W.; Chen, D. Multifunctionalization of Cotton Fabrics with Polyvinylsilsesquioxane/ZnO Composite Coatings. *Carbohydr. Polym.* **2018,** *199*, 516–525.

Martin, J.; Cardamone, J. M.; Irwin, P.; Brown, E. Keratin Capped Silver Nanoparticles— Synthesis and Characterization of a Nanomaterial with Desirable Handling Properties. *Colloids Surf. B: Biointerf.* **2011,** *88* (1), 354–361.

Merdan, N.; Koçak, D.; Şahinbaşkan, B. Y.; Yüksek, M. Effects of UV Absorbers on Cotton Fabrics. *Adv. Environ. Biol.* **2012,** *6* (7), 2151–2157.

Mondal, S.; Hu, J. L. A novel approach to excellent UV Protecting Cotton Fabric with Functionalized MWNT Containing Water Vapor Permeable PU Coating. *J. Appl. Polym. Sci.* **2007,** *103* (5), 3370–3376.

Moshammer, H.; Simic, S.; Haluza, D. UV-Radiation: From Physics to Impacts. *Int. J. Environ. Res. Public Health* **2017,** *14* (2), 200.

Ogulata, R. T., Mavruz, S. Investigation of Porosity and Air Permeability Values of Plain Knitted Fabrics. *Fibres Text East Eur.* **2010,** *82* (5), 71–75.

Ortelli, S.; Costa, A. L.; Dondi, M. TiO$_2$ Nanosols Applied Directly on Textiles Using Different Purification Treatments. *Materials* **2015,** *8*, 7988–7996.

Pailthrope, M. Apparel Textiles and Sun Protection: A Marketing Opportunity or a Quality Control Nightmare? *Mutat. Res.* **1998,** *422*, 175–183.

Pakdel, E.; Daoud, W. A.; Wang, X. Self-Cleaning and Superhydrophilic Wool by TiO$_2$/SiO$_2$ Nanocomposite. *Appl. Surf. Sci.* **2013,** *275*, 397–402.

Passeron, T.; Bouillon, R.; Callender, V.; Cestari, T.; Diepgen, T. L.; Green, A. C.; Van der Pols, J. C.; Bernard, B. A.; Ly, F.; Bernerd, F.; Marrot, L.; Nielsen, M.; Verschoore, M.; Jablonski, N. G.; Young, A. R. Sunscreen Photoprotection and Vitamin D Status. *Br. J. Dermatol.* **2019,** *181* (5), 916–931.

Paul, R.; Bautista, L.; DelaVarga, M.; Botet, J. M.; Casals, E.; Puntes, V.; Marsal, F. Nano-cotton Fabrics with High Ultraviolet Protection. *Text. Res. J.* **2009,** *79* (16).

Pokropivny, V. V.; Skorokhod, V. V. Classification of Nanostructures by Dimensionality and Concept of Surface Forms Engineering in Nanomaterial Science. *Mater. Sci. Eng. C.* **2007,** *27*, 990.

Poole, C. P. Jr.; Owens F. J. *Introduction to Nanotechnology*; John Wiley & Sons: New Delhi, 2006.

Postle, R. Thickness and Bulk Density of Plain-Knitted Fabrics. *J. Textile Inst.* **1971,** *62*, 219–231.

Postle, R. A geometrical Assessment of the Thickness and Bulk Density of Weft-Knitted Fabrics. *J. Text. Inst.* **1974,** *65*, 219–231.

Radetic, M. Functionalization of Textile Materials with TiO$_2$ Nanoparticles. *J. Photoch. Photobio. C.* **2013,** *16*, 62–76.

Reichrath, J. Skin Cancer Prevention and UV-Protection: How to Avoid Vitamin D-Deficiency. *Br. J. Dermatol.* **2009,** *161*, 54–60.

Ruegemer, J.; Schuetz, B.; Hermann, K.; Hein, R.; Ring, J.; Abeck, D. UV-Induced Skin Changes Due to Regular Use of Commercial Sunbeds. *Photodermatol. Photoimmunol. Photomed.* **2002,** *18* (5), 223–227.

Saif, M.; El-Molla, S. A.; Aboul-Fotouh, S. M. K.; Hafez, H. S.; Ibrahim, M. M.; Abdel-Mottaleb, M. S. A.; Ismail, L. F. M. Synthesis of Highly Active Thin Film Based on TiO_2 Nanomaterial for Self-Cleaning Application. *Spectrochim. Acta Part A Mol. Biomol. Spectrosc.* **2013,** *112,* 46–51.

Sample, A.; He, Y. Y. Mechanisms and Prevention of UV-Induced Melanoma. *Photodermatol. Photoimmunol. Photomed.* **2018,** *34* (1), 13–24.

Sayre, R. M.; Dowdy, J. C.; Shepherd, J. G. Variability of Pre-Vitamin D-3 Effectiveness of UV Appliances for Skin Tanning. *J. Steroid Biochem.* **2010,** *121* (1–2), 331–333.

Shi, Y.; Manco, M.; Moyal, D.; Huppert, G.; Araki, H.; Banks, A.; Joshi, H.; Mckenzie, R.; Seewald, A.; Griffin, G.; Sen-Gupta, E.; Wright, J. A.; Ghaffari, R.; Rogers, J.; Balooch, G.; Pielak, R. M. Soft, Stretchable, Epidermal Sensor with Integrated Electronics and Photochemistry for Measuring Personal UV Exposures. *PLoS One* **2018,** *13* (1), 0190233.

Situm, M.; Buljan, M.; Bulat, V.; Mihic, L. L.; Bolanca, Z.; Simic, D. The Role of UV Radiation in the Development of Basal Cell Carcinoma. *Collegium Antropologicum* **2008,** *32,* 167–170.

Sivakumar, A.; Murugan, R.; Sundaresan, K.; Periyasamy, S. UV Protection and Self-Cleaning Finish for Cotton Fabric Using Metal Oxide Nanoparticles. *IJFTR* **2013,** *38,* 285–292.

Sliney, D. H.; Benton, R. E.; Cole, H. M.; Epstein, S. G.; Morin, C. J. Transmission of Potentially Hazardous Actinic Ultraviolet Radiation through Fabrics. *Appl. Indust. Hygiene* **1987,** *12* (1), 36–44.

Scott, R. A., Ed. *Textile for Protection*; Woodhead Publishing Limited: Cambridge, 2005.

Yadav, A.; Prasad, V.; Kathe, A. A.; Raj, S.; Yadav, D.; Chandrasekaran, S.; Nadanathangam, V. Functional Finishing in Cotton Fabrics Using Zinc Oxide Nanoparticles. *Bull. Mater. Sci.* **2006,** *29* (6), 641–645.

Yalcinkaya, F.; Komarek, M.; Lubasova, D.; Sanetrnik, F.; Maryska, J. Preparation of Antibacterial Nanofibre/Nanoparticle Covered Composite Yarns. *J. Nanomater.* **2016,** *7,* 7565972.

Yang, H. Y.; Wang, Y. L.; Liu, K. S.; Liu, X.; Chen, F. X.; Xu, W. L. Facile Fabrication of Ultraviolet-Protective Silk Fabrics via Atomic Layer Deposition of TiO_2 with Subsequent Polyvinylsilsesquioxane Modification. *Text. Res. J.* **2019,** *89* (17), 3529–3538.

Yetisen, A. K.; Qu, H.; Manbachi, A.; Butt, H.; Dokmeci, M. R.; Hinestroza, J. P.; Skorobogatiy, M.; Khademhosseini, A.; Yun, S. H. Nanotechnology in Textiles. *ACS Nano* **2016,** *10* (3), 3042–3068.

Yuranova, T.; Mosteo, R.; Bandra, J.; Laub, D.; Kiwi, J. Self-Cleaning Cotton Textiles Surfaces Modified by Photoactive SiO_2/TiO_2 Coating. *J. Mol. Catal. A Chem.* **2005,** *244,* 160–167.

Wijesena, R. N.; Tissera, N.; Perera, R.; De Silva, K. M. Side Selective Surface Modification of Chition Nanofibers on Anionically Modified Cotton Fabrics. *Carbohydr. Polym.* **2014,** *109,* 56–63.

Wang, L.; Zhang, X.; Li, B.; Sun, P.; Yang, J.; Xu, H.; Liu, Y. Superhydrophobic and Ultraviolet-Blocking Cotton Textiles. *ACS Appl. Mater. Interf.* **2011,** *3* (4), 1277–1281.

Wang, X.; Chen, X. G.; Cowling, S.; Wang, L. B.; Liu, X. Q. Polymer Brushes Tethered ZnO Crystal on Cotton Fiber and the Application on Durable and Washable UV Protective Clothing. *Adv. Mater. Interf.* **2019,** *6* (14).

Williams, J. D.; Bermudez, Y.; Park, S. L.; Stratton, S. P.; Uchida, K.; Hurst, C. A.; Wondrak, G. T. Malondialdehyde-Derived Epitopes in Human Skin Result from Acute Exposure to

Solar UV and Occur in Nonmelanoma Skin Cancer Tissue. *J. Photoch. Photobio. B.* **2014,** *132,* 56–65.

Wong, J. C.; Cowling, I.; Parisi, A. V. Reducing Human Exposure to Solar Ultraviolet Radiation, 1997.

Wong, W.; Lam, J.; Kan, C.; Postle, R. Influence of Knitted Fabric Construction on the Ultraviolet Protection Factor of Greige and Bleached Cotton Fabrics. *Text. Res. J.* **2013,** *83* (7), 683–689.

Wong, W.; Lam, J. K.; Kan, C.; Postle, R. Influence of Reactive Dyes on Ultraviolet Protection of Cotton Knitted Fabrics with Different Fabric Constructions. *Text. Res. J.* **2015,** *86,* 5.

Zhou, Y.; Crews, P. Effect of OBAs and Repeated Launderings on UVR Transmission through Fabrics. *Text. Chem. Color* **1998,** *30* (11), 19–24.

CHAPTER 7

Nanoparticles Modifications of Textiles Using Plasma Technology

HEND M. AHMED[1*], MOHAMED MEHAWED ABDELLATIF[2], and FATEN HASSAN HASSAN ABDELLATIF[3]

[1]*Textile Research Division, Dyeing, Printing and intermediates, National Research Centre, Dokki, Giza, Egypt*

[2]*Chemical Industries Division, Chemistry of Tanning, Materials and Leather Technology, National Research Centre, Dokki, Giza, Egypt*

[3]*Textile Research Division, Pre-treatment and Finishing of Cellulosic Fabric Department; National Research Centre, Dokki, Giza, Egypt*

Corresponding author. E-mail: hend_plasma@yahoo.com

ABSTRACT

The textile industry in developed countries faces many challenges that lead toward highly advanced development of functional and high value-added textiles. Recently, plasma has been used in textile industry as a green alternative to chemical wet processing due to its economical cost and ecological advantages. Plasma technique is a type of clean surface treatment applied simply on the fabric's surface without using solvent in addition to keeping the main properties of the fabrics. Plasma is usually utilized for complete cleaning of the fabric surface and for deposition of nanometer range materials to provide new functionalities. Plasma treatment efficiency is governed by the substrate type and the condition of treatment.

This chapter reviews the different types of fibers, the different steps of wet processing of textile, and the influence of plasma technology in enhancing the textile functionality.

Fundamentals of Nano-Textile Science. Prashansa Sharma, Devsuni Singh & Vivek Dave (Eds.)
© 2022 Apple Academic Press, Inc. Co-published with CRC Press (Taylor & Francis)

7.1 INTRODUCTION

Conventional wet chemical processing techniques are very essential for textile industries , and according to environmental regulations they cause pollution and are considered harmful. Subsequently, plasma technology provides a green replacement for the traditional wet chemical processing. As a new technology, plasma technique is a simple surface treatment used as clean and environment friendly technique. Plasma is mainly used in physical treatment for the textile surface which increases the chances for finishing textile without influencing the bulk properties of the textile fibers. Ablation and deposition are two processes that can be provided by plasma. Ablation is usually used for clearing the textile surface from manufacturing residuals, while deposition process can be governed to nanometer range to create new functions. The efficiency of plasma treatment depends on many factors such as fabric nature and the treatment conditions,discharge power, plasma gas type, exposure time, and device of plasma used.

7.2 TEXTILE FIBERS

Fibers are the basic constituents of textile substrates. Depending on the origin of fibers, they can be classified into natural or synthetic fibers (Ahmed et al., 2019). Figure 7.1 illustrates the different types of fibers. Their availability, quality, and the cost of production are the main factors that decide the development of textile industry.

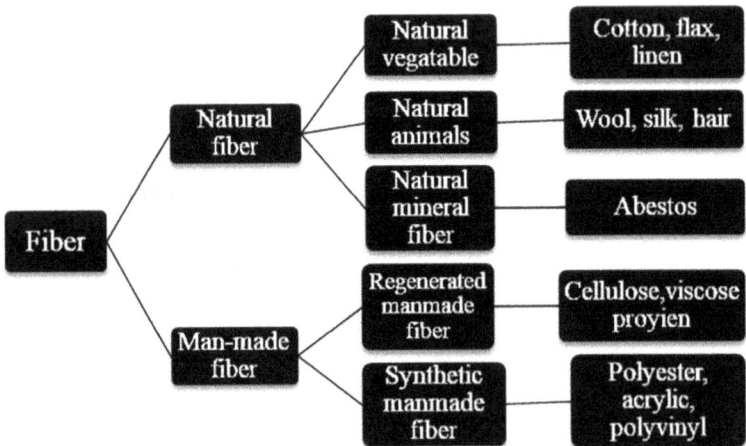

FIGURE 7.1   Classification of textile fibers.

7.2.1 WET TREATMENT OF FABRICS

Textile wet processing consists of different mechanical and chemical processes done on the fabrics to remove the impurities and improve their quality (Fig. 7.2). Wet treatments generally include three stages: pretreatment, coloration (dyeing and printing), and finishing (Madhav et al., 2018).

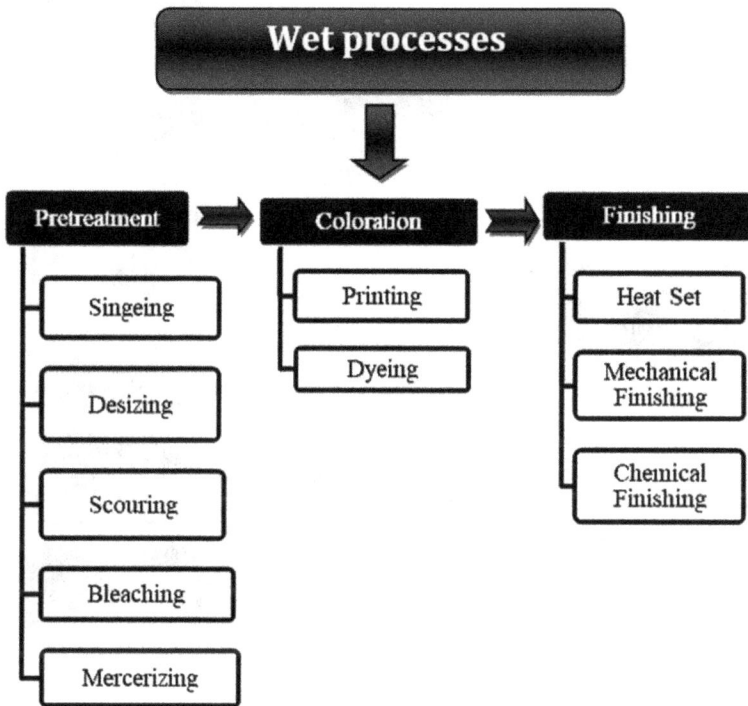

FIGURE 7.2   Textile wet processing steps.

All the above-mentioned processes are explained briefly:

Singeing process is used to improve the fabric surface appearance by removing the ejecting fibers or protruding fibers to improve the fabric smoothness. Singeing process is very important for the success of the next wet processing steps. In the singeing process, one of the following three techniques can be applied (Yasir, 2016):

- Plate singeing,
- Roller singeing,
- Gas singeing.

Sizes, for example, starch, CMC, etc. are necessary to facilitate textile processing. They are applied on the warp yarn of woven fabrics and have to be eliminated before coloration process (Mangesh and Tesfaye, 2016).

Desizing is necessary to eliminate the sizing materials. Desizing process classifications are illustrated in Figure 7.3 (Wadje, 2009).

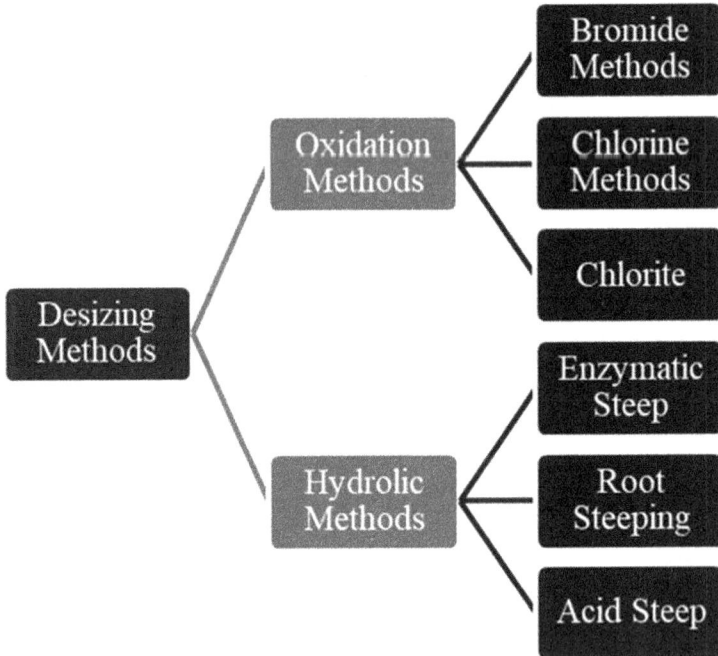

FIGURE 7.3 Classification of desizing processes.

Scouring process is important to remove hydrophobic impurities such as natural wax, oil, fats, and nonfibrous impurities (e.g., seed fragments remains) to improve the wettability of the fabrics. Scouring is done in alkali conditions using sodium hydroxide (Karmakar, 1999; Wadje, 2009).

Bleaching process is done to improve the whiteness of the fabrics by eliminating the remaining impurities and oxidizing natural coloring materials. Sodium hypochlorite diluted or diluted hydrogen peroxide is widely used as a bleaching agent. The level of bleaching depends on the final product. White textile substrate needs the maximum bleaching level whereas dyeing a deep dark substrate will need minor level of bleaching (Ibrahim, 2005).

Mercerization is necessary to improve the appearance of the fabric to become shiny and to enhance the fabric dyeability, strengthen them by improving their dimensional stability and tensile strength. The mercerization treatment includes two methods; caustic mercerization (cold mercerization—hot mercerization) and liquor ammonia mercerization. Mercerizing can be applied on grey or bleached fabric (Kumar and Choudhury, 2017).

7.2.2   DYEING AND PRINTING

Dyeing and printing are two ways to give color to textile substrate and to make them more smart and attractive (Gulrajani 2013; Xiaodong et al., 2018).

Dyeing is a chemical process used to color the textile fabrics using natural dyes or synthetic dyes. This technique consists of three basic steps: (1) retardation, (2) migration, and (3) diffusion. The quality of dyed garments determines the fastness value, homogeneity of the dyeing process, and the color strengths. In the dyeing process, the dye is smoothly spread on the fabric to achieve colored fabric. There are two techniques for dyeing the fabric: batch wise method (where jigger and jet dyeing machine are used) and continuous method where the dye is padded on fabric and then steaming, dry heat treatment, and soaping are done(Khatri et al., 2015; Lewis, 2014).

Printing is a process where color designs are produced on textiles by printing using paste form of pigments or dyes. Printing is used for localization of dyes on specific places on the fabric surface using specific machines. Printing of textile fabrics can be attained by either wet prints by employing thick paste or dry prints. On the face of the printed textile fabrics, there are cut edges in the design portion and the color rarely penetrates completely at the back of the fabric (Ahmed et al., 2017). On a large scale, printing techniques can be classified in these groups:

- Flat bed screen
- Hand screen printing
- The fully automatic flat screen printing
- Rotary screen printing
- Heat transfer printing

Recently, digital ink jet printing machines have been developed. These machines make great advances in textile printing. They have the ability to print fabrics with width up to 2 m by utilizing acid dye, reactive dye, or disperse dye ink sets. In this new printing technique, tiny droplets of printing solutions are ejected on particular location of textile substrates.

This technique is very simple but leads to diverse implementations (Park and Carr, 2006; Xueni et al., 2019). Ink jet printing process can be done in four steps, starting with fabric pretreatment, followed by fabric printing, then fixation, and finally washing off.

7.2.3 FINISHING OF TEXTILE FABRICS

The innovation of functional textile auxiliaries and the development of application techniques are necessary to produce textile substrates with innovative functionality and comfort properties. The different substrate types, quality of the finished product, cost, and available technology are greatly affected by the finishing process of textile substrates (Ibrahim, 2005).

Imparting new functionality in textile substrates can be attained mechanically, chemically, or by using bio-finishes (Horrocks et al., 2016).

Mechanical Finishing:

- **Calendaring:** In this method, the fabric is compressed between two rollers to enhance the compact, smooth, supple, flat, and glaze of the fabrics. Moreover, calendar machines usually contain many rollers that may be heated and change their speed. Furthermore, pressure and sharpening accomplishment can be applied to extend luster (Kumar and Choudhury, 2017; Paul, 2015).
- **Napping and shearing:** This method is usually used to produce uniform heights by cutting the raised naps (Kumar and Sundaresan, 2013).

Chemical Finishing:

Scientists have created numerous types of finishing agents taking into account the concerns of ecological system (Paul, 2015) as well as hygiene which is strongly required for the types of chemical finishing that are listed below:

- **Water repellency:** Water-repellent finishes resist wetting. It finds application in raincoats, unrestricted coats, hats, capes, umbrellas, and shower curtains.
- **Flame retardant:** These finishing materials are applied to flammable materials utilized in children's sleepwear, carpets, and curtains.

- **Resin finishing "anti-crease":** Anti-crease finishes are usually applied to polysaccharide fibers that easily wrinkle, for example, cotton, rayon, and linen. This can be attained by using durable crease-resistant materials that resist wrinkling and conjointly facilitate to take care of pleats and creases.
- **Softening:** It is used to improve the softening properties of the material.
- **Oil and soil repellency:** Prohibit soil, dirt, and stains from being absorbed in materials. These finishes will be used to finish materials employed in consumer goods and article of furniture.
- **Antistatic finishes:** Reduce electricity, which can collect on fibers. The foremost common forms of antistatic finishes are material softened.
- **Antimicrobial finishes:** Antimicrobial finishes act as safeguard for the user and textile substrate itself form microbial attack. Antimicrobial finishing agents can protect the textile substrates from microbes that lead to fabric discoloration and staining in addition to protect the mechanical properties of the fabrics and prevent odor creation. Antimicrobial finishing agents are applied to fabrics using exhaustion, pad dry-cure, spray, coating, and foam techniques. This type of finishing agent is currently used for sports, outside, and aid sector textiles (Ghaly et al., 2014).
- **UV protection:** Modification of fabric using UV absorbers ensures that the clothes do not absorb the harmful UV rays of the sun, which reduces the user's exposure to harmful UV rays and protects the skin from potential hazards.
- **Moth proofing and insect damage:** This type of finishing agent is used for protection of protein fibers, for example, wool from being attacked by mites, carpet beetles, and various insects (Gulrajani, 2013).

7.3 PLASMA IN TEXTILE FIELD

7.3.1 PLASMA BASICS AND FUNDAMENTALS

7.3.1.1 DEFINITION

Plasma is a group of different charges negative and positive which behave collectively (Fig. 7.4). The charges Coulomb's forces influence the forces applied externally and control the effects of collisions between charges themselves and any present neutral gas. "Self-generated" electric fields have a great role in molecules moving (Morent et al., 2008). One of the main results is the plasma ability to detect local density disturbances and form a sheath area between the plasma and the adhered surfaces. The term plasma cannot be used for all ionized gases or discharges (Jacak, 2016). Specific standards must be adhered before ionizing gas to have phenomena associated with plasma collective behavior (Li et al., 2019; Shishoo, 2007). Some examples of natural plasma as shown in Figure 7.5.

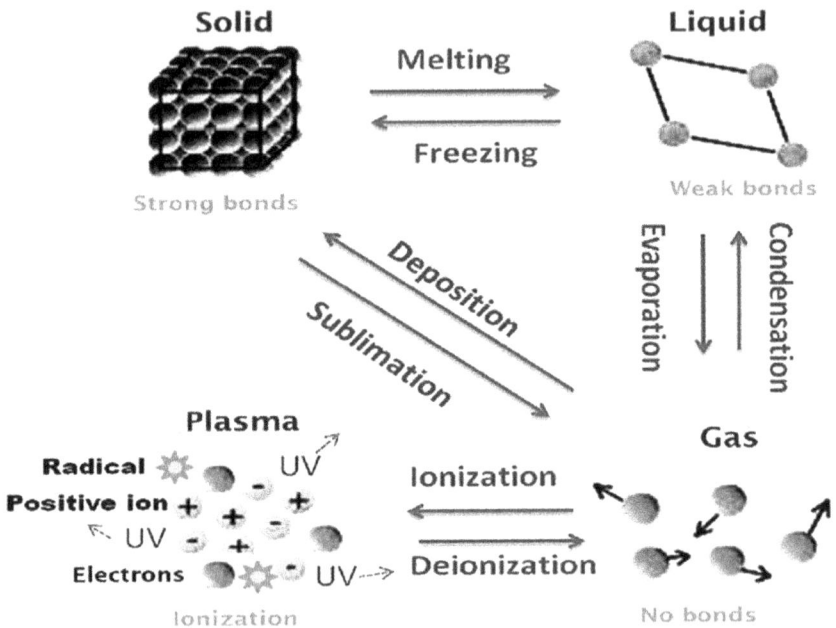

FIGURE 7.4   Fundamental of plasma. (Reprinted from Dave, H.; Ledwani, L.; Nema, S. K.; Shahid ul, I.; Butola, B. S. Chapter 8—Nonthermal Plasma: A Promising Green Technology to Improve Environmental Performance of Textile Industries. In *The Impact and Prospects of Green Chemistry for Textile Technology*; Woodhead Publishing, 2019; pp 199–249). Copyright Elsevier reprinted with permission).

7.3.1.2   GENERATION

Plasma is formed by applying energy to the gas (Freidberg, 2008) to reorganize the electronic structure of atoms and molecules species and the production of ion species. The applying energy can be transportable either by radiation or electromagnetic current, or thermal energy. These sorts are moving under electromagnetic fields. In case of placing textile substrates within plasma or passing through it, this leads to a range of surface processing as well as breaking the links to form active sites, graft chemicals, and functional groups, volatility of materials, removal (etching), disintegration of contaminants/surface layers (etching/ablation), and deposition of corresponding coating (Virendra et al., 2010).

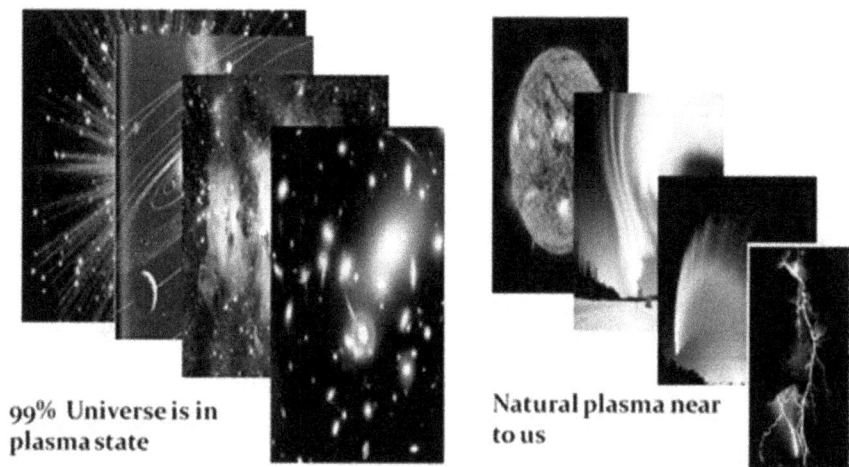

99% Universe is in plasma state

Natural plasma near to us

FIGURE 7.5 Natural plasma examples. (Reprinted from Dave, H.; Ledwani, L.; Nema, S. K.; Shahid ul, I.; Butola, B. S. Chapter 8—Nonthermal Plasma: A Promising Green Technology to Improve Environmental Performance of Textile Industries. In *The Impact and Prospects of Green Chemistry for Textile Technology*; Woodhead Publishing, 2019; pp 199–249. Copyright Elsevier reprinted with permission).

7.3.1.3 CLASSIFICATION

Depending on the temperature, plasma is categorized to two main groups: cold plasma and thermal plasma. In thermal plasma or hot plasma, all components are in equilibrium, that is, temperatures of electrons and different particles are too high whether charged or neutral, that is close to the maximum ionization (approximately 100%) (Klaus et al., 2019). Hot plasma includes plasma jets, electrical arcs, and thermonuclear reaction-generated plasma. On the other hand, corona discharge, low-pressure direct current (DC), discharges from fluorescent (neon) enlightening tubes and dielectric barrier discharge (DBD) are some examples of nonthermal plasmas (Fig. 7.6) (Dave et al., 2019).

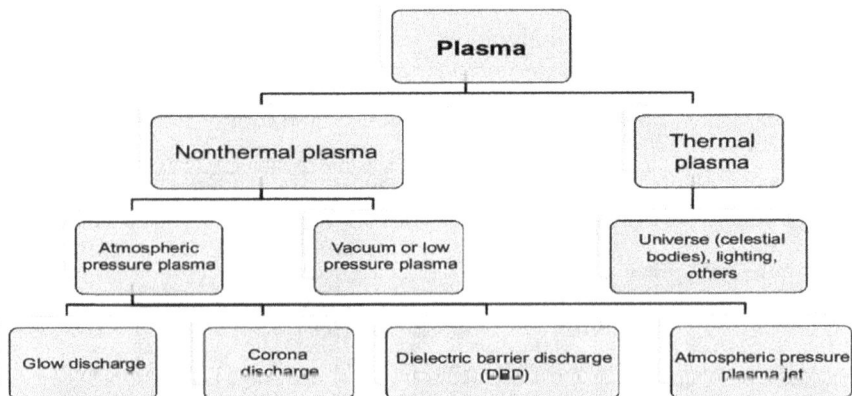

FIGURE 7.6   Different plasma categories. (Reprinted from Dave, H.; Ledwani, L.; Nema, S. K.; Shahid ul, I.; Butola, B. S. Chapter 8—Nonthermal Plasma: A Promising Green Technology to Improve Environmental Performance of Textile Industries. In *The Impact and Prospects of Green Chemistry for Textile Technology*; Woodhead Publishing, 2019; pp 199–249. Copyright Elsevier reprinted with permission).

7.3.2   OPERATING-PRESSURE PLASMA

7.3.2.1   LOW-PRESSURE PLASMA

Low-pressure plasma has been established for the microelectronics sector. However, in many cases, the produced microelectronics are not economically compatible with the treatment of textiles, and therefore, developing of low-pressure reactors can handle many of the textiles in inexpensive and economical technique (Geyter et al., 2006). This is done by using high discharge pumps ranging from 10 to 2 to 10^{-3} m bar. The gas is ionized in these pumps by the aid of high-frequency generator. Low-pressure plasma technique is well controlled and scalable to repeat (Ahmed, 2011).

7.3.2.2   ATMOSPHERIC-PRESSURE PLASMA

Atmospheric-pressure plasma has great productivity advantage compared to low-pressure plasma in terms of the pressure. This can operate in associate open perimeter, allowing direct continuous process with online traditional production line. It was found that the glow discharge apparatus is suitable for treating textiles in a low-pressure system. As for the atmospheric pressure system, there are three suitable devices for treating textiles—the corona, the DBD, and the atmospheric pressure glow discharge (Ahmed et al., 2019).

7.3.2.3 CORONA TREATMENT

Corona generating devices consist of two different electrical electrodes deliverers separated by a space consisting of gas which leads to plasma generation after their connection to high voltage source. The electrodes of electrical engineering are generally similar. They include needle sharp tapered poles, wire thin flat planes, or large-diameter opposed cylinder. It is operated using high, continuous, or vibrated DC or AC voltages. High electrical field around the singular, that is, the needle or wire point, leading to electrical breakdown and ionization of any gas surrounding the individual, and the plasma is generated, that disposes of an atomizer similar to the exit from the point or wire (Akishev et al., 2005).

7.3.2.4 DIELECTRIC BARRIER DISCHARGE

DBD is considered as a type of cold plasma. The discharge in this type of plasma takes a large area and a uniform distance. Plasma generation occurs as a result of the neutralization of the electric field applied between the two electrodes. AC is used to power the DBD and the high voltage power provides administration at frequencies of 1–100 kHz (Khanit and Wongkuan, 2016).

DBD can be in filamentary and homogeneous forms. DBD is also known as glow discharge atmospheric pressure. The general type is the filamentary, showing plasma as a group of "micro" tiny discharge inflections. These are straight parallel filament plasma channels between the electrodes with cross section diameter of ~100 μm, transfer density of the high current of ~100 A/cm², and only a period of a few nanoseconds (Kogelschatz, 2003; Eliasson and Kogelschatz, 1991a, 1991b).

Characteristic little gas warming results due to partial charge transportation in short time and the energy indulgence in each micro discharge. The huge part of the free electron energy of the excited atoms or molecules is triggered within the background gas, thus making the constructors required to initiate surface chemical reactions and/or emission of radiation that energetic photons can also support to motivate surface chemistry. This clarifies the large importance in DBDs for several applications (Fig. 7.7) (Wagner et al., 2003).

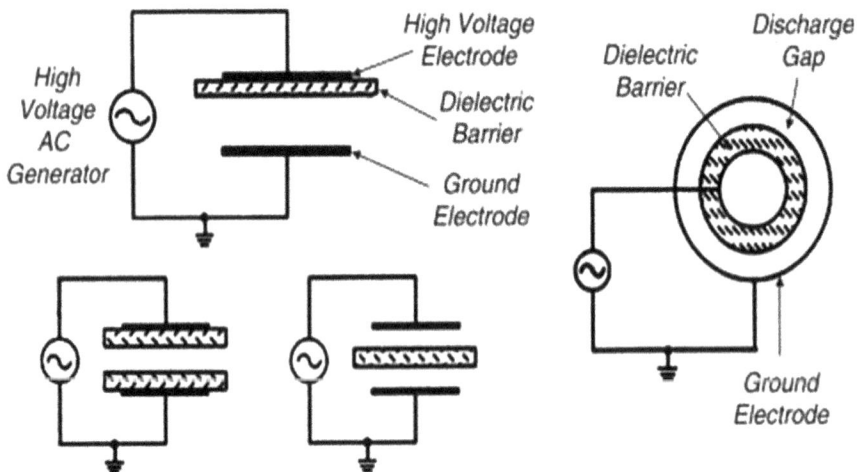

FIGURE 7.7   The schematic diagram of DBD cell used for treatment fabrics (Reprinted from Wagner, H. E.; Brandenburg, R.; Kozlov, K. V.; Sonnenfeld, A.; Michel, P.; Behnke, J. F. (2003). The barrier discharge: basic properties and applications to surface treatment. *Vacuum, 71*, 417–436. Copyright Elsevier reprinted with permission).

7.3.2.5   *ATMOSPHERIC-PRESSURE GLOW DISCHARGE (APGD)*

APGD is another kind of application that can fit with the textile processing requirements, for example, scale and temperature. APGD plasma occurs between the electrodes in the form of uniform, bright, and homogeneous glow. Nonthermal plasma is generated via controlling the gab of the electrode and therefore the driving voltage frequency. This can limit the formation of an extremely undesirable high current density and hot plasma arc. The utilization of ~99% of helium as a generation gas leads to the prevention of arcing and generation of cold plasma. This is considered as a supply that is rich of chemical species required to use in textile development (Xinxin et al., 2003; Justyna et al., 2018).

7.4   USING OF PLASMA TREATMENT IN TEXTILES FIELD

7.4.1   *PRINCIPLES OF PLASMA PROCESS*

Great chemical and physical transformations occurred for textile materials by plasma treatments. These transformations include chemical and structural changes of surface layers and their physical properties (Chinta et al., 2012; Zohuri, 2017). During plasma treatments, electron collisions and photochemical processes cause molecules disassociating and formation of high density of free radicals. These lead to chemical bonds disruption within the surface of fiber polymer that ends up in the formation of recent chemical species. The chemistry of the surface and its topography are affected and also the specific surface area of fibers is significantly enlarged. New functional groups like —OH, —C=O, and —COOH are produced which influence wettability of the fabric and in addition also facilitate graft polymerization. Plasma is applied on the fabric for fine cleaning, activation and etching of fabric surface, cross-linking and coating deposition that is governed by the plasma type and the used gases (Deshmukh and Bhat, 2011; Jacobs et al., 2013).

7.4.2 APPLICATIONS OF PLASMA TREATMENT IN TECHNICAL TEXTILES

Investigation in the field of plasma treatment of textile is quite extensive. Few of the ideal related applications of plasma treatment in technical textile are summarized below (Kale and Desai, 2011):

- **Mechanical properties improvement:**
  Oxygen plasma treatment is useful for softening cotton and other cellulosic polymers. Felting of wool can also be reduced whereas the treatment by dipping in DMSO and then N_2 plasma can impart high resistance in different fabric types such as cotton, silk, and wool.
- **Electrical properties:** Use of chloromethyl dimethylsilane in plasma treatment led to fabrics with antistatic properties.
- **Wetting:** The synthetic polymers (PA, PE, PP, PET PTFE) wettability can be enhanced via treatment in oxygen, air, or NH_3-plasma. The treatment with siloxan or per fluorocarbon plasma can impart the hydrophobic nature for cotton, cotton/polymer polyethylene (PET). Oleo phobic character can be acquired for cotton/polyester by grafting process.
- **Dyeing and printing:** The treatment in O_2-plasma enhances the capillarity properties in wool and cotton rather than improvement of polyester dyeing with $SiCl^{4-}$ plasma and Ar-plasma for polyamide.
- **Metal-coated organic polymers:** Many applications require coating of organic polymers with metals. The plasma pretreatment of the polymer is used for stronger metal adhering and to fulfill its function.

- **Composites and laminates:** Plasma treatments can be very helpful for good adhesion of fibers. Various laminates relay on the surface properties of the fibers in the layers and the reactions that occur at the interface.

7.4.3   ACTIVE PLASMA SPECIES INTERACTION WITH TEXTILE SURFACE

Plasma treatment aims to introduce new functional groups to the textile surfaces and change their chemical compositions without influencing the bulk properties of fabrics. The surface treatments usually occur in an intensity of few hundred astronomy (Fridman, 2008). The treatments take place because of their interaction with the plasma material. Plasma leads to formation of charged particles, for example, ions and electrons. Moreover, atoms/molecules, met stables, and free radicals are present in the region of active plasma and photons because of the generation of UV light. These particles react differently with the substrate, resulting in many various surface processes. With the exception of photons, the depth of the effected substrate is nearly 10 nm which proves that plasma only affects the outer layer of the substrate. Subsequently, plasma is considered as a vital technique for modifying the surface and not affecting the bulk properties of the substrate. Nevertheless, the contamination on the surface can be harmful to plasma process. Plasma processing can change hydrophobic nature of the surface to hydrophilic and vice versa depending on the gas/monomer type used in generating plasma and whether the chemicals have been converted/rounded or shifted on the surface of the textiles. Similarly, compatibility, adhesion, resistance to wear and tear, the rate and depth of dyeing, fiber cleaning surfaces, and desizing can be improved.

7.4.4   PLASMA CLEANING AND ETCHING

Materials production is subjected to solvents, grease, and various volatile substances, etc. These undesirable substances will be absorbed and accumulated on the materials surface with time, leading to change the surface and nonproliferation with low likely performance of the product.

When the fabric is used as a Gersaih, this results in reduced surface contamination, bacterial adhesion, proliferation, and growth (Larsson Wexell et al., 2013; Morra and Cassinelli, 1997). The contaminants can be removed fast with nonthermal plasma treatment (Akishev et al., 2005). The long time plasma treatment will lead to the drilling of the upper layers of the material surface rather than pollution removal (Akishev et al., 2005). The density of the subjected material and hardness needed inflections more intense and/or long exposure to get a noticeable influence. Because in most cases, textiles (biomedical) are made of soft material relatively, it cannot supervise the impact of drilling and will present certain roughness in nanoscale on the fiber surface (Washburn et al., 2004; Khorasani et al., 2008; Persin et al., 2013) (Fig. 7.8).

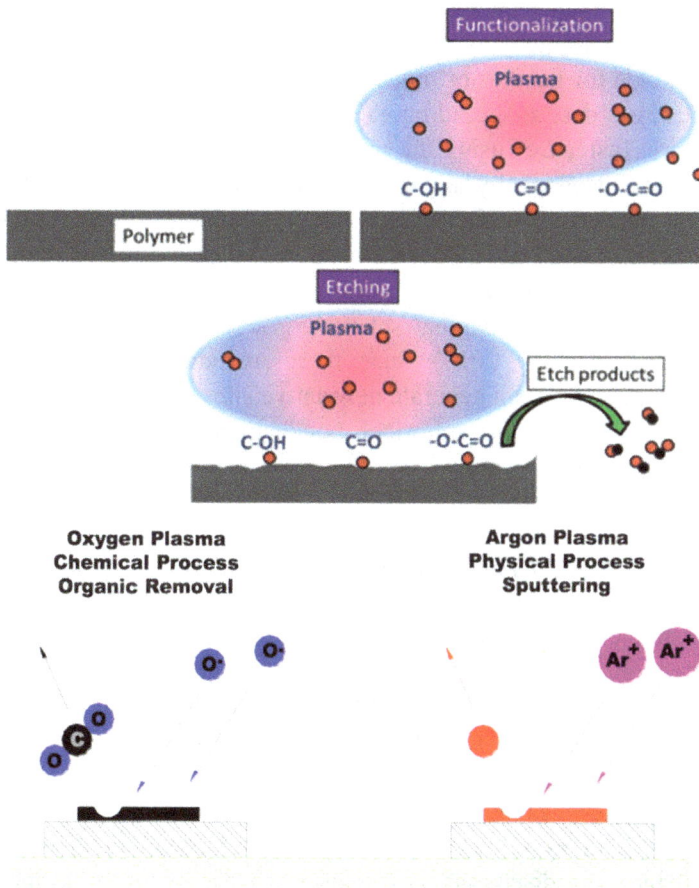

FIGURE 7.8 (a) Plasma cleaning processes, (b) Plasma etching processes (Reprinted from. Puliyalil, H.; FilipiÄ, G.; Cvelbar, U.; Thomas, S.; MozetiÄ, M.; Cvelbar, U. et al. Chapter 9—Selective Plasma Etching of Polymers and Polymer Matrix Composites. In *Nonthermal Plasma Technology for Polymeric Materials*; Elsevier, 2019; pp 241–259. Copyright Elsevier reprinted with permission).

7.4.5 ACTIVATION USING PLASMA

Plasma activation is a surface change by the reactive molecules formed in the plasma. The combination will result from the interactive molecules to integrate radical sites on the substrate surface for a depth up to 10 nm. These sites will interact with other existing radicals, and combine different functional groups according to the gas used. The new reactive sites influence the surface properties significantly like wetting ability and the surface free energy that support the interactions of physical materials and cells. In most cases, the follow-up increases in water resistance to improve the performance of textile. In many applications like the surface of the heart valves or internal shields for needles or industrial stents, the cells and proteins adhesion is strongly desirable leading to a blockage and medical devices vital early failure. Instead of using the usual feed gas to the plasma processing (noble gases, nitrogen, dry air, oxygen, etc.), the use of fluorinated gases such as CF_4 that create surfaces possess hydrophobic nature with high water contact angles reaches to 150 degree or more. These hydrophobic interfaces prohibit cells and proteins from sticking strongly and thus ensuring the best performance of the implant substrate (Geyter et al., 2008; Mattioli et al., 2012; Jacobs et al., 2013). Sure, there are other methods rather than plasma activation that can generate new active sites on the substrate surface; however, plasma is nonmetaphorical, nonchemical, and it ensures that the most sensitive structures are preserved.

7.4.6 PLASMA GRAFTING AND POLYMERIZATION

Cold plasma can be used in other application such as starting radical polymerization, which leads to thin layers' deposition on wide ranges. To improve the bonding between thin films and biological materials, this process takes place in plasma which offers radical sites and allows covalent polymer in conjunction with the surface of the substrate. The polymers can deposit on the surface placing in between the two electrodes. Polymerization process can occur by ionic or radical process which is initiated by plasma generated by glow discharge. For plasma polymerization, plasma is used as a method to initiate and remain active during the whole polymerization reaction. This results in the formation of initiation sites on both surfaces of the substrate and the monomer. Rather than chemical initiation, the plasma can be used for any functional groups of polymeric precursors to initiate the chain reaction. Highly cross-linked, hole free, and amorphous thin film is usually obtained by using the plasma initiation that differs from its traditional equivalent and adheres to a wide range of surfaces. Changing the discharge power leads to a control over the formed functionalities on the surface of the films. The density of functional group plays essential function in the biomedical applications regarding the proliferation and growth of cells that depend on the used cells.

7.4.7   PLASMA COPOLYMERIZATION AND NANO COATING OF TEXTILE

The most popular technique of nanoparticles deposition on textile materials surfaces is pad-dry-cure technology. Dip coating, electrochemical roads, and layer-by-layer depositions are alternative techniques for nanoparticles deposition. However, these techniques have shown several limitations on deposits of nanoparticles (Yang and Yao, 2017; Liu and Abbasi, 2017; Surmeneva and Sharonova, 2017).

Plasma is used for etching of the fiber which increases irregularity of the surface. Therefore, the adhesion properties of substrates toward nanoparticles can be enhanced (Bozzi et al., 2009; Cunko and Varga, 2006; Gorensek et al., 2010; Gorenšek et al., 2010; Marija et al., 2013). Adhesion of titanium nitride to polypropylene and polycarbonate was obtained with modifying the materials with low-pressure argon plasma (Pedrosa et al., 2010). The irregularity of substrates surfaces is enhanced after treatment with plasma in the range of 15–17 nm for polypropylene and 12–30 nm for the polycarbonate. Subsequently, the contact angles reduced from 95 to 59 degree for the polycarbonate and 87° to 35° for polypropylene. Modifying the surface of the PET with oxygen and nitrogen plasma was also attained at different exposure times (Junkar et al., 2009). Polyester polymer surface was improved to optimize the attachment of materials to fucoidan that is biologically active materials with anticlotting properties. The adhesion of fucoidan to materials surfaces was increased by oxygen plasma treatment, particularly, due to the increase in surface roughness. The adhesion action, surface polarity, and surface energy of the polyamide six fibers were improved by DBD using helium gas at atmospheric pressure.

UV protection and self-cleaning properties of polyester fiber were enhanced dramatically after adjusting fiber with oxygen plasma and loaded with TiO_2 by sol-gel process (Borcia and Dumitrascu, 2006).

Cotton fabric indicated self-cleaning functionality after plasma and RF processing TiO_2 (Qi et al., 2007). TiO_2 deposition on textile materials can be used in biomedical applications to enhance the efficiency of antimicrobial textiles (Mejia et al., 2009). Utilizing radio-frequency oxygen plasma at the top of the input energy led to increasing fiber roughness as well as TiO_2 adhesion on the modified fabric.

The treatment of polyamide and polyester using corona plasma enhanced their adhesion to silver, that imparts the antifungal resistance for the materials (Ilic et al., 2009a). The amount of silver loaded on the textile-treated plasma is threefold greater than the untreated fabric.

Effective antimicrobial textiles are used for medical, leisure, military, technical textiles, bedding, and sports. Various nano-materials are used in textiles fictionalization, for example, silver nanoparticles, silver chloride (Ag Cl), composite of silver, and titanium dioxide (Ag-TiO_2) (Gorjanc et al., 2010b; Gavriliu et al., 2010; Choi et al., 2008; Marija et al., 2013). The efficiency of antimicrobial will increase by increasing the surface layer of silver nanoparticles on TiO_2 which increases the amount of particles per unit area compared with the employment of an equal part of pure silver (Rimai et al., 2000; Dastjerdi et al., 2008).

There are various ways for coating silver nanoparticles on synthetic and natural textiles substrates. Sol-gel and dip coating are two examples for loading different textile fabrics with nanoparticles (Marija et al., 2013; Moazami et al., 2010). The exhaustion method is another example for nanoparticles deposition on fabric surface. Exhaustion method led to uniform distribution of nanoparticles that is especially appropriate for simultaneous application of nanoparticles and dyeing the fabric is achieved by effectively dyed material with antimicrobial activity (Gorjanc et al., 2010a).

The required fabric function can determine if the treatment bath could have dye only, dye and silver nanoparticles, or just silver nanoparticles. Exhaustion technique was used to load silver nanoparticles on silk fabric (Moazami et al., 2010). When using different concentrations of colloidal silver (10, 25, 50, and 100 ppm), it has been found that the pH of the medium influences the deposited nanoparticles quantity on the surface of fabric. The effectiveness of antimicrobial fiber was practically the best for samples with high silver concentration and samples treated with a low pH. The addition of salt (NaCl) provided silver particles distributed homogenously on the surface of the fabrics that improved the effectiveness of antimicrobial activity of the materials.

The impact of pad-dry-cure and exhaustion methods on adhesion activity and antimicrobial materials using silver nanoparticles and technical organic–inorganic binder was investigated (Tomšič et al., 2009). The results displayed that using the pad-dry-cure method for loading the fabric with silver nanoparticles led to deposition of low amount of the nanoparticles compared to the amount deposited on the fabric using exhaustion technique. Application of silver nanoparticles on textile substrates can be obtained by plasma polymerization method where the fabric surface run using silver microfilm through sputtering (Jiang et al., 2010). Modification of textile using silver can affect the color of the fabric and the presence of silver nanoparticles in the solution or on the substrate of the fabric can be controlled or detected by using UV/Vis reflectance spectroscopy, which is considered a direct measure of silver on the top layer of the fabric surface. Where the device analyzes the light reflected from the sample treated with silver nanoparticles and results in simple absorption of some of the electrons in the silver particles that appear in the form of an electronic cloud. The light that falls on these electrons gives rise to collective oscillations, which is known as surface plasma. The resonance condition is created once the frequency of light photons becomes suitable to the natural frequency of surface electrons oscillatory against the restoring force of positive nuclei. Various standard tools are based on surface plasma resonance for mensuration adsorption of fabric on to two-dimensional metal (usually gold and silver) or on the surface of metal nanoparticles. This is often the basic principle for several of the applications biosensor based on color. As a result of particle growth, the presence of intense optical phenomenon of 400–415 nm was discovered resulting from the collective excitement of all the free electrons during a particle (Malinsky et al., 2001). The growth in nanoparticles diameter of 1–100 nm causes a shift within the surface plasma absorption wavelength bar to high (Mock et al., 2000). This means that the particle size of the nanoparticles are determined through visual response, and thus by the seen color (Ilic et al., 2009b). The colloid of silver nanoparticles loaded on to bleached cotton fabric can produce yellow coloration of the fabrics with a maximum absorption of 370 nm (Yuranova et al., 2003). This evaluates the color changes to the textile that have been modified by silver nanoparticles in color CIELAB space (Ilic et al., 2009b). Deposited silver nanoparticles colloidal concentration of 10 ppm in six dyeing environment friendly fabrics and finishing a very small inferred, and is insensitive to the eye, and color change (ΔE^* <1) on the cloth. When the loaded silver nanoparticles colloidal concentration was 50 ppm, it was a clear change in the cloth (* E * = color 15.09). Silver nanoparticles deposited on fabric during the dyeing process can alter the color but the change is not as wide as is the case with white fabric. Loading colloidal silver before dyeing led to color change of $\Delta E^* = 1.44$ while loading after dyeing led to the color change of $\Delta E^* = 2.73$. The difference in the color can occur in textiles by plasma processing. The ozone plasma can be used for bleaching of raw cotton fabric (Navarro and Bautista, 2005). The whiteness index of the cloth enhanced after processing the plasma better than after peroxide bleaching. Since modified materials by plasma enhance the adhesion of metal nanoparticles, DBD plasma-supported deposition of AgNPs has shown to enhance the NPs adhesion prepared in water distributions. DBD plasma is environment friendly dry technique used to modify the surfaces of the fabric evenly without traditional chemical procedures (Zanini and Citterio et al., 2018). Active plasma processing of DBD fabric surface is done by introducing new functional groups, for example, carboxyl and carbonyl, ether, amine and hydroxyl depending on the used gas (such as air, helium, nitrogen or oxygen) (Zanini and Citterio et al., 2018). These polar groups have the ability to enhance the surface energy and therefore, exchange the wettability of fabric. Also, it can enhance the cross-linking and etching without affecting the material bulk properties (Zanini and Citterio et al., 2018). In addition, plasma processing for the deposition AgNPs is attractive due to the formation of roughness minutes linked to chemical changes on the surface that improves installation AgNPs. The deposit AgNPs on various parts of the textile material is inaccessible mechanically by physical attraction such as Van der Walls or electrostatic forces (Gorjanc and Gorenšek, 2013).

Furthermore, plasma allows incorporating of nanoparticles in the polymer surrounding substance. The nanoparticles/plasma metal in this procedure freezes the nanoparticles on site, which is possible during plasma deposition, providing a first-step coating process that eliminates the necessity of directly dealing with the nanoparticles (Balazs et al., 2005; Favia et al., 2000). Moreover, the quantity of metal materials can be decreased considerably compared to electrochemical techniques and thus have a significant cost impact, especially when metals are precious, for example, silver or gold. This tool allows the control of the amount of metal materials that are integrated into the paint through the changes in particle size. Deposition of silver nanoparticles on PP fabrics using plasma are illustrated in (Fig. 7.9) (Dong, 2019).

7.5   FUTURE TRENDS

Plasma processes with low pressure are achieved by invasions even when economic limitations are slight, since the plasma processes has unique and benign potential of environmentally sound. Subsequently, plasma can be an alternative to many traditional processes, especially those wet chemicals. The approach taken in the deposit of nano-coatings measured on textiles is of

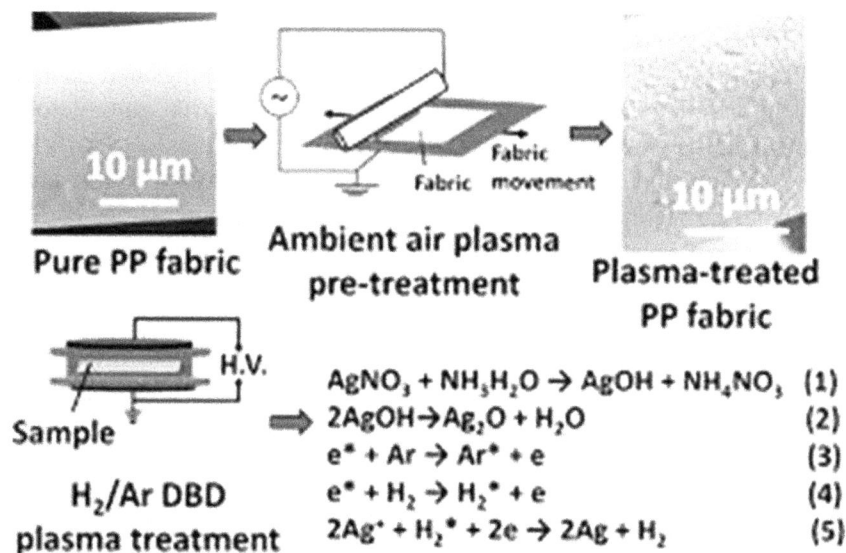

FIGURE 7.9 Deposition of silver nanoparticles on PP fabrics using plasma. (Reprinted from Dong, P.; Nie, X.; Jin, Z.; Huang, Z.; Wang, X.; Zhang, X. Dual Dielectric Barrier Discharge Plasma Treatments for Synthesis of Agâ€"TiO₂ Functionalized Polypropylene Fabrics. *Indust. Eng. Chem. Res.* **2019,** *58* (19), 7734–7741. Copyright (2019), American Chemical Society).

particular importance, where the bulk of fabric properties are still uninfluenced. For example, the fiber can be spun metalized under the same conditions of untreated fiber. The incorporation of electronics in the clothing concept opens up a full range of multifunction electric textiles which can be worn for the functions of sensor/body control, and the provision of communication facilities, data transfer, and control of the individual in the environment, and various new applications (Tao, 2005). Medication distribution controlled significant growth in biotechnology allows plasma processing to produce new substrates with tight control linkage and mutual release rates of polymer when distension. The fabrication of multilayer systems, the gradient, and the reservoir system, which is not easy to attain by wet chemical techniques but it is enhanced by the plasma treatment. This forms new smart and technical textile that promotes wound healing, prohibit infection, and leads to vital improvements of products for humans. Other finishing such as hydrophobic and oleophobic coatings for textile are still of great attention, where wet chemical treatments indicate a deficiency of sufficient washing fastness. Thus, we should also study the plasma layers of cross-linking for fabrics. The same applies to permanent aqueous coatings as new types of functional nano-volatile plasma coatings that appear to reduce aging and enhance washing stability are currently being investigated. Thus, the relevant trends in the textile processors plasma are mainly dealing with coatings hydrophobic/ hydrophobic, functional waterproof, static and conductive, antimicrobial and medical, as well as multipurpose surfaces. Another tendency will lead to a heavy-duty and established textile of heat and high corrosion resistance, where plasma technology also contributes. Besides the function, comfort functional textiles have become more and more user focused.

KEYWORDS

- **textile**
- **plasma**
- **plasma in textiles**
- **plasma copolymerization**
- **nanocomposite coating**

REFERENCES

Ahmed, H. A.; El-Halwagy, A. A.; Abdel-Aaty, A.; Garamoon, A. A. Plasma Application in Textiles. *J. Text. Color Polym. Sci.* **2019,** *16* (1), 35–52.

Ahmed, H. M. Studies on the Effect of Atmospheric Glow Discharge Plasma on Some Natural and Synthetic Fabrics to Improve Their Printability, Faculty of Science of Chemistry Department, Al-Azhar University Cairo, 2011; pp 40–42.

Ahmed, H. M.; Ahmed, K. A.; Mashaly, H. M.; El-Halwagy, A. A. Treatment of Cotton Fabric with Dielectric Barrier Discharge (DBD) Plasma and Printing with Cochineal Natural Dye. *Indian J. Sci. Technol.* **2017,** *10* (10), 1–10.

Akishev, Y.; Grushin, M.; Kochetov, I.; Karal'nik, V.; Napartovich, A.; Trushkin, N. Negative Corona, Glow and Spark Discharges in Ambient Air and Transitions Between them. *Plasma Sources Sci. T.* **2005,** *14* (2), 18–25.

Balazs, D. J.; Hollenstein, C.; Mathieu, H. J. Fluoropolymer Coating of Medical Grade Poly(Vinyl Chloride) by Plasma-Enhanced Chemical Vapor Deposition Techniques. *Plasma Process Polym.* **2005,** *2*, 104–111.

Borcia, C.; Dumitrascu, N. Adhesion Properties of Polyamide-6 Fibres Treated by Dielectric Barrier Discharge. *Surf. Coat. Tech.* **2006,** *201*, 1117–1123.

Bozzi, A.; Yuranova, T.; Guasaquillo, I.; Laub, D.; Kiwi, J. X. Self-Cleaning of Modified Cotton Textiles by TiO_2 at Low Temperatures under Daylight Irradiation. *J. Photoch. Photobio. A* **2009,** *174*, 156–164.

Chinta, S. K.; Landage, S. M. Plasma Technology & Its Application in Textile Wet Processing. *IJERT* **2010,** *1* (5).

Choi, O.; Deng, K.; Kim, N.; Ross, L.; Surampalli, R.; Hu, Z. The Inhibitory Effects of Silver Nanoparticles, Silver Ions and Silver Chloride Colloids on Microbial Growth. *Water Res.* **2008,** *42* (12), 3066–3074.

Cunko, R.; Varga, K. Application of Ceramics for the Production of High-Performance Textiles. *TEKSTIL* **2006,** *55* (6), 267–278.

Dastjerdi, R.; Mojtahedi, M. R. M.; Shoshtari, A. M. Investigating the Effect of Various Blend Ratios of Prepared Masterbatch Containing Ag/TiO_2 Nanocomposite on the Properties of Bioactive Continuous Filament Yarns. *Fiber Polym.* **2008,** *9* (6), 727–734.

Dave, H.; Ledwani†, L.; Nema, S. K. Nonthermal Plasma: A Promising Green Technology to improve Environmental Performance of Textile Industries. *Imp. Pro. G. Chem. Tex. Tech.* **2019,** 199–249.

Deshmukh, R. R.; Bhat, N. V. Pretreatments of Textiles Prior to Dyeing: Plasma Processing. In *Textile Dyeing*; 2011; pp 30–53.

Eliasson, B.; Kogelschatz, U. Plasma Science. *IEEE Transac.* **1991a,** *19* (6), 1063–1077.

Eliasson, B.; Kogelschatz, U. Plasma Science. *IEEE Transac.* **1991b,** *19* (2), 309–323.

Favia, P.; Vulpio, M.; Marino, R.; d'Agostino, R.; Mota, R. P.; Catalano, M. Plasma-Deposition of Ag-Containing Polyethyleneoxide-Like Coatings. *Plasmas Polym.* **2000,** *5*, 1–14.

Freidberg, P. J. *Plasma Physics and Fusion Energy*; Cambridge University Press, 2008.

Fridman, A. *Plasma Chemistry and Plasma Processing*; Cambridge University Press, 2008.

Gavriliu, S.; Lungu, M.; Elena, E.; G. L. C. Composite Nanopowder for Antibacterial Textiles. *Industria Textila.* **2010,** *61* (2), 86–90.

Geyter, N. D.; Morent, R.; Gengembre, L.; Leys, C.; Payen, E.; Vlierberghe, S. V.; Schacht, E. Increasing the Hydrophobicity of a PP Film Using a Helium/CF4 DBD Treatment at Atmospheric Pressure. *Plasma Chem. Plasma P.* **2008,** *28* (2), 289–298.

Geyter, N. D.; Morent, R.; Leys, C. Penetration of a Dielectric Barrier Discharge Plasma into Textile Structures at Medium Pressure. *Plasma Sources Sci. Technol.* **2006,** *15*, 78–84.

Ghaly, A. E.; Ananthashankar, R.; Alhattab, M.; Ramakrishnan, V. V. Production, Characterization and Treatment of Textile Effluents: A Critical Review. *Chem. Eng. Technol.* **2014,** *5* (1), 1–18.

Gorensek, M.; Gorjanc, M.; Bukosek, V.; Kovac, J.; Jovancic, P.; Mihailovic, D. Functionalization of PET Fabrics by Corona and Nano Silver. *Tex. Res. J.* **2010,** *80* (3), 253–262.

Gorenšek, M.; Gorjanc, M.; Bukošek, V.; Kovač, J.; Petrović, Z.; Puač, N. Functionalization of Polyester Fabric by Ar/N2 Plasma and Silver. *Text. Res. J.* **2010,** *8* (16), 1633–1642.

Gorjanc, M.; Bukošek, V.; Gorenšek, M.; Mozetič, M. CF4 Plasma and Silver Functionalized Cotton. *Text. Res. J.* **2010a,** *80* (20), 2204–2213.

Gorjanc, M.; Bukošek, V.; Gorensek, M.; Vesel, A. The Influence of Water Vapor Plasma Treatment on Specific Properties of Bleached and Mercerized Cotton Fabric. *Text. Res. J.* **2010b,** *80* (6), 557–567.

Gorjanc, M.; Gorenšek, M. Eco-Friendly Textile Dyeing and Finishing. University of Engineering & Management India, 2013.

Gulrajani, M. L. *Advances in the Dyeing and Finishing of Technical Textiles*; Woodhead Publishing Limited: Cambridge, UK, 2013.

Horrocks, A. R.; Anand, S. C. *Handbook of Technical Textiles*; Elsevier, 2016; p 394.

Ibrahim, N., et al. Effect of Cellulase Treatment on the Extent of Post-Finishing and Dyeing of Cotton Fabrics. *J. Mater. Process Technol.* **2005,** *160* (1), 99–106.

Ilic, V.; Saponjic, Z.; Vodnik, V.; Molina, R.; Dimitrijevic, S.; Jovancic, P.; Nedeljkovic, J.; Radetic, M. Antifungal Efficiency of Corona Pretreated Polyester and Polyamide Fabrics Loaded with Ag Nanoparticles. *J. M. Sci. Mater. Med.* **2009a,** *44* (15), 3983–3990.

Ilic, V.; Saponjic, Z.; Vodnik, V.; Potkonjak, B.; Jovancic, P.; Nedeljkovic, J. M.; Radetic, M. The Influence of Silver Content on Antimicrobial Activity and Color of Cotton Fabrics Functionalized with Ag Nanoparticles. *Carbohyd. Polym.* **2009b,** *78* (3), 564–569.

Jacak, W. A. Plasmons in Finite Spherical Electrolyte Systems: RPA Effective Jellium Model for Ionic Plasma Excitations. *Plasmonics* **2016,** *11* (2), 637–651.

Jacobs, T.; Declercq, H.; Geyter, N. D.; Cornelissen, R.; Dubruel, P.; Leys, C.; Beaurain, A.; Payen, E.; Morent, R. Plasma Surface Modification of Polylactic Acid to Promote Interaction with Fibroblasts. *J. Mater. Sci. Mater. Med.* **2013,** *24* (2), 469–478.

Jiang, S. X.; Qin, W. F.; Zhang, L. Surface Functionalization of Nanostructured Silver-Coated Polyester Fabric by Magnetron Sputtering. *Surf. Coat. Technol.* **2010,** *204* (21–22), 3662–3667.

Junkar, I.; Vesel, A.; Cvelbar, U.; Mozetič, M.; Strnad, S. Influence of Oxygen and Nitrogen Plasma Treatment on Polyethylene Terephthalate (PET) Polymers. *Vacuum* **2009,** *84* (1), 83–85.

Justyna, S.; Wiesława, U. D.; Waldemar, M.; Henryk, W.; Karolina, Ł.; Beata, G. Low Temperature Plasma for Textiles Disinfection. *Int. Bio. B.* **2018,** *131*, 97–106.

Kale, K. H.; Desai, A. N. Atmospheric Pressure Plasma Treatment of Textiles Using Non-Polymerising Gases. *Ind. J. Fibre Textile Res.* **2011,** *36*, 289–299.

Karmakar, S. R. Combined Pre-Treatment Processes Of Textiles. *JTST* **1999,** *12*, 336–343.

Khanit, M.; Sommavan, W. Non-Thermal Dielectric Barrier Discharge Generator. *Pro. Comp. Sci.* **2016,** *86*, 313–316.

Khatri, A.; Peerzada, M.; Mohsin, M. A. Review on Developments in Dyeing Cotton Fabrics with Reactive Dyes for Reducing Effluent Pollution. *J Clean Prod.* **2015,** *87*, 50–57.

Khorasani, M. T.; Mirzadeh, H.; Irani, S. Plasma Surface Modification of Poly (l-Lactic Acid) and Poly (Lactic-Co-Glycolic Acid) Films for Improvement of Nerve Cells Adhesion. *Radiat Phys. Chem.* **2008,** *77* (3), 280–287.

Klaus, D. W.; Juergen, F. K.; Marcin, H.; Dirk, U.; Milan, Š.; Kostya, O.; Satoshi, H.; Uroš, C.; Mirko, Č.; Alexander, F.; Pietro, F.; Kurt, B. The Future for Plasma Science and Technology. *Plasma Process Polym.* **2019,** *16* (1), 1–29.

Kogelschatz, U. Plasma Chemistry and Plasma Processing, **2003,** *23* (1), 1–46.

Kumar, A.; Choudhury, R. Principles of Textile Finishing, 2017.

Kumar, R. S.; Sundaresan, S. Mechanical Finishing Techniques for Technical Textiles; Woodhead Publishing Limited, 2013.

Labay, C.; Canal, C.; RodrÃguez, C.; Caballero, G.; Canal, J. M. Plasma Surface Functionalization and Dyeing Kinetics of Pan-Pmma Copolymers. *Appl. Surf. Sci.* **2013,** *283*, 269–275.

Larsson Wexell, C.; Thomsen, P.; Aronsson, B. O.; Tengvall, P.; Rodahl, M.; Lausmaa, J.; Kasemo, B.; Ericson, L. Bone Response to Surface-Modified Titanium Implants: Studies on the Early Tissue Response to Implants with Different Surface Characteristics. *Int. J. Biomat.* **2013**.

Lewis, D. M. Developments in the Chemistry of Reactive Dyes and their Application Processes. *Color Technol.* **2014,** *130*, 382–412.

Li, J.; Ma, C.; Zhu, S.; OrcID, F. Y.; Dai, B.; Yang, D. A Review of Recent Advances of Dielectric Barrier Discharge Plasma in Catalysis. *Nanomaterials* **2019,** *9* (10), 1–34.

Liu, K. G.; Abbasi, A. Deposition of Silver Nanoparticles on Polyester Fiber Under Ultrasound Irradiations. *Ultrason Sonochem.* **2017,** *34*, 13–18.

Madhav, S.; Ahamad, A.; Singh, P.; Mishra, P. K. A Review of Textile Industry: Wet Processing, Environmental Impacts, and Effluent Treatment Methods. *Manage. Environ. Qual.* **2008,** *27* (3), 31–41.

Malinsky, M. D.; Kelly, K. L.; Schatz, G. C.; Van, D. R. P. Chain Length Dependence and Sensing Capabilities of the Localized Surface Plasmon Resonance of Silver Nanoparticles Chemically Modified with Alkanethiol self-Assembled Monolayers. *J. Am. Chem. Soc.* **2001,** *123,* 1471–1482.

Mangesh, D. T.; Tesfaye, T. A. Short and Efficient Desizing and Scouring Process of Cotton Textile Materials. *IJETT* **2016,** *35* (6), 256–269.

Marija, G.; Marija, G.; Petar, J.; Miran, M. *Multifunctional Textiles—Modification by Plasma, Dyeing and Nanoparticles*; Licensee InTech, Supported by Slovenian Research Agency, 2013.

Mattioli, S.; Kenny, J.; Armentano, I. Plasma Surface Modification of Porous PLLA Films: Analysis of Surface Properties and in vitro Hydrolytic Degradation. *J. Appl. Polym. Sci.* **2012,** *125* (S2), E239–E247.

Mejia, M. I.; Marin, J. M.; Restrepo, G.; Pulgarin, C.; Mielczarski, E.; Mielczarski, J.; Arroyo, Y.; Lavanchy, J. C.; Kiwi, J. Self-Cleaning Modified TiO2-Cotton Pretreated by UVC-Light (185 nm) and RF-Plasma in Vacuum and also Under Atmospheric Pressure. *Appl. Catal. B. Environ.* **2009,** *91* (1–2), 481–488.

Moazami, A.; Montazer, M.; Rashidi, A.; Rahimi, M. K. Antibacterial Properties of Raw and Degummed Silk with Nanosilver in Various Conditions. *J. Appl. Polym. Sci.* **2010,** *118* (1), 253–258.

Mock, J. J.; Barbic, M.; Smith, D. R.; Schultz, D. A.; Schultz, S. Shape Effects in Plasmon Resonance of Individual Colloidal Silver Nanoparticles. *Int. J. Chem. Phys.* **2000,** *116* (15), 6755–6759.

Morent, R.; Geyter, N. D.; Verschuren, J.; Clercke, K. D.; Kiekens, P.; Leys, C. Non-Thermal Plasma Treatment of Textiles. *Surf. Coat. Technol.* **2008,** *14* (202), 3427–3449.

Morra, M.; Cassinelli, C. Evaluation of Surface Contamination of Titanium Dental Implants by LV-SEM: Comparison with XPS Measurements. *Surf. Interf. Anal.* **1997,** *25* (13), 983–988.

Navarro, A.; Bautista, L. *Surface Modification and Characterization in Cotton Fabric Bleaching*, Proceedings of 5th World Textile Conference AUTEX 2005, Maribor.

Park, H.; Carr, W. W.; OK, H.; Park, S. Image Quality of Inkjet Printing on Polyester Fabrics. *Text. Res. J.* **2006,** *76* (9), 720–728.

Paul, e. b. R. *Functional Finishes for Textiles: An Overview*; Woodhead Publishing is an Impert of Elsevier, UK, 2015.

Pedrosa, P.; Chappe, J. M.; Fonseca, C.; Machado, A. V.; Nobrega, J. M.; Vaz, F. Plasma Surface Modification of Polycarbonate and Poly(propylene) Substrates for Biomedical Electrodes. *Plasma Proc. Polym.* **2010,** *7* (8), 676–686.

Persin, Z.; Mozetic, M.; Vesel, A.; Maver, T.; Maver, U.; Kleinschek, K. S. Plasma Induced Hydrophilic Cellulose Wound Dressing, Medical, Pharmaceutical and Electronic Applications, 2013.

Qi, K.; Xin, J. H.; Daoud, W. A.; Mak, C. L. Functionalizing Polyester Fiber with a Selfcleaning Property Using Anatase TiO$_2$ and Low-Temperature Plasma Treatment. *Int. J. Appl. Ceram. Technol.* **2007,** *4* (6), 554–563.

Rimai, D. S.; Quesnel, D. J.; Busnaina, A. A. The Adhesion of Dry Particles in the Nanometer to Micrometer-size Range. *Colloids Surf. Physicochem. Eng. Aspects* **2000,** *165* (1–3), 3–10.

Shishoo, R. *Plasma Technology in the Textile*; Woodhead Publishing Limited in Association with The Textile Institute: Cambridge, England, 2007.

Surmeneva, M.; Sharonova, A., et al. Incorporation of Silver Nanoparticles into Magnetron-Sputtered Calcium Phosphate Layers on Titanium as an Antibacterial Coating. *Colloids Surf. B. Biointerf.* **2017,** *156,* 104–113.

Tomšič, B.; Simončič, B.; Orel, B.; Žerjav, M.; Schroers, H. J.; Simončič, A. Antimicrobial Activity of AgCl Embedded in a Silica Matrix on Cotton Fabric. *Carbohydr. Polym.* **2009,** *75* (4), 618–626.

Virendra, K.; Jerome, P.; Hubert, R.; Ilaria, M.; Francois, R.; Farzaneh, A. K. Fluorocarbon Coatings Via Plasma Enhanced Chemical Vapor Deposition of 1H,1H,2H,2H perfluorodecyl Acrylate 2, Morphology, Wettability and Antifouling Characterization. *Plasma Process Polym.* **2010,** *7* (11), 926–938.

Wadje, P. R. Textile – Fibre to Fabric Processing, *IEI J.* **2009,** 90, 28–36.

Wagner, H. E.; Brandenburg, R.; Kozlov, K. V.; Sonnenfeld, A.; Michel, P.; Behnke, J. F. The Barrier Discharge: Basic Properties and Applications to Surface Treatment. *Vacuum* **2003,** *71*, 417–436.

Washburn, N. R.; Yamada, K. M.; Simon, C. G.; Kennedy, S. B.; Amis, E. J. High Throughput Investigation of Osteoblast Response to Polymer Crystallinity: Influence of Nanometer-Scale Roughness on Proliferation. *Biomaterials* **2004,** *25* (7), 1215–1224.

Xiaodong, M.; Zhong, Y.; Hong, X.; Linping, Z.; Xiaofeng, S.; Zhiping, M. A Novel Low Add-On Technology of Dyeing Cotton Fabric with Reactive Dyestuff. *Text. Res. J.* **2018,** *5* (2), 376–765

Xinxin, W.; Chengrong, L.; Mingze, L.; Yikang, P. Study on an Atmospheric Pressure Glow Discharge. *Plasma Sources Sci. Technol.* **2003,** *12*, 358–361.

Xueni, H; Guoqiang, C; Tieling, X; Zhenzhen, W. Reactive Ink formulated with Various Alcohols for Improved Properties and Printing Quality Onto Cotton Fabrics. *J. Eng. Fiber Fabr.* **2019,** *14*, 1–11.

Yang, C.; Yao, Y. Silk Inverse Opals: Modulation of Multiscale 3D Lattices through Conformational Control: Painting Silk Inverse Opals with Water and Light. *Adv. Mater.* **2017,** *29* (38).

Yasir, N. Textile Engineering: An introduction, illustrated ed. Walter de Gruyter GmbH & Co KG, 2016.

Yuranova, T.; Rincon, A. G.; Bozzi, A.; Parra, S.; Pulgarin, C.; Albers, P.; Kiwi, J. Antibacterial Textiles Prepared by RF-Plasma and Vacuum-UV Mediated Deposition of Silver. *J. Photoch. Photobio. A.* **2003,** *161* (1), 27–34.

Zanini, S.; Citterio, A. Efficient Silver Nanoparticles Deposition Method on DBD plasma-Treated Polyamide 6,6 for Antimicrobial Textiles. *Mater. Sci. Eng.* **2018,** *427* (1), 90–96.

Zohuri, B. *Magnetic Confinement Fusion Driven Thermonuclear Energy*, 1st ed.; Springer Publishing Company, University of New Mexico, 2017.

CHAPTER 8

Nanoparticle Application by Layer-by-Layer Deposition Technique to Produce Polyelectrolyte Multilayers on Fabrics

CHET RAM MEENA

National Institute of Fashion Technology, Jodhpur 342037, Rajasthan, India

E-mail: chetram.meena@nift.ac.in

ABSTRACT

Few decades back, a "nano" word with big promises has been precipitously implying itself into the world's realization and associated with everyday life, economics, and globally consequences. Nanotechnology may deliver better performances to textile materials due to high surface area and energy. Further, nanoparticles can be applied on the fabrics by coating method without affecting the comfort and feel of the fabrics. It improves various properties like ultraviolet protection, antibacterial and stain proof, and so on. Layer-by-layer (LBL) techniques are used to produce a thin polymeric film in a controlled manner on a surface of fabrics by using different sizes of molecular weights and charges of polyelectrolytes. The unique feature of this technique is that it forms a very thin layer on fabric surface (1–10 nm) as compared to other available techniques; thus, there is no deterioration of surface properties of the substrate on which they are deposited. Nano-TiO_2 and ZnO particles along with polyelectrolytes produce polyelectrolyte multilayers on the Nylon 66 substrate by using LBL technique.

Fundamentals of Nano-Textile Science. Prashansa Sharma, Devsuni Singh & Vivek Dave (Eds.)
© 2022 Apple Academic Press, Inc. Co-published with CRC Press (Taylor & Francis)

8.1   INTRODUCTION

Nanotechnology exists to engineer materials. It usually involves the design, characterization, production, and application of materials, and the scope has nowadays been extended to also include electronic gadgets rather than just materials. Standard unit is nanometer dimensions [0.000000001 (10^{-9}) m] scale of particles (Ramsden, 2005). Some interesting properties of nanoparticles used in textile applications (Table 8.1) (Zhu and Narh, 2004).

TABLE 8.1   Nanoparticles and Potential Textile Applications (Zhu and Narh, 2004).

Sr. No.	Nanoparticles	Properties
1	Silver	Antimicrobial
2	Fe	Conductive magnetic
3	ZnO and TiO$_2$	UV resistant, antimicrobial
4	MgO	Chemical and biological resistant, self-cleaning
5	SiO$_2$ or Al$_2$O$_3$ nanoparticles with PP or PE coating	High water repellency
6	Indium–tin oxide	EM/IR resistant property
7	Ceramic	Improving the abrasion resistance
8	Carbon black	Enhance the abrasion, chemical resistance, and electrical conductivity
9	Clay	Impart the electrical, heat, and chemical resistance
10	Cellulose nanowhiskers	Crease resistant, antistain, and water resistant

8.1.1   WHAT ARE POLYELECTROLYTES?

Polyelectrolyte is defined as any polymeric substance in which the monomer units of its constituent's macromolecules hold ionizable groups. Electrochemically, a polyelectrolyte can be classified as either as polyacid, a polybase, or a polyampholyte, depending upon the nature of ionization in the aqueous solution (Fig. 8.1).

The usual anionic polyelectrolytes are homopolymers and copolymers of Na salt of acrylic acid with acrylamide.

Cationic polyelectrolytes are photopolymers or copolymers with acryl amide as major cationic monomer (Fig. 8.2).

Nowadays polyelectrolytes are used in various industries, such as paper manufacturing, oil processing, treatment of water, and so on. Polyelectrolytes are responsible for the formation of a thin polymeric film on the surface of textile materials and offer new innovations in unconventional textiles.

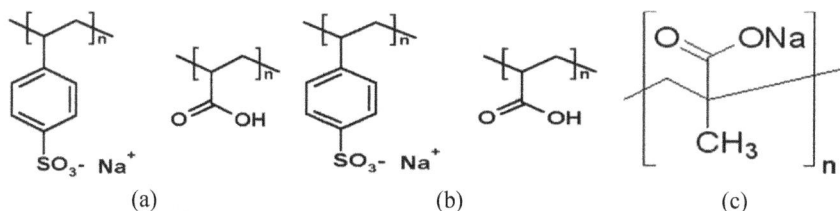

(a)   (b)   (c)

FIGURE 8.1 (a) Poly(sodium 4-styrene sulfonate) (PSS), (b) poly(acrylic acid) (PAA), and (c) poly(methacrylic acid sodium salt).

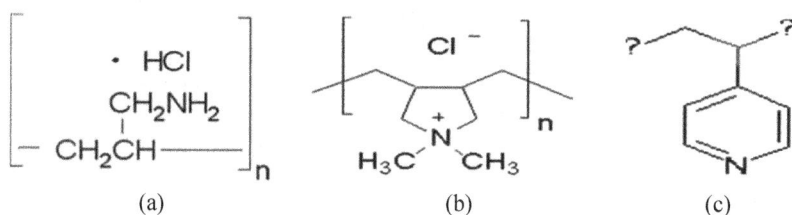

(a)   (b)   (c)

FIGURE 8.2 (a) Poly(allyl amine hydrochloride) (PAH), (b) poly(diallyl dimethyl ammonium chloride), and (c) poly(4-vinylpyridine).

8.1.2 SEM PHOTOGRAPH OF PEM FILM

The polyelectrolyte layer-by-layer (LBL) deposition technique has mainly been applied to deposit films onto planar surfaces or spheres. Progressive adsorption of oppositely charged colloids extended to the preparation of multilayers of polycations and phosphonate ions as well as to the layering of polyelectrolytes.

The electrostatic self-assembly process is based on the alternating adsorption and self-organization of charged cationic and anionic species. The SEM photograph of PEM film shown in (Fig. 8.3).

LBL thin films can also be fabricated by exchanging charged species such as nanoparticles in place of or in addition to one of the polyelectrolytes (Fig. 8.4) (Decher, 1997).

LBL technique is suitable for natural and synthetic fibers in form of woven, knitted, and nonwoven structure. A significant feature of LBL technique is that it can be used for optimizing the performance by controlling the number of layers of polyelectrolyte multilayers (PEMs). Other potentials of PEMs in the various fields are air filters, nanocapsules, conducting films for solid-state electrolytes, tubular titanium nanostructures, nano-electromechanical systems, and micro-electromechanical systems, patterning of PEM, and so on (Kale and Meena, 2008).

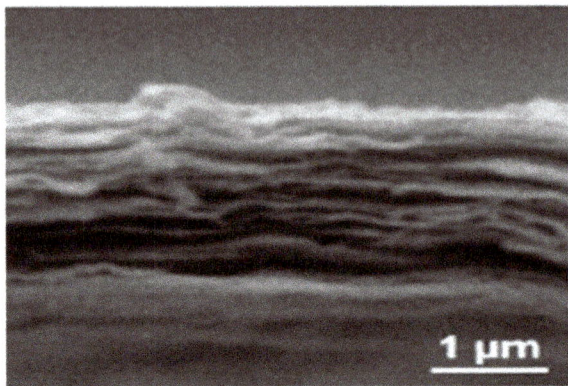

FIGURE 8.3   SEM photograph of PEM film.

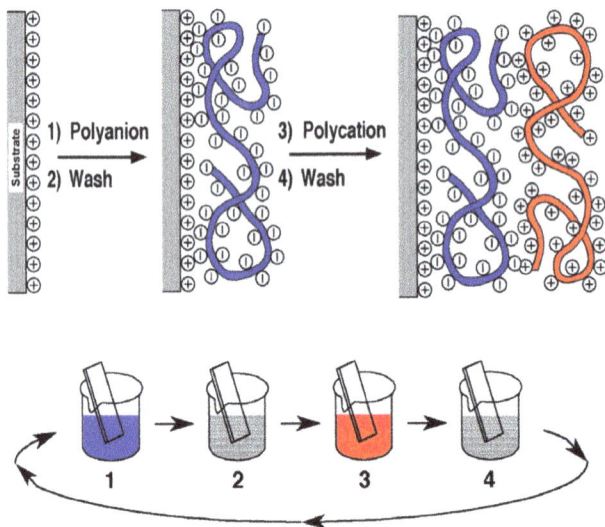

FIGURE 8.4   Simplified molecular picture of the first two adsorption steps, depicting film deposition starting with a negatively charged substrate. Scheme of the LBL film-deposition process using glass slides and beakers. Steps 1 and 3 represent the adsorption of a polyanion and polycation, respectively, and steps 2 and 4 are washing steps (Decher, 1997).

8.1.3 NECESSITY OF MODIFICATION OF FIBERS

The "modification" term is itself self-explanatory that to define a deliberate modification in assembly or structure leading to improvements in some fiber properties like affinity for cationic dyes, nonflammability, as high strength, soft hand, stretch resistance, and so on (Shenai, 1994). For more than three decades, various types of modifications of fibers have been systematically studied. The demand of smart materials in daily life has of late abruptly increased in the area of modification. Technologies that comprise engineering to convert cheap materials into exclusive finished merchandises have become more important in the current situation (Desai Shrojal and Singh, 2004).

The constraint of alteration for a fiber is dependent on its characteristics and the end applications. Textile finishes are generally classified according to purpose or end uses. These classifications are standard, wet, or chemical finishes and decorative or mechanical finishes. Individuals concerned with end merchandises (designers, merchandisers, and sales personnel) usually categorize finishes as esthetic finishes and functional finishes. The earlier modifies the appearance and/or handle (feel) of fabrics, while the later improves the performance of a fabric under specific end-use conditions (Bajaj, 2001). Functional modifications of fibers are to enhance its performance in specific field. The surface properties of the textile fibers manage the macroscopic properties of a textile such as absorbency and visual appearance to a large extent (Knittel and Schollmeyer, 2000). Modification of fiber's surface plays a significant role in textile finishing. There are few selected methods to modify textile surfaces. Various such finishes add more than one property to a textile. Some chemical finishes are added to the surface of the textile substrates without penetration inside the fibers (Hall, 1957). Though modification of fibers is not an easy task preparation as an effort is made to improve certain properties, there is inclination for other useful properties to be negatively disturbed (Vaidya, 1986).

8.2 SYNTHESIS OF NANOPARTICLES

8.2.1 NANO TITANIUM DIOXIDE (TiO₂)

Tetraisopropyl orthotitanate of 2.84 ml added in a solution of 0.5 ml methanol and 5 ml ethanol with molar ratio (1:1:10) and refluxed for 6 h at 60°C. Distilled water (30 ml) added drop wise in the above solution at 60°C under the continuous stirring condition. The precipitate isolated by filtration and washed 2–3 times with warm water. Precipitate dried at 130°C for 12 h and then subjected to calcinations at 550°C for 10 h in muffle furnace. The white crystalline powder of nano-TiO_2 was obtained at the end of the process (Kale and Meena, 2012).

8.2.2   NANO ZINC OXIDE (ZnO)

Zinc acetate dihydrate (10.98 g) treated with ethanol (300 ml) for 30 min at 60°C. Oxalic acid dihydrate (12.6 g) dissolved in ethanol (200 ml) at 50°C. The oxalic acid solution added slowly under the stirring conditions to the warm ethanolic solution of zinc acetate. A high viscous white gel was formed, which was dried at 80°C for 20 h and calcinations carried out in muffle furnace at 600°C for 2 h to yield ZnO nanoparticles (Kale and Meena, 2011). There are some functional properties of Zinc Oxide applied on the textile materials as shown in Table 8.2 (Verbič et al., 2019).

TABLE 8.2   The Functional Properties of Zinc Oxide on Textiles (Verbič et al., 2019).

Functional properties of zinc oxide (ZnO)	Antibacterial property
	UV absorption
	Flame resistance
	Hydrophobicity
	Air and water purification
	Self-cleaning
	Breathable finish
	Electrical conductivity
	Thermal resistance

8.3   PREPARATION OF PEM FORMATION

Poly(ethylene imine) (PEI) (0.1 wt%) solution was prepared by adding HNO_3 (1 M) to maintain the slightly acidic pH. Polyacrylic acid (PAA) (0.1 wt%) solution was prepared the same way as PEI to achieve pH 5. Addition of nano-ZnO and TiO_2 to the PEI solution was always done just prior to the beginning the self-assembly procedure.

Fabric strips should be fixed on sample holder in an open condition. Then the fabric should be dipped into a solution of positively charged polyelectrolyte (polyanion) results in adsorption of first layer. A washing step is needed to remove unfixed particles. Now the surface potential is positive and a polyanion is adsorbed and thus one bilayer is formed. Subsequently again, a washing step is required in the end. The initial layer for each fabric strip was deposited by dipping the fabric into each of the ionic solutions for 2 min, beginning with the cationic solution. Subsequent layers were also deposited with 2 min dips. Each dip was followed by rinsing with normal water for 2 min. Succeeding the deposition of required number of layers, fabric strips were dried in air to remove remaining water.

8.4 ANTIMICROBIAL ACTIVITY

Currently, there are two essential aspects of textiles such as comfort and safety. Users are more concerned about health that encouraged a need for fabrics that can prevent the growth of bacteria and other microorganisms which can cause offensive odors, skin irritation, visual deterioration, and disfiguring spots.

Limitations of antimicrobial agents are harmful to human being and problem of degradation in the environment. The textile industry has constant to search for eco-friendly processes as replacements for toxic chemicals and processes. It may be possibly fixed by ecofriendly and biodegrade antimicrobial agents (Shalini and Anitha, 2016). Antimicrobial activity of metal and metal oxide based nanoparticles as shown below in Table 8.3 (Karwowska, 2016).

TABLE 8.3 Antimicrobial Activity of Nanoparticles (Karwowska, 2016).

Antimicrobial activity of nanoparticles				
Abolishing and malfunction of cell membrane (Ag, ZnO, Au, TiO$_2$)	Disruption of electron transport and energy production (Ag)	Formulation of reactive oxygen species (TiO$_2$)	Renewing of cell enzymes	Prohibition of DNA reproduction (Ag)

8.4.1 NECESSITY OF ANTIMICROBIAL TREATMENT

Antimicrobial treatment for textile materials is necessary to fulfill the following objectives:

• To prevent cross-contamination by pathogenic microorganisms;
• To regulate the infestation by microbes;

- To capture metabolism in microbes in order to diminish the growth of odor; and
- To safeguard the textile products from stain, mark, and quality decline (Hausam, 1962).

The functionalization of textile polymers can be achieved by using LBL technique by adding the active component such as nanoparticles into one of the polyelectrolyte solutions. Nylon being an amphoteric fiber has both carboxyl and amine end groups. PEI and PAA are used and they have amine and acidic groups, respectively, in its structure, which is capable of forming ionic bonds with nylon as well as with each other. Therefore, PEM that is built onto the nylon is strongly bonded with the fabric. AATCC 100-2004 method is used to evaluate the antibacterial activity of the treated samples.

8.5 ULTRAVIOLET PROTECTION FACTOR

Source of ultraviolet radiation is sun, which is invisible and cannot be felt. It is a high-energy radiation which depends on its wavelength that causes sunburns, eye disorders, skin aging, cancer, and so on (Table 8.4). Ultraviolet protection factor (UPF) depends on various parameters of fabrics such as fiber composition, yarns, and construction of fabric. Treated fabric has shown the UV protection to satisfactory level against the solar radiations compared to unfinished fabrics.

TABLE 8.4   Effect of UV Rays on Different Types of Skin (Saravanan, 2007).

Skin type	Critical dose (mJ/cm^2)	Self-protection time (min)	Risk level
White	15–30	5–10	Burns easily, skin aging, and skin cancer
White	25–35	8–12	Burn and remarkably tan
Brownish	30–50	10–15	Tan and irregularly burn
Brown	45–60	15–20	Tan and irregularly burn
Brown	60–100	20–35	Sufficient levels of melanin and remarkably burns, easily tan
Dark Brown-Black	100–200	35–70	Sufficient levels of melanin pigment provide protection. Very rarely burns, easily tan

8.5.1 NECESSITY OF AN IDEAL UV ABSORBER

- An efficient UV absorber has to be absorbed throughout the spectrum to remain stable against UVR and then to disperse the absorbed energy to avoid ruin of fabric or loss of color value.
- Maximum absorbency in the ultraviolet region (290–340 nm) and no absorbency in the visible region.
- Should be heat stable and harmonious with other additives in the finish invention.
- Should be nontoxic and nonskin irritation (Xin et al., 2004).

The UPF or sun protection factor terms are commonly used as a measuring unit for ultraviolet radiation. Spectrophotometer was used to measure the percent transmission at wavelength intervals up to 5 nm in the 290–400-nm spectral span. The UPF is examined by the following equation (Singh and Singh, 2013):

$$\text{UPF} = \frac{\sum_{280 \text{ nm}}^{400 \text{ nm}} E_\lambda S_\lambda \Delta_\lambda}{\sum_{280 \text{ nm}}^{400 \text{ nm}} E_\lambda S_\lambda T_\lambda \Delta_\lambda} \tag{8.1}$$

where E_λ is the spectral weighting function of erythemal action spectra, S_λ is the spectral irradiation for appropriate solar radiation spectrum (W m^{-2} nm^{-1}), T_λ is the spectral transmittance through specimen, and Δ_λ is the appropriate wavelength measuring interval (nm).

8.6 RESULTS AND DISCUSSIONS

8.6.1 ANTIMICROBIAL ACTIVITY AGAINST NANO TiO$_2$

Nano-TiO$_2$ particles are added into the PEI solution so that it gets entrapped into the PEM as the number of layers is built up onto the nylon fabric.

In Figures 8.5 and 8.6, results of antimicrobial activity on nylon fabric versus number of layers deposited and amount of nano-TiO$_2$ are shown. The percentage reduction in number of colonies of *Staphylococcus aureus* and *Escherichia coli* for TiO$_2$ increases from 1st to 25th PEM. As the concentration of TiO$_2$ increases from 0.05, 0.1 to 0.2 g/l, the percentage reduction in number of colonies increases from 61% to 71% to 95%, respectively, for the 25th layer.

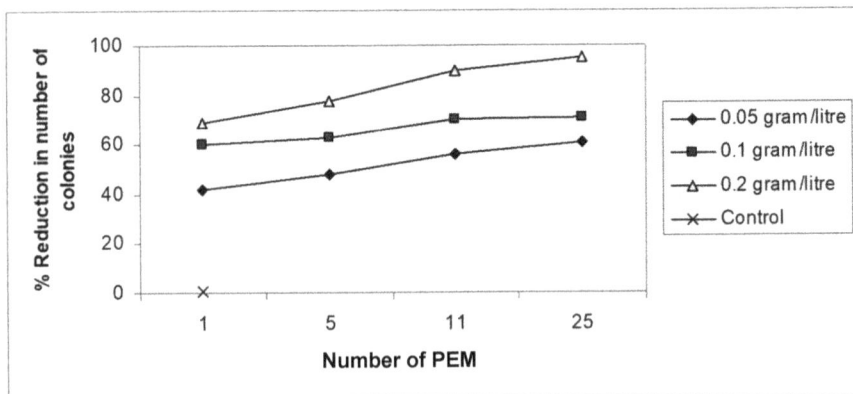

FIGURE 8.5   Relation between different amount of synthesized TiO$_2$ and layers of PEM on percentage reduction in number of *Staphylococcus aureus* colonies on nylon fabric.

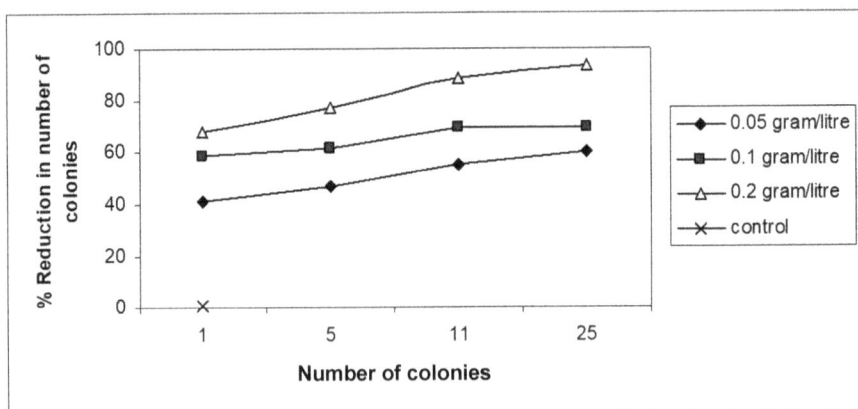

FIGURE 8.6   Relation between different amount of synthesized TiO$_2$ and layers of PEM on percentage reduction in number of *Escherichia coli* colonies on nylon fabric.

Thus, if we increase the concentration of nano-TiO$_2$ in the polyelectrolyte solution, the required effect can be obtained at less number of layers, thereby saving time and chemicals. Further, also as the number PEM increased from 1st to 25th PEM, there is an additional improvement in antimicrobial activity because as the number of layers is built up, there is a proportionate increase in the concentration of TiO$_2$ onto the fabric.

The washing durability of nano-TiO$_2$ treated fabric as shown in Table 8.5. It can be noticed that there is a negligible drop in antimicrobial activity from 69% to 64.5% for the 1st layer and from 95% to 93.5%

for the 25th layer. Results verify that nano-TiO_2 particles deposited on the fabric are strongly bonded to the fabric. It is clear that synthesized nano-TiO_2 either before and after washing shows more or less similar antimicrobial property.

TABLE 8.5 Effect of Washing Treatment on Synthesized Nano-TiO_2 and Number of PEM on Antimicrobial Activity of Nylon Fabric for *S. aureus*.

Number of layers	Amount of nano-TiO_2, 0.2 g/l					
	Before wash			After wash		
	Number of colonies after		% Reduction in number of colonies	Number of colonies after		% Reduction in number of colonies
	0 h	24 h		0 h	24 h	
#Control	786	780	0.76	786	780	0.76
1	99	31	69	105	38	64.5
5	72	16	78	79	19	76
11	55	6	90	61	8	88
25	17	1	95	23	2	93.5

Control = Nylon fabric treated with only PEM (up to 25 layers).

8.6.2 ANTIMICROBIAL ACTIVITY AGAINST NANO-ZnO

The treated fabric shows that antimicrobial activity is because of the presence of nano-ZnO on the fabric which is a very efficient antimicrobial compound. Also as the number of layers increases from 1 to 25, there is further improvement in antimicrobial activity because as the number of layers is built up, there is a proportionate increase in the concentration of ZnO on the fabric. Also as the concentration of ZnO added into the PEI solution when increased from 0.05 to 0.1 to 0.2 g/l, the amount of nano-ZnO that gets deposited onto the fabric increases and this results in the enhanced antimicrobial activity for the same number of layer but having different concentration.

Here also, the percentage reduction in a number of colonies increased from 39% for the 1st layer to 62% for the 25th layer for 0.05 g/l concentration of nano-ZnO for *S. aureus* and from 38% to 61% for *E. coli*. The antimicrobial activity showed an improvement up to 96% for *S. aureus* and 94.5% for *E. coli* when the concentration of ZnO increased to 0.2 g/l when one considers the 25th layer as shown in Figures 8.7 and 8.8.

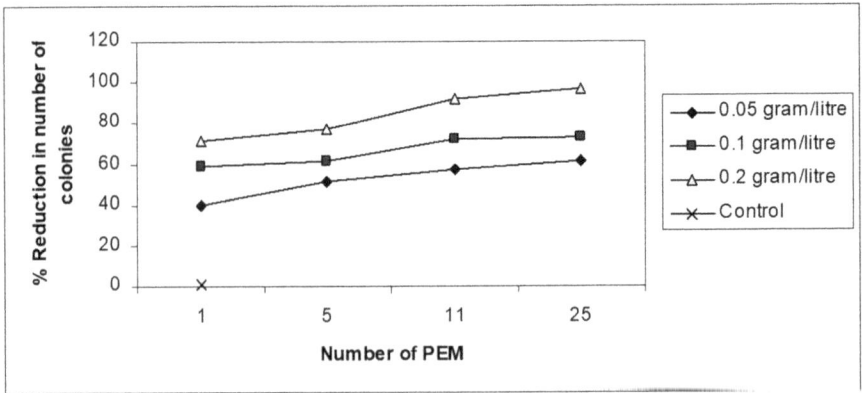

FIGURE 8.7   Relation between amount of synthesized ZnO and layers of PEM on percentage reduction in number of *S. aureus* colonies on nylon fabric.

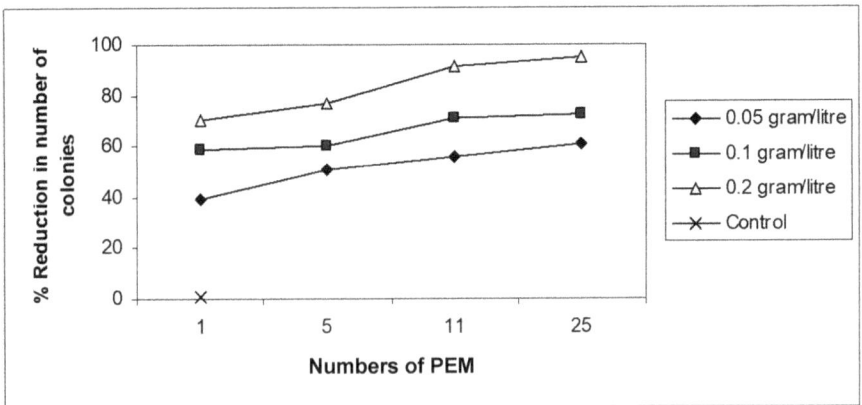

FIGURE 8.8   Relation between amount of synthesized ZnO and layers of PEM on percentage reduction in number of *E. coli* colonies on nylon fabric.

8.6.2.1   *COMPARISON BETWEEN SYNTHESIZED NANO-ZNO AND FABSHIELD AEM 5700 (ALDRICH-SIGMA) FOR ANTIMICROBIAL ACTIVITY*

Fabshield AEM 5700 (Aldrich-Sigma), a commercial antimicrobial sample based on silver, is also tested for the comparison, using 0.05 g/l concentration to assess the efficacy of synthesized nano-ZnO particles for *S. aureus* as shown in Table 8.6.

The percentage reduction in the number of colonies was 39.5% and 56% for synthesized ZnO and Fabshield AEM 5700, respectively, for the 1st layer.

The correspondingly values were 61.5% and 96% for the 25th layer. These values indicate that 0.05 g/l concentration of nano-ZnO is not sufficient to get the comparable results with the Fabshield but 0.2 g/l concentration is to be used for which the values are 96.5% and 96%.

TABLE 8.6 Effect of Synthesized Nano-ZnO and Fabshield AEM 5700 and Number of PEM on Antimicrobial Activity of Nylon Fabric for *S. aureus*.

Number of layers	Concentration of nano-ZnO and Fabshield AEM 5700, 0.05 g/l					
	Nano-ZnO			Fabshield AEM 5700		
	Number of colonies after		% Reduction in number of colonies	Number of colonies after		% Reduction in number of colonies
	0 h	24 h		0 h	24 h	
#Control	786	780	0.76	786	780	0.76
1	158	96	39.5	90	40	56
5	145	72	51	62	21	67
11	98	43	57	33	8	78
25	81	32	61.5	21	1	96

8.6.3   ULTRAVIOLET PROTECTION AGAINST NANO-ZNO

The ultraviolet protection of treated fabric versus number of layers deposited and concentration of synthesized nano-ZnO and commercial nano-ZnO is shown in Table 8.7. UPF of treated fabric progressively increases from 167 nm for the 1st layer to 271 nm for the 25th layer when the concentration of nano-ZnO in the solution is 0.05 g/l.

It is also observed that as the concentration of ZnO increased from 0.05 g/l to 0.2 g/l, the ultraviolet protection factor of nylon fabric enhanced further from 271 nm for 0.05 g/l to 295 nm for 0.1 g/l and up to 325 nm for 0.2 g/l with the number of layers being the 25th for all three concentration. This fact is further underlined by the fact that in case of 0.2 g/l of concentration for the first layer itself, the ultraviolet protection factor 235 nm is almost equal to that of the 25th layer of 0.05 g/l concentration 271 nm. Thus, as the concentration of nanoparticles in the polyelectrolyte solution increases, the desired effect can be achieved at less number of layers, thereby saving time and chemicals.

The comparative study of UPF of the nano-ZnO obtained from Sigma-Aldrich with that of synthesized nano-ZnO synthesized as shown in Table 8.7. Here again, the same trend in results is observed for nano-ZnO particles from

Aldrich-Sigma where used. For the 1st layer, when concentration is 0.05 g/l, the UPF values are 167 nm and 185 nm, respectively, for synthesized ZnO and ZnO obtained from Aldrich-Sigma. For the 25th layer and same concentration, the values are 271 nm and 268 nm. Similar behavior is observed for the other two concentrations. Thus, the synthesized nanoparticles are equivalent to the commercial nano-ZnO particles.

TABLE 8.7 Effect of Concentration of Nano-synthesized ZnO, Aldrich-Sigma and Washing Treatment and Number of PEM on Ultraviolet Protection Factor of Nylon Fabric.

Concentration of nano-ZnO (g/l)	No. of layers	UPF (nm)		
		Synthesized (ZnO)	Aldrich-Sigma (ZnO)	After one wash
Untreated (nylon fabric)			74	
Control (nylon fabric treated with only PEM, up to 25 layers)			75	
0.05	1	167	185	164
	5	225	221	219
	11	244	243	236
	25	271	268	256
0.1	1	226	224	215
	5	231	228	222
	11	235	241	227
	25	295	289	277
0.2	1	235	232	225
	5	259	255	244
	11	298	297	282
	25	325	319	306

It would be worthwhile to see whether the nanoparticles deposited via LBL technique are resistant to washing treatment. To assess this, nylon fabric activated by synthesized nano-ZnO particles was subjected to one washing cycle and then again evaluated for their UPF. Presented in Table 8.7, it can be seen that UPF has marginally dropped due to washing effect. It is 167 nm for unwashed sample and 164 nm after first washing cycle for the 1st layer for 0.05 g/l concentration. For the same concentration, the values for the 25th layer are 271 nm for unwashed sample and 256 nm for washed sample. Thus, the UPF did not deteriorate too much after washing. The only possible reason for this is the ability nano-ZnO particles hold onto the fabric even during washing. The scanning electron microscope (SEM) pictures of unwashed and washed nylon fabric are shown in Figure 8.9. We can see the presence of nanoparticles on the washed samples, and this is the reason why even after washing, nanoparticles are firmly bound to the fabric and hence show UPF property.

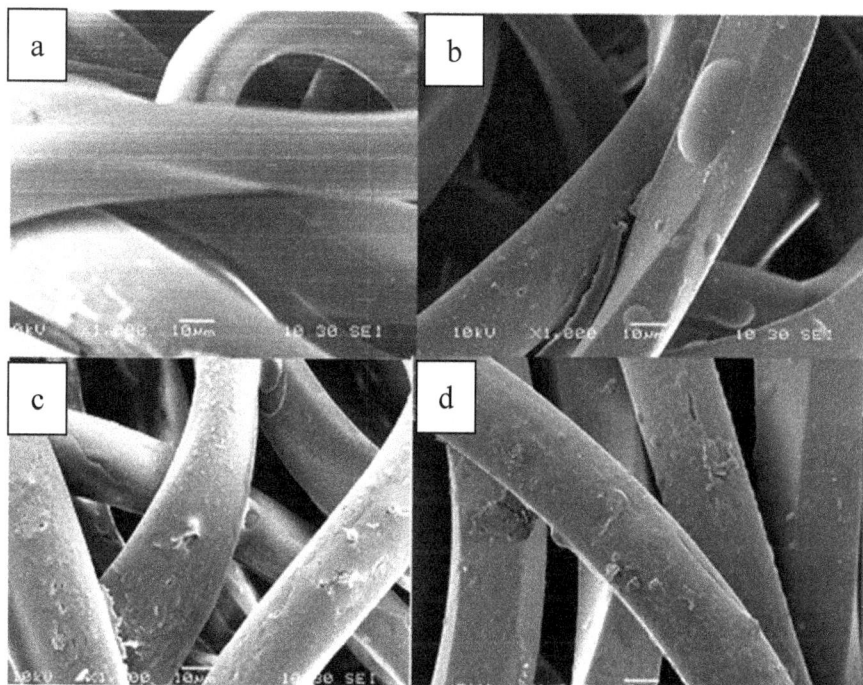

FIGURE 8.9   Scanning electron microscope (SEM) image of nylon surface: (a) untreated, (b) nano-ZnO, (c) Aldrich-Sigma, and (d) nano-ZnO (after washing).

8.6.4   ULTRAVIOLET PROTECTION AGAINST NANO-TIO$_2$

The UPF value for nano-TiO$_2$ increases from 198 nm for the 1st layer to 298 nm for the 25th layer for 0.05 g/l concentration. It is clear from Table 8.8 that as the concentration of TiO$_2$ increases from 0.05 to 0.1 to 0.2 g/l, the UPF increases from 298 nm to 311 nm to 391 nm, respectively, for the 25th layer.

The washing fastness of TiO$_2$ deposited on nylon for single washing cycle is assessed. The UPF is 198 nm and 188 nm for unwashed and washed samples, respectively, for concentration 0.05 g/l. The values for the 25th layer are 298 nm and 278 nm for the same concentration. Thus, the UPF decreased after single washing but to a lower extent because some of the nanoparticles washed.

TABLE 8.8 Effect of Different Concentration of Nano-TiO$_2$, Washing Treatment, and Number of PEM on Ultraviolet Protection Factor of Nylon Fabric.

Concentration of nano-TiO2 (g/l)	No. of layers	UPF (nm)	
		Synthesized	After one wash
Untreated (nylon fabric)		74	
Control (nylon fabric treated with only PEM, up to 25 layers)		75	
0.05	1	198	188
	5	231	222
	11	271	259
	25	298	278
0.1	1	235	230
	5	256	246
	11	298	287
	25	311	300
0.2	1	255	249
	5	296	282
	11	351	346
	25	391	380

8.7   CONCLUSION

Worldwide nanoparticles have the potential market. It has numerous areas to improve the performance of textiles and other materials. The most challenging area is the application techniques of nanoparticles on textiles and to get the optimized results. There are few conventional methods/techniques for application of nanoparticles by exhaust, padding, coating, and so on. One of the unconventional techniques is LBL technique. It has additional significant benefits that are controlled and effective over the conventional techniques and allow more versatility on different surfaces. Polyelectrolytes are the basic composition along with functional elements for LBL techniques. LBL technique may be used on textiles, optical, magnetic, chemical, electrical, and so on. The functionality of the treated fabric has increased with increasing quantity of TiO$_2$ and ZnO nanoparticles and the number of layers. Therefore, higher quantity of nanoparticles can improve the antimicrobial property and UPF property of treated fabric with less number of PEMs. The effectiveness of the antimicrobial activity and UPF is comparable with the commercially available nanoparticles/products in the market.

KEYWORDS

- **nanoparticles**
- **LBL**
- **PEM**
- **TiO$_2$**
- **ZnO nanoparticles**
- **antimicrobial activity**
- **ultraviolet protection**

REFERENCES

Bajaj, P. Ecofriendly Finishes for Textiles. *Indian J. Fbr. Text. Res.* **2001,** *26* (1), 162–186.

Decher, G. Fuzzy Nanoassemblies: Towards Layered Polymeric Multicomposites. *Science* **1997,** *277*, 1232–1237.

Desai Shrojal, M.; Singh, R. P. Surface Modification of Polyethylene. *Adv. Polym. Sci.* **2004,** *169*, 231–293.

Hall, A. J. *A Handbook of Textile Finishing*; The National Trade Press Ltd.: London, 1957.

Hausam, W.; Milleiand Textilber. *Am. Dyestuff Rep.* **1962,** *43*, 277.

Kale, R. D.; Meena, C. R. Application of Polyelectolyte Multilayers on Textile Polymers. *Int. Dyer* **2008,** *193* (10), 26–29.

Kale, R. D.; Meena, C. R. Synthesis and Application of Zinc Oxide Nanoparticles on Nylon. *Int. J. Basic. Appl. Chem. Sci.* **2011,** *1* (1), 1–8.

Kale, R. D.; Meena, C. R. Synthesis of Titanium Dioxide Nanoparticles and Application on Nylon Fabric Using Layer by Layer Technique for Antimicrobial Property. *Adv. Appl. Sci. Res.* **2012,** *3* (5), 3073–3080.

Karwowska, E. Antibacterial Potential of Nanocomposite-Based Materials: A Short Review. *Nanotechnol. Rev.* **2016,** *6* (2).

Knittel, D.; Schollmeyer, E. Technologies for a New Century: Surface Modification of Fibres. *J. Text. Inst.* **2000,** *91* (3), 151–165.

Ramsden, J. J. What Is Nanotechnology? *Nanotechnol. Percept.* **2005,** *1*, 3–17.

Saravanan, D. UV Protection Textile Materials. *AUTEX Res. J.* **2007,** *7* (1), 53–62.

Shalini, G.; Anitha, D. A Review: Antimicrobial Property of Textiles. *Int. J. Sci. Res.* **2016,** *5* (10), 766–768.

Shenai, V. A. *Textile Fibers*; Sevak Publications: Bombay, 1994; Vol. 1, pp 11–15.

Singh, M. K.; Singh, A. Ultraviolet Protection by Fabric Engineering. *J. Text.* **2013,** *2013,* 1–6.

Vaidya, A. A. *Production of Synthetic Fibres*, 1st ed.; Prentice-Hall of India Private Limited: New Delhi, 1986; pp 1–10, 280–285.

Verbič, A.; Gorjanc, M.; Simončič, B. Zinc Oxide for Functional Textile Coatings: Recent Advances. *Coatings* **2019,** *9* (9), 550.

Xin, J. H.; Daoud, W. A.; Kong, Y. Y. A New Approach to UV-Blocking Treatment for Cotton Fabrics. *Text. Res. J.* **2004,** *74* (2), 97–100.

Zhu, L.; Narh, K. A. Numerical Simulation of the Tensile Modulus of Nanoclay-Filled Polymer Composites. *J. Polym. Sci. B: Polym. Phys.* **2004,** *42* (12), 2391–2406.

CHAPTER 9

Nanotechnology in Sports Clothing

M. PARTHIBAN[1*] and K. SARAVANAN[2]

[1]*Department of Fashion Technology, PSG College of Technology, Coimbatore, Tamil Nadu, India*

[2]*Department of Fashion Technology, Bannari Amman Institute of Technology, Sathymangalam, Tamil Nadu, India*

Corresponding author. E-mail: parthi111180@gmail.com

ABSTRACT

Nanotechnology is a term that has advanced around sci-fi for quite a long time. The equivalent word of nanotech is straightforward: minor, frequently infinitesimal machines are structured and modified to do a basic assignment, something that would be in some way or other difficult to achieve by a human or machine. In other words, this would just be a sort of nanomaterial made for a specific reason, and typically a reason it can satisfy considerably more proficiently than everything else available. The innovation behind this has been in progress for a considerable length of time, yet, even during a time where researchers can create molecularly measured capacitors in microchips, these are still amazingly essential machines, and where nanotech would be somewhat further developed.

Nanoengineered functional textiles are about to demonstrate a vital role in the futuristic development of the clothing industry. The basic potential of this technology in the development of new textile materials is highly considerable. The application of nanotechnology in the sportswear sector is increasing gradually. The properties, such as mechanical protection, chemical protection, physical and chemical behavior, thermal behavior apart from improving water-repellency, oil recovery, and antifouling functions are the major functions of nanotechnology. Sports textiles require the use of nanofibers, nanocomposites, nanofinished fabrics to impart multifunctional properties to sportswear.

Fundamentals of Nano-Textile Science. Prashansa Sharma, Devsuni Singh & Vivek Dave (Eds.)
© 2022 Apple Academic Press, Inc. Co-published with CRC Press (Taylor & Francis)

Most of the textile companies have utilized nanotechnology in their apparel supply chain. Sport products are developed using nanotechnology which could enhance good comfort, improved air permeability, improved wind and water resistance, and enhance self-cleaning properties.

9.1 INTRODUCTION

Sportswear or dynamic wear is apparel, as well as footwear, worn for sport or work up. Game express attire is worn for many games and work up for viable, comfort, or security reasons. Everyday game outfit consists of "shorts, tracksuits, and T-shirts." Certain games, such as swimming, surfing, skiing, and vaulting etc. need wet suits. It also includes some clothing, for example, the athletic supporter and sports bra. Nowadays, sportswear is also used as easygoing style attire. For most of the sports, the participants wear a combination of various things, for example, sports shoes, jeans, and shirts. In certain games, defensive rigging might be required to be worn, for example, head protectors or American football body armor.

To provide comfort to the user, sportswear is normally expected to remain weightless. The suitable athletic wear for certain types of activity, for example, cycling, is a tight fit, because it will not drag air and slow you down while riding. Then again, sportswear should be adequately free so as not to limit agility. Certain games have obvious style necessities, for instance, the "'Keikogi'" utilized in karate. Different truly risky games, such as for hockey, soccer, wrestling, fencing, etc. require defensive kit. Standardized sportswear may also be worn as a uniform. In team sports, the opponent sides are generally made identifiable by the exclusively designed dress, while team associates can be made noticeable by a number on their jersey back. Spandex is the favored material for perfectly sized sportswear, for example, it is utilized in wrestling, track and field, move, vaulting, and swimming (Milenkovic et al., 1999).

9.2 CRITICAL REQUIREMENTS AND CHARACTERISTICS FOR SPORTS CLOTHING

Sportswear configuration essentially considers protecting the wearer from hot or cold weather. In hot circumstances, sportswear ought to permit the wearer to remain cool; while in cold circumstances, it should assist the user to stay warm. Sportswear needs to have an option to move sweat away from the skin through the surface of the fabric. Spandex is a well-known material utilized as base layer to absorb sweat. When exercising, the workout wear needs to be comfortable which allows the wearer to move freely and also to keep the person cool during workouts (Das and Alagirusamy, 2010).

Exercise has been very well known and sports garment makers have rushed to react to the customer's request. Each game has its particular attire prerequisites. However, one of the most significant needs is that sports garments should be breathable, and they should permit the body to inhale regularly and to stretch (Oğulata, 2007).

Cotton T-shirts, high-cut elasticated shorts, and one-piece leotards with tights which fit cozily over the legs are well known and satisfy the essential prerequisites. It is additionally essential to wear a tracksuit toward the beginning and end of an activity which needs warming of the body. If the temperature is extremely low, the most ideal approach to keep warm is by wearing at least two layers of garments. The inward layer ought to permit sweat to get away with the objective that wearer will not' get wet. The external layer must be wind-safe, waterproof, and preferably must allow perspiration to get away (Das et al., 2007).

Up to 40% of body's heat can be lost through the head so that the person needs to put on a cap in a cold climate. They should no longer need to overdress or they may feel overheat in the course of exercise. It is a clever thought to wear an external layer which can be zipped open for snappy chilling. Gloves are a really useful add-on for some games. Weight lifters make use of fingerless gloves to improve hold very close and to prevent rankles framing; cyclists utilize cushioned gloves to reduce the chance of harming the fragile structures with the palm, at the same time as preventing handlebars for large stretches, and sprinters frequently put on narrow woolen gloves to hold their hand's heat in cold climate (Das et al., 2007).

They have enough stretch; however, still have those super soft woolly senses. A wide scope of footwear is accessible. However, it is essential to get shoes explicitly intended for your movement and which suit your prerequisites. Those practicing as a major aspect of a weight decrease program ought to keep themselves away from rubber-treated or plastic suits as they are expected to retain heat and can be extremely risky (Das et al., 2009).

9.3 APPLICATION OF SPORTS CLOTHING

Sportswear is the most adaptable and fastest growing part of the exhibition apparel market. Emotional way of life changes, a quickly maturing populace, and expanding sports interest advertisement well-being awareness have made an enormous interest in useful games clothing. In this manner, the scope of significant worth included apparel with properties, such as temperature guideline, the dampness the executives, stretch and smell decrease, and lightweight is accessible in the market for ordinary games wear. While these garments may have multipractical properties, they do not fall in the classification of utilitarian apparel (Das A et al., 2007).

Sports utilitarian attire as a class speaks about those presentation-improving garments that help athletes contend at forefront execution. Two standards essentially oversee the structure and building of this sportswear, for example, use of pressure on explicit muscles to build bloodstream and utilization of standards of optimal design to diminish wind or air haul in fast games. The two standards can be utilized in a blend or independently, contingent upon the prerequisites. In contrast to different classes of utilitarian dress, style is likewise a significant structure basis in this classification (Das et al., 2007).

It has ordinarily concurred that textures with damp wicking properties will direct internal heat level, improve muscle performance, and hold over fatigue. Whereas traditional strands, as an example, cotton could be cheap for the dress worn for low degrees of movement, factory-made textures made from nylon or polyester are a lot of qualified materials for high levels of activity. They ingest considerably less water than cotton, but will in any case wick the damp quickly through the texture.

The important characteristic required for comfort and utility properties are given below:

- Water-repellent, breathability and comfort
- Humidity/perspiration transport
- Warmness/heat control
- Easy-care maintenance
- Smart and practical design

9.4   MOISTURE MANAGEMENT TECHNOLOGY

The part that carries dampness in materials closely resembles the wicking of fluid in vessels. Two fundamental properties of the vessels are chosen for surface pressure:

(1) They reach across and (2) its inner face surface vitality.

If little the circulation or more the strength of the air, the more is a fluid's tendency to travel up. In material structures, vessels are formed viably by the gaps between the filaments. Consequently, the smaller the spaces between these strands, the more noteworthy the intensity of the material to wick dampness. Texture developments, which are successfully structure-restricted vessels, gain dampness without any problem. Such developments incorporate textures made up of small-scale strands, which are pressed intently together. In any case, surface strain stops when all parts of the garment are similarly wet (Ishtiaque et al., 2003).

The surface vitality of a material system is essentially chosen through the chemical structure of the 'fiber's exposed surface as follows: Hydrophilic filaments require a high surface vitality. Thus, they can achieve added prompt humidity than hydrophobic filaments or fibers. Hydrophobic filaments on the contrary, have poor surface vitality and are repulsive for humidity. Specific completion procedures are routinely acclimatized to increase the differentiation in surface vitality between the nature of a material and the 'back of the material to assist its wickability. There are few variables in a texture of fabric which affect the transport of humidity. These include the nature of the fiber, fabric weave structure, weight, or thickness of the material, and finishing treatment given to the material.

Crafted strands may have either wetting (hydrophilic) or nonwetting (hydrophobic) surfaces. Similarly, they have a degree of mass porousness, which suppliers periodically discover and check the relationship, as the rate of clamminess regained by weight. Generally speaking, for garments worn as a base layer, the created surfaces are seen as the best option. It is because they can give the fabric a good mix of sogginess, fragile consistency, and health. Although most surfaces, both ordinary and built, are capable of wicking the soddenness away from the skin. These surfaces are not speedy drying, not permeable to air. Two factors are affecting cooling and perceived comfort. Forefront designed surfaces are lightweight, are fit for moving away clamminess efficiently and dry reasonably quickly (Ishtiaque et al., 2003).

9.5 WATER-REPELLENT BREATHABLE FABRICS FOR SPORTSWEAR USING NANOMEMBRANES

Water-repellent breathable material is intended to be utilized in clothing articles and assures natural variables, such as wind, downpour, and body heat loss. Waterproof structure forestalls fluid water penetration and retention. The word breathable indicates that there is appropriate ventilation of the material. Latently, breathable textures permit water fumes to disperse over them but stop the fluid water from entering. High functional textures enhance versatile sportswear with a sense that puts high ability almost as much solace on it.

Eventually, products with both warming and cooling properties have recently pulled out in light of a clear business concern. Each of these materials does 'not seek after a solitary capacity; however, unique practical properties are consolidated on a more significant level (Babus'Haq et al., 1996).

Textures that can pass on water fume from the body, sweat out through the material while staying impenetrable to outside fluids, for example, water is generally utilized in casualwear and relative applications. Water-safe and dampness porous materials might be partitioned into three principle classifications—high-thickness textures, tar-covered, and film-coated fabric that are chosen by makers as per the completed article of clothing necessities in easygoing, sports, ski, or outside attire.

9.5.1 COMPACTLY WEAVED WATER-REPELLENT FABRICS

The breathable, water-repellent fabric is thickly woven and consists of cotton or microfilament yarns with condensed weave structure. One of the popular water-repellent fabrics known as VENTILE has been manufactured by using long-staple cotton with less space between the fibers. Fluffy yarns are typically weaved which allow water to penetrate, correspondingly without pores. The cotton filaments swell at the point at which the textured surface is wetted by water, transversely decreasing the size of pores in the texture and requiring high strain to induce infiltration; hence, waterproof shall be given for completion without the use of any water-repellent treatment. Thickly woven textures can also be produced from miniaturized denier size manufactured fiber yarns. The individual fibers in these yarns have a size of less than 10 μ, hence the texture can be made with extremely small pores (Das et al., 2008).

9.5.2 LAMINATED BREATHABLE, WATERPROOF FABRICS

Covered breathable waterproof textures are produced by using layers of films into material pieces. Such layers of films are slim, made from polymeric materials. They provide a high level of protection against water penetration and simultaneously allow water to fume. The 'maximum thickness of the filmy layer is 10 μ. There are of two kinds of films: (1) micropermeable films and (2) hydrophilic films.

The micropermeable layers of miniaturized scale have moderate openings on their surface that are smaller than a raindrop but larger than the atom of water vapor. The films are made using PTFE polymer, PVDF, etc.

The hydrophilic layers are thin, chemically modified polyester or polyure-thane films. These polymers are altered by a polyfuse. The poly (ethylene oxide) forms the hydrophilic portion of the layer by forming shapeless locale within the simple polymer system. This shapeless area acts as intermolecular pores that allow water fumes to pass through but prevent the entry of fluid water due to the film's robust concept (Frydrych et al., 2009).

9.5.3 COATED WATER-REPELLENT BREATHABLE FABRICS

Covered surfaces of water-repellent breathable texture consist of polymeric content that is applied to one layer. Polyurethane is used as the base for the coating. The coatings are of two kinds: (1) permeable microlayers, (2) hydrophilic layers.

The covering in a microporous layer is made up of fine interconnected channels that are much smaller than the fine raindrop but larger than the particles of water fumes. Hydrophilic coatings are similar to hydrophilic films, but the difference between microporous and hydrophilic material is that in the former, water fume travels through the permanent air-porous network, while the latter transmits gases through mechanisms, such as adsorption and desorption.

The qualities of leisurewear are:

- Ideal warmth besides humidity guidance.
- Strong air and water smoking porousness.
- Rapid humidity retention and movement cap.
- Absence of clamminess.
- Rapid drying to avoid dropping with bugs.
- Low water retention of the attire layer just placed on the skin.
- Dimensionally firm even in damp condition.
- Robust.
- Easy to care.
- Less weight.
- A soft and pleasing touch.

It is beyond the imagination to presume to attain these properties in a clear system of any single fiber or its combination. The two-layer structure has a near-skin layer of wicking type, engineered filaments, such as the miniaturized denier polyester, and the usually assimilating and disappearing outer layer of cotton or rayon's. Miniaturized denier polyester is suitable for wicking away the skin sweat. The use of superfine or microfiber yarn enables the development of thick textures that stimulate the slim activity and give the best wicking properties.

No single fiber or mixture of different filaments can produce perfect sportswear. The right kind of fiber should be in the right place. The mixing of filaments does not affect the texture of multiple layers. The wicking effect of the texture depends to a large extent on the moisture properties of its base filaments (Morton and Hearle, 2008).

9.5.4   TRANSPORT MECHANISM FOR MOISTURE

The method by which the moisture is transported in the materials is like wicking the fluid in the vessels. Narrow operation is determined by two slim core properties:

(1) Its breadth and (2) surface strength of the inner face.

The narrower the gap over or the more notable the vitality of the air, the more prominent a fluid's inclination to ascend the vessels. For material systems, vessels are effectively formed by gaps between the filaments. The narrower the gaps between these threads, therefore, the more pronounced is the material's potential for wicking moisture. Texture structuring, which properly structures tight vessels have no problem of having dampness. These texture constructions integrate textures that are created using smaller scale filaments, which are pressed together intently. The wicking however ends when all parts of clothing item are equally wet. Within the textile structure, the surface vitality is determined by the chemical structure of the exposed surface of the fiber and is expressed as follows.

- Hydrophilic fibers are of extraordinary vitality on the surface. And they get more prompt humidity than hydrophobic strands.
- Hydrophobic strands, on the other hand, have poor surface viability and exhibit repulsive behavior for humidity.

Specific completion procedures may be used to expand the difference between the nature of texture and the rear of the texture in surface vitality to boost its wickability.

9.5.5   FACTORS INFLUENCING DAMPNESS TRANSPORT

There are a few elements, which influence dampness carrying in a surface. The best significant elements areas follows:

- Fiber type
- Cloth development or weave
- Weight or thickness of the material
- Existence of blend treatments

Engineered strands may have surfaces that are either hydrophilic (wetting) or hydrophobic (nonwetting). They also have a mass retention array, usually reported by providers and test associations as the rate of dampness regain by weight. Engineered fabrics are generally considered to be the best alternative to clothing worn as a base coat. This is because they can give the board a good mix of moisture, nonabrasiveness, and insulation (Farnworth, 1983).

While most textures, both common and manufactured, can transport the moisture away from the skin by wicking, which is not quick drying and air-tightening components affect cooling and snipping comfort. Cutting edge fabric textures are lightweight, efficient, and generally quick to dry for shipping dampness.

It is widely accepted that materials with wicking properties of dampness can control body temperature, improve muscle execution, and delay depletion. For example, while common strands may be appropriate for wearing, low-motion cotton clothing, engineered nylon, or polyester fabrics are more suitable for substantial movements. They assimilate considerably less water than cotton, but even now, they can quickly dampen the wick through the texture.

The key comfort and relevance criteria are:

- Water and wind verification, breathability, and consolation
- Execution of moisture
- Perspiration
- Temperature control
- Easy care
- Smart and efficient design

9.6 LATEST NOVELTIES IN LEISUREWEAR

The 1980s period stayed a time of profoundly productive development in leisurewear pieces of clothing. Some sensibly straightforward microfibers and covered textures were created; variations of which have addressed the requirements of numerous game articles of clothing. The development of new materials and articles of clothing was fruitful to the point that in numerous games, the essential execution necessities have been distinguished and to a great extent fulfilled. These days, from exceptionally basic microfibers to substantially more mind-boggling textures are successfully utilized in dynamic sportswear. The most recent materials substantially have more capacity for satisfying explicit needs in various games exercises (Woo et al., 1994).

9.6.1 ABSORPTION OF SWEAT AND A FAST-DRYING PROPERTY

Moisture control of material properties during genuine physical exercise was seen as the key problem in the execution of comfort. As a general rule, the comfort perspective on clothing pieces is determined by the surface's wetness or dryness and warm estimates resulting from the exchange of surface-related properties.

It should have (a) sweat management and sweat releasing property to the environment for the fabric that is worn next to the skin, and (b) snappy drying property for incremental material relaxation.

The friction force required for the surface movement against sweating of the skin (due to physical activity, high temperature, and natural component moisture) is much higher than that for dry skin improvement. As a consequence of its remaining habit, the wet surface will give the wearer an additional concern.

Some textile manufacturers ensure that the fiber's lightness is enormous by eliminating the liquid sweat from the skin, and therefore, cotton or gooey is an essential component for the near-skin sportswear. While some claim that the fibers in such pieces of clothing do not retain moisture, it is intended that sweat is wicked away from the skin to the outer layers of clothing products from which it can spread to the air. An impossible arrangement would, in any case, depend on the degree of development considered. Believe it or not, the synthetics will certainly be sponsored in interactive games clothing pieces because they do not retain wetness and this has the advantage of holding clothing pieces lighter than cotton when it is wet. Similarly, crafted fibers have some extra room piece of rapid dry and extraordinary property for form upkeep. By far, most products used in the front line follow the simple concept of restricted action for sweat absorption and snappy drying (Yoon et al., 1984).

9.6.2 APPLICATION OF HIGH-END FIBERS FOR SPORTSWEAR

HYGRA: A sheath-core thread made of a hydrophilic polymer and nylon fiber, respectively was launched by Unitika Limited. The water-absorbing polymer has a special network structure that absorbs its own water weight 35 times and offers rapid release properties that conventional water-absorbing polymers cannot achieve. Nylon, on the other hand, provides strong tensile strength and dimensional stability to the core. Hygra also has excellent anti-static properties even in low-humidity settings. The main products produced using Hygra are fitness apparel, skiwear, golf apparel, etc.

LUMIA: Lumia is a Unitika product too. This is a series of polyester filaments with varying fineness and uneven cross-sections (0.5–2.0 denier per filament). The Hygra–Lumia mixed knitted fabric is quite common among Japanese top athletes.

DRYARN: This is the new filament produced by Aquafil. It is a polypropylene microfiber and is fully recyclable. Dryarn fabric is brand new and very lightweight and is used in various sports. It also has a soft handle and strong thermoregulatory capabilities and dries fast as well. Bacteria cannot flourish on the smooth surface of the fiber eliminating the unpleasant odor associated with bacterial decomposition.

KILLAT N: It is a . nylon hollow filament produced by Kanebo Ltd. The production technique of Killat N is entertaining. In the bicomponent filament, yarn nylon as the outer part is spun with soluble polyester copolymer as the core part. Then the soluble polyester copolymer of the bicomponent filament will be dissolved by giving alkali weight-loss treatment and a large hollow portion (over 30% of the cross-section) is formed, as a result of which strong water absorption and warmth retention properties are gained by the material.

TRIACTOR: Toyoba Co., Ltd has developed Triactor. It is a fast-absorbing/quick-drying polyester filament. Polyester is hydrophobic and does not absorb moisture, but by modification of the filaments to the Y-shaped cross-section, rapid absorption of the suction through capillary action will take place. The hydrophobic structure and large polyester filament surface lead to fast exposure to air and stimulating effect.

LYCRA: Lycra, a completely organic long-chain nylon fabric made of at least 85% segmented polyurethane, has a wide range of end-use applications, such as swimwear, recreational sportswear, floor gym thanks to its flexibility and fitness. Adding Lycra to a cloth gives it stretch and recovery, particularly in gymnastics and swimwear where body skin flexing and stretching is unavoidable.

ROICA AND LEOFEEL: Roica is a dry spinned spandex of polyether-type, and Leofeel is a smooth nylon-66 filament yarn. A knitted tricot fabric, Roica and Leofeel blend offers incredibly smooth touch and stretch. It is particularly used to manufacture swimwear.

9.7   MULTILAYERED STRUCTURES FOR SPORTSWEAR TECHNOLOGY

Mixtures in the shape of fiber layers are capable of giving each of the best properties. In a single layer, man-made filaments like polyester or polypropylene can be blended with wool for better wicking and insulation properties (Das et al., 2007).

Push–pull fabrics are bicomponent materials consist of an absorbent hydrophilic layer on the outside and nonabsorbent hydrophobic layer on the inside. The absorbent hydrophilic fiber is mostly nylon and the hydrophobic material is normally polyester.

Sports wool from The Wool Company is a fabric designed to withstand humidity. Created in 1994 by scientists, it is a composite material consisting of a fine insulating sublayer of Merino wool and an exterior polyester that draws moisture from the layer of wool into the air.

Besides the skin, the wool fiber collects unwanted molecules of vapor and disperses them into the atmosphere so that they can condense into the air. The fabric has drawn attention of the top Australian rivals and Manchester United football team. The biggest downside, however, is that because of the consistency of the wool, it takes longer time to dry.

Dri-release is a wicking yarn developed by US-based Optimer, a company set up by a group of former DuPont scientists. This patented fabric is an intimate 85%–90% mixture of low moisture-absorbing hydrophobic staple fibers such as polyester and 10%–15% hydrophilic wicking staple fiber such as cotton (Cooke, 2011).

Dri-release finds applications in athletic equipment, in footwear and underwear. It 'is used in clothing by several major brands, including Nike, Fila, etc. Dri-release combines the wicking qualities of cotton and the flexible touch properties along with the nonabsorbent property of polyester.

During the manufacturing cycle, chemicals may be applied and combined with polyester in small quantities to avoid the formation of the body odor for which polyester is known in the finished product. During the spin, the inclusion of cotton results in an intimate blend that locks the wicking cycle into the structure. And the impact is permanent.

Over time, the cotton's soft ends are more visible on the surface of the fabric. That improves the wiping action and provides a softer touch to the cloth. Dri-release cotton differs from the process used in other synthetic polymers, where topical silicone finishes make the fabrics wettable.

Tests performed by Optimer and its customers show that its 85/15 copolyester/cotton blend wicks and sends away moisture faster than a 100% polyester fiber tissue whose entire surface has been transformed into hydrophilic hydroxyls. Post-release of moisture was shown that it is four times faster than cotton in tests and as fast as other polyester performance, especially after multiple washes.

Dri-release also requires a New Weapon finish. This neutralizes the odors that textiles have stored in the fabric throughout their entire life span. Optimer is looking into the possibility of combining other hydrophilic fibers with polyester to create moisturizing fabrics. One of those fibers is wool.

"Entrant Dermizax EV" is a weightless cloth with a feather-soft texture and long-lasting water-repellence. It is an outstanding and first outdoor sportswear product with waterproof/moisture permeability of the world top quality as well as water-repellence that is highly durable.

9.8 OUTFIT, GLOVES, RAINWEAR, AND PROTECTIVE GEAR

Outfit's primary objective remains utilitarian, by way of defense against the elements. The garment has significant social and cultural roles as well. For example, a uniform may recognize a team or a company. For economic, health, or safety considerations, the bulk of sports and athletic activity are done with protective clothes (Nihat et al., 2008). Popular garments in sportswear include special coats, ties, trouser suits, and salopettes *(trousers with a high waist and shoulder straps)*.

9.9 MOTORBIKE LEATHER OUTFITS

Typically, one-piece or two-piece suits are leather jackets with a grouping of jackets and trousers worn mainly by bikers for safety in a collision. The thicker, more flexible and much tougher leather will be used to make leather suits. The leather suit is formally tested for the resistance to ergonomics and symptoms of scratch, cut, slit, and blast. They need 'not be too thick.

The motorcycle leathers are of two key styles: The first style is the closely tailored, at times; colorful one- or two-piece suits are manufactured using leathers and are usually black and frequently ornamented with metal pins and fringes on leather pants and jackets. The second style is often worn by people who like the style especially the jackets but those are nonriders. The new leather suits have armors on the ground in large-effect areas, such as arms, elbows, hips, knees, and back. They are designed to spread impact weight and shear stresses to avoid and decrease the level of damage to wound and disability.

For both men and women, the SuperMoto and Motorbike Leather Suits by Befit Sports are made of 1.3 mm superior leather with go through cooling panels. CE approved waterproof coatings on the Elbow, Knee, and Shoulder parts, replaceable PU Knee Sliders and silicone filling on the arms, back, and bump are provided (Shahbaz et al., 2005).

9.9.1   *LEATHER JACKETS FOR MOTORCYCLE*

A jacket is a kind of hip or waist-length sleeved upper-body dress. Most coats remain trendy and there are some that double as shielding outfit. Motorbike riders wear a specially made jacket by padding on the knees and ankles, arms, and rear for full-body safety. Motorcycle jackets are protective devices, besides remain sometimes thicker and frequently fitted with armor as a very practical piece of outfit (Au, 2011).

9.9.2   *LEATHER PANTS FOR MOTORCYCLE*

The pants are a bit of the outfit worn on the lower, some portion of the body from the midsection to the lower leg that spread the two legs independently. Motorcycle leather pants are designed for comfort and durability, made from heavy-duty milled cowhide. The driver will be protected from abrasion but will also be able to ride comfortably without the clothing being restricted (Mukherjee et al., 2009).

Befit Sports is a dealer of top-notch Leather Pants, Motorbike Leather Pants, Men's Leather Pants, Biker's Leather Pants, Motorcycle Leather Pants, Off-Road Pants, and Race Trousers made of cowhide 1.2 mm genuine leather with 100% polyester lining where well-being and comfort factor is significant. Most of the Bikers will have these extreme premium quality leather trousers that are well fitted with their thighs, waist and back with CE Protectors on ankles (Ishtiaque, 2001).

9.9.3   *TEXTILE JACKETS FOR MOTORCYCLE*

Owing to their enhanced environmental protection from the sun, wind, and rain, motorcyclists gradually prefer protective clothing constructed from man-made textiles, and the added versatility these fabrics appear to have in terms of pockets and vents. Popular fabrics include 600–1000 Denier Corduroy high quality or Kevlar, Corduroy, and Lycra combinations, which also have lightweight liners made with fabrics such as Gore-Tex. Some motorcyclists say that those synthetic fabrics are extra comfortable (Gagge et al., 1969).

As manufacturers and exporters, Befit Sports has delivered a wide variety of Waterproof Textile Apparels, inclusive of Motorcycle Jackets, Long Jackets, Men's/Women's Jackets with Removable CE (French phrase "'Conformité Européene'" which means 'European Conformity'). Prevention is more vital than defense, so Befit Sports used in the jackets reflective piping. With Befit Sports Textile Clothing, street bikers and riders feel outstanding joy and luxury on long trips (Pavko-Čuden et al., 2010).

9.9.4   MOTOCROSS AND MOTORCYCLE PANTS

Protection through pants is down to three factors:

- What material is used?
- Do they have any kind of security builds?
- Are they will compensate for extra protection?

Man-made materials usually deliver greater safety against extreme climate conditions. Crucial performance essentials include strength; protective clothing, in particular pants, necessity to retain reliability in the event of a crash, abrasion resistance, and heat resistance; while road roughness may result in adequate heat to melt many man-made materials, elongating ability, and breathing ease (Spencer, 2001). The bottom wears available in the market are made with different fabrics, but the Befit Sports cloth bottoms are made of Corduroy 600 D with a combination of 1600 D and synthetic leather on the knee, particularly for biker pants. In several clothes, additional protection is provided using knee or hip protectors. Befit Sports goods are the best when you 'are looking for defense. Protection is one of the most critical considerations when choosing MX pants.

Befit Sports has suffered as OEM, designers, and vendors in developing new versions of Cloth Clothing, Cloth Shorts, MX Pants, Supercross Pants, Dirt-bike Pants, Street-bike Pants, Enduro Shorts, and Race Shoes. Befit Sports highly recommends its Motocross Gear and MX clothes aimed at the health, and then the security of the passionate Biker and Riders. Befit Sports also sells tailor-made models (Araujo and Fangueiro, 2011).

9.9.5   TRANSFER PRINTED T-SHIRTS

The shirts with transfer printing effects are the most common in biker culture nowadays. The material used in those pullovers with 140–200 GSM is a water-repellent polyester material. With increased absorption and heat resistance, unfading sublimated logos, set-in sleeves, and a large nonrestrictive collar made from the same fabric as the rest of the jersey, Befit Sports pays special attention to making the jerseys (Matusiak, 2010). The distinctive collection of Befit Sports are Sublimation hats, Sublimated tops, MX tops mountain bike jerseys, and stylish-looking T-shirts.

9.9.6 *RAIN WEAR*

Rainwear protects rider from poor weather. A raincoat is an impermeable or waterproof gown used to protect the body of the wearer from rain (Gagge et al., 1969). Recent raincoats are often made of blended fabrics which are breathable and durable, such as Gore-Tex, nylon, polyester, etc. They allow air to pass in these fabrics, allowing the garment to "'breathe" so that sweat vapor can escape from the body. Innovative air vent fabric stretches and fits best for the body offering complete wind and rain protection (Wang et al., 2002).

9.10   CONCLUSION

The use of nanotechnology in sports apparel is infinite. The new trend in fabrics and garments is still a challenge for the industry to improve further. People are health conscious and exercise, yoga, and other physical sports are the everyday necessities of keeping a person away from real stress. As a result, the sports market is expected to grow in the coming years and for the production of apparel, manufacturers will focus on emerging technologies, such as nano, plasma, microencapsulation, etc., to meet the standard and demand of the customer if and when necessary.

KEYWORDS

- **nanotechnology**
- **nanofibers**
- **nanocomposites**
- **nanofinishes**
- **sportswear**

REFERENCES

Araujo, M. D.; Fangueiro, R. Weft-Knitted Structures for Industrial Applications. In *Advances in Knitting Technology*; Au, K. F., Ed.; Woodhead Publishing Ltd: Cambridge, 2011; pp 136–170.

Au, K. F. Quality Control in the Knitting Process and Common Knitting Faults. In *Advances in Knitting Technology*; Au, K. F., Ed.; Woodhead Publishing Ltd: Cambridge, 2011; pp 213–232.

Babus'Haq, R.F; Hiasat, M. A. A; Probert, S. D. Thermally Insulating Behaviour of Single and Multilayers of Textiles under Wind Assault. *Appl. Energy* **1996**, *54* (4), 375–391.

Celik, N.; Coruh, E. Investigation of Performance and Structural Properties of Single Jersey Fabrics Made from Open End Rotor Spun Yarns. *Tekstil ve Konfeksiyon* **2008**, *18* (4), 268–277.

Cooke, B. The Physical Properties of Weft Knitted Structures. In *Advances in Knitting Technology*; Au, K. F., Ed.; Wood Head Publishing Ltd.: Cambridge, 2011; pp 37–47.

Das, A.; Alagirusamy, R. *Sciences in Clothing Comfort*; Woodhead Publishing: New Delhi, India, **2010**; p 87.

Das, A.; Alagirusamy, R.; Banerjee, B. Study on Needle-Punched Non-Woven Fabrics made from Shrinkable and Non-Shrinkable Acrylic Blends. Part 2: Transmission Behaviour. *J. Text. Inst.* **2009**, *100* (4), 350–357.

Das, A.; Kothari, V. K.; Balaji, M. *J. Text. Inst.* **2007**, *98* (3), 363–375.

Das, A.; Kothari, V. K.; Balaji, M. Studies on Cotton-Acrylic Bulked Yarns and Fabrics. Part I: Yarn Characteristics. *J. Text. Inst.* **2007**, *98* (3), 261–267.

Das, A.; Kothari, V. K.; Makhija, S.; Avyaya, K. Development of High-Absorbent Light-Weight Sanitary Napkin. *J. Appl. Polym. Sci.* **2008**, *107* (3), 1466–1470.

Das, B.; Das, A.; Kothari, V. K.; Fangueiro, R.; Araújo, M. Moisture Transmission through Textiles: Part I: Process Involved in Moisture Transmission and the Factors at Play. *AUTEX Res. J.* **2007**, *7* (2), 100–110.

Das, B.; Das, A.; Kothari, V. K.; Fanguiero, R.; De Araújo, M. Effect of Fibre Diameter and Cross-Sectional Shape on Moisture Transmission through Fabrics. *Fibers Polym.* **2008**, *9* (2), 225–231.

Farnworth, B. Mechanisms of Heat Flow through Clothing Insulation. *Text. Res. J.* **1983**, *53* (12), 717–725.

Frydrych, I.; Sybilska, W.; Wajzsczyk, M. *Fibres Text. Eastern Eur.* **2009**, *17*, 50–56.

Gagge, A. P.; Stolwijk, J. A.; Saltin, B. Comfort and Thermal Sensations and Associated Physiological Responses during Exercise at Various Ambient Temperatures. *Environ. Res.* **1969**, *2* (3), 209–229.

Ishtiaque, S. M. Engineering Comfort. *Asian Text. J.-Bombay* **2001**, *10* (11), 36–39.

Ishtiaque, S. M.; Das, A.; Chaudhari, S. Influence of Fiber Openness on Processibility of Cotton and Yarn Quality. Part I: Effect of Blow Room Parameters. *Indian J. Fibre Text. Res.* **2003**, *28* (4), 399–404.

Ishtiaque, S. M.; Das, A.; Chaudhari, S. Influence of Fiber Openness on Processibility of Cotton and Yarn Quality. Part I: Effect of Carding Parameters. *Indian J. Fibre Text. Res.* **2003**, *28* (4), 405–410.

Matusiak, M. Thermal Comfort Index as a Method of Assessing the Thermal Comfort of Textile Materials. *Fibres Text. Eastern Eur.* **2010**, *18* (2), 45–50.

Milenkovic, L.; Skundric, P.; Sokolovic, R.; Nikolic, T. Comfort Properties of Defense Protective Clothings. *Work. Liv. Environ. Prot.* **1999**, *1* (4), 101–106.

Morton, W. E.; Hearle, J. W. S. *Physical Properties of Textile Fibres*; The Textile Institute, CRC and Woodhead Publishing Limited, Cambridge, **2008**, *175*,163–167.

Mukherjee, S.; Punj, S. K.; Ray, S. C. Dimensional Properties of Single Jersey Fabrics-a Critical Review. *Asian Text. J.* **2009**, 52–58.

Oğulata, R. T. The Effect of the Thermal Insulation of Clothing on Human Thermal Comfort. *Fibres Text. Eastern Eur.* **2007**, *15* (2), 67–72.

Pavko-Čuden, A.; Elesini, U. S. Elastane Addition Impact on Structural and Transfer Properties of Viscose and Polyacrylonitrile Knits. *Acta Chim. Slov.* **2010**, *57* (4), 957–962.

Shahbaz, B.; Jamil, N. A.; Farooq, A.; Saleem, F. Comparative Study of Quality Parameters of Knitted Fabric from Air-jet and Ring Spun Yarn. *J. Appl. Sci.* **2005**, *5* (2), 277–280.

Spencer, D. J. *Knitting Technology*; Woodhead Publishing Ltd.: Cambridge, **2001**.

Wang, Z.; Li, Y.; Kowk, Y. L.; Yeung, C. Y. Mathematical Simulation of the Perception of Fabric Thermal and Moisture Sensations. *Text. Res. J.* **2002**, *72* (4), 327–334.

Woo, S. S.; Shalev, I.; Barker, R. L. Heat and Moisture Transfer through Nonwoven Fabrics: Part I: Heat Transfer. *Text. Res. J.* **1994**, *64* (3), 149–162.

Yoon, H. N.; Buckley, A. Improved Comfort Polyester: Part I: Transport Properties and Thermal Comfort of Polyester/Cotton Blend Fabrics. *Text. Res. J.* **1984**, *54*, 289–298.

CHAPTER 10

Sustainable Physiological Adaptation of Humans to Diverse Environment Conditions Using Smart NanoTextiles

RENU BALA YADAV[1], VINAY KUMAR YADAV[2], DHARAM PAL PATHAK[3], and RAJESH ARORA[1*]

[1]*Defence Institute of Physiology and Allied Sciences (DIPAS), DRDO, Timarpur, Lucknow Road, Delhi, India*

[2]*Department of Computer Science and Engineering, Institute of Engineering and Technology (IET), Dr. Rammanohar Lohiya Avadh University, Ayodhya, Uttar Pradesh, India*

[3]*Delhi Institute of Pharmaceutical Sciences and Research, M.B. Road, PushpVihar, New Delhi, India*

Corresponding author. E-mail: rajesharoratejas@gmail.com

ABSTRACT

Soldiers operate in diverse environmental conditions and their operational environments are very challenging in terms of human physiology as they have to operate in extremely cold and hot environmental conditions. In order to protect them from the vagaries of nature, unique kinds of textile materials are needed to protect the human body from the external environment. Every special task assigned to the soldiers is quite challenging as compared to conventional jobs. Smart nanotextile material is an advanced material which can be used to design textile modules for smart clothing such as functionalized smart textiles with antimicrobial agents, use of smart textiles in diagnosis and monitoring of various diseases, RT-ST (real-time smart textile) for management of various cardiovascular and respiratory diseases, smart

Fundamentals of Nano-Textile Science. Prashansa Sharma, Devsuni Singh & Vivek Dave (Eds.)
© 2022 Apple Academic Press, Inc. Co-published with CRC Press (Taylor & Francis)

textiles in hazard management, smart helmet subsystem, fabricated gloves and socks, water repellent shoes, and gloves. Besides the listed advantageous features, advanced nanotextiles materials can be useful for combating situations like biological and chemical warfare attacks, protection against bio-terror agents, resistance to extreme cold/hot environment, act as an antimicrobial agent, etc. Nanofabrication tools such as electrospinning, split assembling, self-assembling of polymer, natural/biological/microbiological synthesis of nanoparticles, etc., are used for preparing nanotextile materials with a few unique blend of designs and texture, which are helpful in the designing of advanced material with many specific and targeted approach for sustainable physiological adaptation to the extreme and exhaustive environmental conditions. Some such latest trends and developments in the area of smart textiles are highlighted in this chapter.

10.1 INTRODUCTION

Introduction of smart approach in textile materials is aimed to achieve many benefits in military scenarios such as to give protection from extreme environmental conditions such as extreme heat in hot and dry area and extreme cold in wet and humid area. Other than these benefits, smart textile materials provide casting in a desired form such as: (1) Sensor-responsive textile material (Castano and Flatau, 2010), (2) Carbon nanotube (CNT)-incorporated nanotextile for resistance to heat/cold shock (Kiu et al., 2007; Panhuis et al., 2007; Liu et al., 2008), (3) Nuclear and biological hazards protective smart nanotextile materials (4) Chemical hazards protective nanotextile materials (Turaga et al., 2012), (5) Nanotextile clothing materials for high altitude (Jhala, 2006), (6). Smart textile and antimicrobial agent incorporated gloves (Gugliuzza and Drioli, 2013), etc. Thus, an integrated approach in designing and fabrication of textile materials can give advanced solution in textile engineering for ultimate physiological adaptation.

Smart nanotextile materials are those materials that can sense the external stimuli and can provide comfort or response accordingly. Responses of smart fabrics are dependent upon chemical, physical, mechanical, magnetic, electrical, biological, and environmental stimuli. Examples of smart nanotextile materials includes nanoparticles incorporated fabric materials for protection against microbial infections, smart nanotextiles which control muscular vibration and are aligned to normal physiology, smart nanotextiles which can manage and regulate core body temperature, etc. (Syduzzaman et al., 2015). Thus, by incorporating miniature electronic or sensing device into nanotextile/fabrics makes smart nanotextiles which can be further used as diagnostic, therapeutic, managing, protective, and detective agents.

Various smart nanotextile materials are being used in the military and a few of them are as: "smart carpet", that is, piezoelectric material installed in advanced textile material used as a security system; "smart suit" especially designed for monitoring the physiological condition such as hypoxia, pulmonary threats, freezing cold, neurophysiologic status, etc. (Karl, 2018); "chameleon suit" is designed to liberate and explore working gesture in space (Hodgson, 2003).

Functionalization of nanotextile materials are greatly influenced by the technique used in processing and material used for selected purpose. A large number of advanced textile materials can be achieved by manipulating the technique to get a desired one. In the late 20th century, the first ICD+ outerwear was introduced and was found successful in monitoring altered physiological conditions. Use of functional fibers such as optical fiber, conductive fiber, flame retardant fiber, and antimicrobial fibers are current research topics in nanotextile research and development. Other than these materials, smart textile fibers such as fibers capable in energy harvesting, energy storage smart textile fibers, color-changing fibers, shape deformable fibers, high-performance fibers in extremes, etc., are being used in novel source in textile research (Shi et al., 2019).

Artificial intelligence (AI) operating system impregnated textiles is one step further forward system for managing and controlling operations and monitoring physiological system balanced just through one systematically designed operating system. Highly validated technologies with affordable sensing devices can cut the cost of many extra devices. This usage is basically a closed-loop system to sense, communicate, and respond to a data-based system. The possibility of using conductive coating with a specific designed purpose of application can in build a large function and can transform a common textile into intelligent and smart textile (Javaid et al., 2018).

Amplification of fluorescent polymer in nanotextile used as protective materials which detect hazardous substances as well as have the capability of detoxifying such agents with extremely high efficiency. Such materials are being used in CBRN (Chemical, Biological, Radiological and Nuclear hazards) protective clothing. Since soldiers live in very uncertain conditions so along with an active training and capability development program, a protective wearable is also required (Potter et al., 2017). Thus, this modulated research in smart textile material provides suitability for mission-oriented and mission-specific requirements such as acid and heat resistant wearable, water-resistant clothing and wearable, optic devices, insulated weapons, etc. A future research in smart nanotextile material suggests that optimization of an ideal military clothing and wearable with a less physiological burden which signify compliances.

10.2 APPLICATION OF SMART NANOTEXTILE IN MILITARY

Smart nanotextile engineering has been introduced to develop a user-specified textile material by monitoring physiology and other environmental conditions.

10.2.1 SMART NANOTEXTILE FOR SOLDIERS

Some examples of smart nanotextiles for improving the physiological adaptation to diverse extreme environments are enumerated below:

- Real-time monitoring of physiological states
- Wearables with ergonomically active design
- Compliance with fitness and duty
- Cognitive status and neurophysiological monitoring system
- High-energy expenditure monitoring by maintenance of metabolic heat production
- Assessment of muscle fatigue and limits any injury occurring due to muscular stress
- Sweat monitoring and electrolyte imbalance restoration in extreme environments
- Pulmonary stress monitoring and exposures limit performances
- Early detection and management of CBRN (Chemical, Biological, Radiological, and Nuclear) hazards
- Thermoregulation monitoring
- Electrocardiogram (ECG) monitoring smart nanotextiles
- Telemedicine
- Biometric identification
- Superior gas mask for armed forces
- Chemo/bio-sensors
- Artificial gills-advanced re-breathers for indefinite length underwater for military personnel.

10.2.2 SMART NANOTEXTILE FOR MEDICAL APPLICATIONS

Smart nanotextiles can have a multitude of medical applications. Some areas of application include, but not limited to only these are mentioned below:

- Optimization between individual health and fitness through unique algorithms.
- Protection from hypoxia, cold injury, pulmonary stress, etc.
- Monitoring of real-time glucose monitoring by smart fluorescent nanotextile materials such as "Smart Tattoos."
- Antimicrobial and breathable polymeric nanotextile to prevent any dermal etiology.

A future direction along with smart nanotextile material by incorporating knowledge of some basic treatment agent, for example, a quick sealant system fabrication in the fluorescent nanotextile or pH-dependent nanotextile can give benefit in assessing and treatment of any pellet injuries in soldiers.

An application-based use of smart nanotextile materials, their features, and function for sustainable adaptation to extreme environmental as well as health monitoring conditions are shown in Figure 10.1.

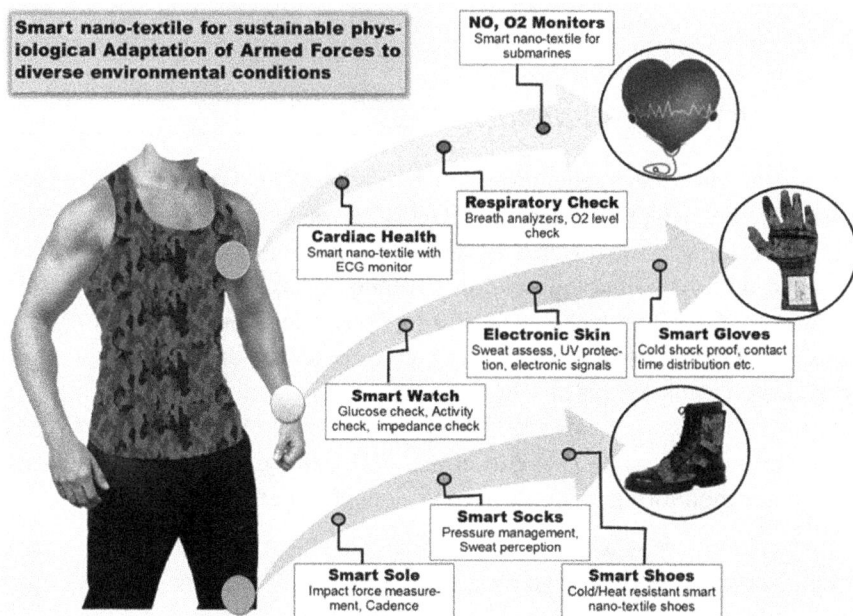

FIGURE 10.1 Smart nanotextile approach in physiological adaptation and real-time (RT) health management.

10.3   CLASSIFICATIONS OF SMART NANOTEXTILES MATERIAL

10.3.1   PASSIVE SMART NANOTEXTILES

Passive smart textile is the first-generation textile material which is subjected to "sense" the environmental condition but do not able to modify the fabrics according to the changes. For example, an insulating overcoat used in military for various purposes is able to provide insulation but does not resist any change in textile material. Other smart nanotextile materials are silver nanoparticles-incorporated fabrics used as an antimicrobial fabric especially designed for soldiers posted in remote area (Zhang et al., 2014), anti-static jackets for soldiers, bulletproof jackets, etc., are also examples of passive smart nanotextiles (Abdullaeva, 2017).

10.3.2   ACTIVE SMART NANOTEXTILES

Active smart nanotextile material is the second-generation nanofabrics which are designed to "sense" as well as "respond" to the given stimuli. Well-known examples of active smart nanofabrics are shape memory, Ni–Ti alloy (Lah et al., 2019), graphene nanotubes (with variant applications) (Cai et al., 2017), GZE shirt, etc. (Vagott et al., 2018).

10.3.3   ULTRA-SMART NANOTEXTILES

These are the third-generation smart nanotextile which can "sense," "respond," and adapt the changes as per the given stimuli. Latest example of ultra-smart nanotextile material is Jacquard™ introduced by Google which is designed with AI technology and this technology is implemented by Levi's Commuter™ Trucker Jacket, which allows user to manage their calls, text massages, GPS, Google maps, and other Google's in-built active functions without using smart phones (Vagott et al., 2018). Few other examples of ultra-smart nanotextile materials are Hexoskin™ (Hexoskin & Carre Technology Inc.), DuPont (Intexar™) (Vagott et al., 2018), Integration of Piezoresistive sensors for monitoring respiratory condition, etc. (Paradiso et al., 2005).

10.3.4   FUTURISTIC SMART NANOTEXTILES

Futuristic approach for smart nanotextile would be in combinations. It will be given auto response, act and modulate without aid of any external energy.

And fabrication will be possible on synthetic skin as well as on breathable and biodegradable polymers in some specific cases.

10.4 TYPES OF MATERIAL REQUIRED IN TEXTILE ENGINEERING TO BUILD SMART FABRICS/TEXTILES

Textile material science is a broad area of research and categories on the basis of their nature, applicability, durability, and functions. Smart nano-textile materials are being aimed to design with intelligence system and intellectualized with respect to external stimuli. These functional materials are prefabricated or synthesized by advanced technology. On the basis of their capability to act "smart," smart textile materials are categorized as follows:

10.4.1 BIOPOLYMERS FOR ADVANCED TEXTILE MATERIALS

Polymers are universally acceptable materials in the field of manufacturing and production. Textile industry always aims to generate new, user friendly, and cost-effective material; so this becomes necessary to choose a versatile material in the case of smart nanotextiles Biopolymers are those polymers which are produced from the biomass, for example, microorganism, plants, and animals and precursor for producing biopolymers in textile industry is mostly used from plant sources as starch, corn, fats, etc. Sustainable production of smart nanotextile material can be achieved by employing some modern techniques such as electrospinning, non-weaving web forming, hot-air bonding, etc. (Biresaw et al., 2004). Based upon the industrial application of biopolymers in smart textile material and research, many natural and advanced materials have been manipulated according to the necessity as well as cost effectiveness. Examples of such materials are cellulose, starch, alginates, PCL (polycaprolactone), PTT (polytrimethylene terephthalate), PLA (polylactic acid), PHA (polyhydroxyalkanoates), PBA (polybutylene succinate), PU (polyurethanes), PHB (polyhydroxy butyrate), etc., frequently used to prepare smart nanotextile materials (Crank et al., 2005; Potter et al., 2013). Modifications of biopolymer are based on application and strength of material. Few famous examples of smart biopolymers are green nano-mesh widely used in the area of health and medicine; PLA-taped Dura cover, etc.

10.4.2   INTRINSICALLY CONDUCTIVE POLYMERS

Intrinsic conductive polymers (ICP) are named so for their inbuilt intrinsic functional properties. In this category, stimuli-responsive polymers such as electroactive biomaterials play a vital role in the generation of smart nanotextile materials for various uses. Examples being ICPs are: PPy (polypyrrole); PEDOT (poly (3,4-ethylenedioxythiophene); PANI (polyaniline), etc. (Grancaric et al., 2017). PPy is the material that possesses high electrical conductivity as well as environmental stability and modification of textiles in the form of smart textile materials are done with the help of surface modification technique. Over ease of preparation, these kinds of smart textiles possess a problem of high fragility which reduces the shelf-life of the textile materials. In the case of PEDOT, these materials are transparent and highly stable with environmental changes with good electrochemical stability. PEDOT is very less soluble in water; hence this characteristic gives a better and reliable smart textile material with the scope of nano-incorporation to fictionalizations of material (Balint et al., 2014; Tamburri et al., 2009; Ding et al., 2010). In the context of ICPs material PANI, this is also highly electroactive material used various textile material to functionalized as smart textile for ammonia sensing, neural probes, and application in many drug delivery systems (Tamburri et al., 2009). PANI is available in various diverse structural forms, and hence can be used as modified engineering textile material, but their non-biodegradable nature lacks an eco-friendly approach but due to their low cost is used in many textile applications.

10.4.3   METALLIC FIBERS

Nano/macro metallic threads along with textile threads are woven together to obtain a smart nanotextile material with a wide range of application such as sensing material, diagnostic material, conducting material, etc. Diameter of these metallic wires is lies in between the range of 2–40 μm (Matteo et al., 2014). Metallic fibers have an advantages over others fibers that they possess a good mechanical strength thus can be used for many military appliances, wide range of composition such as iron, nickel, aluminum, silver, gold, platinum, etc., biological inertness and ease of availability. Metallic fibers are inert in nature and due to this they are insensitive to external damages such as sweating and washing. Metallic fibers are used in recording and monitoring of physiological activity such as ECG signals.

10.4.4 AEROGEL-INCORPORATED SMART TEXTILES

Aerogel-incorporated smart textiles are the textile used in space technology for making space suits and space shuttle, cryogenic material, protection suits for high altitude regions, etc. Aerogels are the material having the lightest weight on the Earth and have much exceptional physical properties which make these gels a unique gel. They aligned with covalent bonds and are nanometer in size (1–10 nm diameters) arranged in a space of three-dimensional structures. Aerogel shows extremely low thermal conductivity, low dielectric constant, and high optical transparency physical properties (Bheekhun et al., 2013). Aerogel having low density are compressible and possess good elasticity thus can be used in high range of military appliances such as non-thermal face mask for extremely high temperature range areas, ergonomically designed wearable soles for high altitude trekking, heavy exercises, running, and training, etc.

10.4.5 CONDUCTIVE INK

Conductive ink is a precise conductive inks used for printing conductive material on a woven fabric for customization of fabric material for the purpose of smart textile uses. The metal precursors used to produce conductivity on the nanofabrics are silver, copper, gold nanoparticles, etc. The ink materials are especially hydrophilic in nature; so after the incorporation of nanofabrics, a protective layer or shield is needed to provide over the printed area (Matteo et al., 2014). Integration of textile material and screen printing with conductive ink along with the systematic arrangement of electronics device gives a smart fabric used in various applications such as military security and threat detecting textiles and devices, etc.

Following needs to be fulfilled by a fabric material to have a capability for conductive printing on textile materials as:

- Fabric material should free from particle aggregations while printing on it.
- Fabric material should have antioxidant property or should not be pH modulated types.
- Fabric material should have ease of drying ability.
- Fabric material should have good adhesive property with the conductive ink or material should be pre-coated with some of the conductive polymers or metals.

- Fabric material should possess defined viscosity and surface tension for conductive ink printing.

A subsequent addition of nanoparticle-based ink system with conductive ink printing is required for the execution of smart nanotextile fabrics (Eugene, 2008). To achieve conductivity upon a nanotextile material for smart wearables, sintering of nanoparticle-based ink is required for a continuous connectivity. For a good smart nanotextile material, screen printing is appropriately used for fabrication of electronic material on textile due to their good ability to produce a specified pattern whichever requires on the wearables. An example of screen printing for smart nanotextile material used in medical and healthcare system is printing of polyurethane paste with the conductive ink of silver nanoparticles which provides a uniform track for wearable health monitoring devices (Matteo et al., 2014; Bheekhun et al., 2013).

10.4.6 OPTICAL FIBER INTEGRATED SMART NANOTEXTILES

Integration of optic fibers into smart nanotextiles for military use gives a wide application range used in confidential communications, threats sensing, direction monitoring, etc., as well as optic fibers-integrated smart nanotextiles are widely used in the field of healthcare and monitoring system (Eugene, 2008).

Personalization of optic fiber-integrated smart nanotextiles is possible due to their soft and curved fitting to the body shape of an individual, thus do not give any such bulgy or suffocating appearances (Zindan et al., 2019). Integration of optic fibers is achieved in fabrics by simple weaving with the yarn of standard textiles fabric and after which a smart approach is created onto it by modulating the intensity and thus signals sensing system of optic fibers.

Polymeric optic fibers (POFs) play an advanced role in making many wearables used in military services. POFs are suitably used due to its biocompatibility, flexibility in nature, lightweight, durability, and electromagnetic interference immune susceptibility and very cost-effective which makes POFs an advanced material to generate smart nanotextiles (Li et al., 2018; Tao, 2015; Tao, 2001). In healthcare and medicine system, POFs play a vital role such as a smart textile wearable for neonatal jaundice treatment was created by Quandt et al. (2015) which is a kind of homogenous luminous textile customized with breathable and comfortable fabrics used with an inbuilt channel of phototherapy used for jaundice treatment at home.

Another successful example of optical fiber-integrated POFs is a textile light diffuser system designed by Cochrane et al. (2013) which stimulates the drug from a core polymeric matrix and diffuses to the skin in a pulsatile manner which offers advanced treatment of dermal diseases.

10.4.7 CHROMIC MATERIAL IN SMART NANO TEXTILES

Chromic materials are defined as the materials which changes distinct color upon exposure of external stimuli such as light, heat, electric current, and change in pH (ionic movement) which is reversible and controllable in nature. Chromic phenomena such as thermochromism, photochromism, electrochromism, and ionochromism, etc., are used for smart textile wearables (Christle, 2013).

Armed forces are posted in extreme to extreme and rushed environmental conditions; the clothing material provided to them must be suitable and comfortable for their living and exercised condition. For military exercise purposes, Karpagam et al. (2017) have developed a chameleon based on thermochromic mechanism which changes color upon changing external environment temperature, for example, jungle color (classic green) in dry and humid temperature and sand brown color in hot and dry temperature. Such innovations are very helpful in camouflaging of the soldiers in the terrain in which they operate.

10.4.8 BIOTECHNOLOGY-ORIENTED SMART NANO TEXTILES

Biotechnology-oriented smart nanotextiles are based example for the uses of natural resources to convert normal fabric into smart and advanced fabric by use of microorganisms such as algae, fungi, and cellulosic materials. A high class nanofabric can be developed for medical use such as wound dressing material, drug-releasing bandages, long-term moisture providing fabrics like gloves and clothing for defense uses in high altitude, antibacterial, and antiviral clothing for defense personnel uses in very hot environment such as dessert. Use of biotechnologically oriented textiles is successfully introduced into the diagnosis and treatment approaches such as use of tissue-engineered implants in growth of cartilaginous cells (for arthritic patients), biotech absorbable tissue fabrication into smart textile material for diabetic foot wound healing (Kelvin and Heather, 2010).

10.5   RESEARCH FOCUSES ON SMART NANOTEXTILE WEARABLE SPECIFICALLY DESIGNED FOR ARMED FORCES

10.5.1   *MANAGEMENT OF PHYSIOLOGICAL CHANGES BY SMART NANOTEXTILES*

Alteration in physiological conditions plays a key role in the soldier's active performance, thus this becomes necessary to maintain a high output throughout the process. Maintaining such conditions by the introduction of devices like Bluetooth is not permissible, hence a medically designed algorithm system is introduced into nanotextile material to assess and maintain any altered physiological conditions.

Potter et al. (2017) described a mathematical model for the prediction of core body temperature influenced from environment, physical activity, and clothing material by the Heat Strain Decision Aid model. This mathematical model used a combination of empirically derived equations to calculate rise in core body temperature (Tc) or thermal tolerance limit with which in concern of physical activity and exercises with a function of extreme environmental conditions and clothing material and others anthropometrics activity. This model was basically introduced to assess a sustained work output in soldiers but this model can be used further in ergonomics and physiological challenges and their management (Biresaw and Carriere, 2004).

Physiological conditions such as total daily energy expenditure, metabolic rate, pulmonary output, cardiac output are the critical parameters to which a smart approach is required.

Soldiers are posted anywhere and thus their field workload and metabolic energy requirements are needed to be measure for stagnant and continuous healthy performances.

10.5.2   *SMART TEXTILES USED IN FITNESS AND ALERTNESS MODEL DURING DUTY ASSESSMENT*

Fitness and alertness during duty given to soldiers are mandated for security and others defense parameters. But sleep is a natural phenomenon and can hinder the alertness during strict duties and exercises. A general perception can be made for non-alertness and decrease in performances of soldiers during duties due to combination of circadian cycles into sleep process and performance model of exercise physiology (Redmond and Hegge, 1985). Situation-based alertness is not only a motive for soldiers as well as this requires also for a general public for example, a driver who drives vehicle in low light, quick reaction to unwanted accidents, decrease performances of soldiers during any emergency exercises due to deprived sleep, and tiredness, etc., are the situations which required a smart measure of alertness and fitness assembly which can be worn easily and provide time-dependent functions when it requires.

Many neurophysiologist researchers have shown their interest in EEG monitoring of alertness via sleep cycles but this has many limitations on actual ground such as unavailability of electrical power for electrodes during war, it is also not feasible to apply electrodes during driving a vehicle and aircraft, thus the concept of alertness and activeness monitoring by EEG is not given a successful tool. Thus, to overcome these hypothetical problems, a concept of smart textile having many inbuilt sensors is created by researchers to provide a solution for such critical monitoring.

Hexoskin, developed by Montreal, QC, Canada, is a smart textile wearables knitted with 73% micro polyamide and 27% elastane equipped with the system to measure heart rate (HR), variability in HR (VHR) as tachycardia and bradycardia, respiratory volume and breath rate, gesture and posture control, steps counting, activity measurement, calories burn measurement, sleep quality and activeness measurement, cognitive functions, etc., are the measure function utilized by Hexoskin (Hexoskin Smart Garments Specifications (online); Abdallah et al., 2017). Few more examples of smart textile which serves the purpose of activeness and alertness are: Bioman+ manufactured by AiQ (AiQBioman+ [online]), SKIIN Textile Computing™ by Myant Inc. (SKIIN Textile Computing™ [online]), and Neuronaute® technology by Bioserenity, etc. (Neuronate® Bioserenity [online]).

10.5.3   CBRN PROTECTIVE WEARABLE SMART NANO TEXTILES

Chemical, biological, radiological, and nuclear (CBRN) threats are a major threat and there is a need to develop CBRN protective wearable smart nanotextiles. Protective measures should be available during attacks and after attack because the resistance time upon the exposed area is more than any other weapons, thus a proper shield and coverage in the form of smart clothing and wearable is required (Turaga et al., 2012).

Chemical hazards material can be classified as those materials which are hazardous in nature and do not occurs in body naturally, on the other hand, biological hazards can be classified as a live threat in the form of bacteria such as *Bacillus anthracis (Anthrax), Ebola virus, Clostridium botulinum (botulism), and Ricinus communis (Ricin toxin)* are few examples of biological hazards which have been tested during World War-II as bioweapons. Nuclear hazards are considered as most dangerous hazards as compared to chemical and biological hazards and one of the biggest examples of a nuclear explosion is Chernobyl, Ukraine (1986) was the world's largest nuclear disaster ever took place (Dickson, 2013).

Using smart technologies in textile and engineering protective clothing and wearable for such agents can be introduced to combat any sudden CBRN warfare threats. Defence Research and Development Organization has developed "Mk V suits," Boot Mk II/Overboot Mk II, CBRN Respirator Mk II/Integrated Hood Mask Mk II, CBRN gloves Mk II and Haversack Mk II for Indian Armed Forces, which are durable and breathable clothing and wearable.

Other than these few of private companies such as DUPONT (DuPont™ Tychem®, DuPont™ Tyvek® and DuPont™ Nomex®) has been also manufactured some protective clothing for the protection from liquid and dry chemical splashes, aerosols mist, hazardous chemical particles, and toxins (DUPONT, [online]). Defence Research and Development Organization (DRDO) has developed PROTECTON™ initially as protective clothing for medical responders who deal with patients contaminated with radiological hazards every day during experimentation or treatment. This protective clothing is specially designed with nanotextile material which does not allow any radiation to pass away due to its highly specialized design and very less pore size. Stability of exertion has been tested as per the ASTM guidelines of textile material and it was found to be durable and possess same quality up to 100 laundry washes. Spread of COVID-19 necessitated the demand of such protective PPEs kits that allow a health-worker to safe and protective PPEs wearables which can meet the criteria of protection of a person from SARS-CoV2 viruses by not allowing its passage from protective PPEs. To meet such criteria, PROTECTON™ was further modified in terms of pore size, textile flexibility, wash durability, resistivity, etc., and rechristened as Bio-Suit PPEs kits for human-to-human transferable viruses such as SARS-CoV2, MERS, H1N1, Ebola (Fig. 10.2).

10.5.4   *EXTREME ENVIRONMENT-RESISTANT SMART NANOTEXTILES*

Soldiers face many challenges due to environmental conditions, which are either too hot or too cold and both are unbearable to those soldiers who posted in these areas. Both physiological and psychological factors are affected by these extreme environmental conditions. Such conditions can be continuously monitored by the real-time physiological status monitoring (RT-PSM) (Karl, 2018), but sometimes status monitoring for a large number of soldiers may not be feasible. Thus, a proper coverage by smart textile inbuilt with advanced technology can be used to protect army personnel from extreme environments.

FIGURE 10.2 PROTECTON™ designed and developed by Defence Research and Development Organization (DRDO), India

Researchers have developed new smart nano textile 3D-printed fabrics on a-BN/PVA (poly-vinyl alcohol) composite nanofibers which exert extreme cooling than cotton as well as highly stretchable (355 MPa) and durable than the cotton material (Tingting et al., 2017). For extreme cold condition, high content (15%) of CNT-incorporated glycerol–water hydrogel with a different range of polymers such as poly (3,4-ethylenedioxythiophene), PANI, and PPy

(Lu et al., 2018), along with elastane or spandex is used to achieve a good textile material which can resist the perception of cold wind or cold temperature.

10.5.5   *MEDICALLY ACTIVE SMART NANO TEXTILE MATERIAL*

Nowadays with the help of nanoparticles and nanomedicines, many smart textiles have been designed which possess antimicrobial, antifungal, wound healer, etc., kind of smart textiles.

Quartinello et al. (2019) have successfully developed a smart textile material functionalized with nanocapsules of human serum albumin/silk fibrin was incorporated in cotton/PET beds and tested for antimicrobial properties. Andrea et al. (2017) developed a RT-ST (real-time smart textile) used for measuring the real-time monitoring of diabetic foot pressure offloading. This system is integrated with a web dashboard and artificially monitored by pressure loads signals produced by pressure wound (diabetic foot) and analyzed by medical specialist so that a proper cure can be done.

Some of the smart wearables are also introduced to diagnose and monitor the disease condition in humans before they occur. Examples are: "Smart shirt" consists of a wearable motherboard developed by Georgia Institute of Technology, Washington, DC, USA under the brand name Sensatex. This shirt is able to monitor human physiology and alertness, used by medical monitoring of disease and military athletics as well as provides battlefield combat management system also. Another example is "Life-shirt," first introduced by VivoMetrics, Inc, Ventura, CA, USA a Southern California-based health information collection and monitoring company. This shirt is embedded with a personal digital assistant system which is able to record more than 30 physiological functions such as heart rate, respiratory volume, pulse rate, body temperature, diseases responses, etc., at a time. A short-listed data can be collected by data card and then information can be sent to the physician to conclude the results, thus a successful monitoring of disease in medical intervention of real world can be made possible by these smart nanotextile approaches (Keri and Stephen, 2007) (Fig. 10.3).

10.5.6   *FEATURING ARTIFICIAL INTELLIGENCE-OPERATING SYSTEM (AI-OS) FOR SMART TEXTILES*

A nano textile or composite textile can be made as smart as a computer by using an AI operating system (AI-OS). Using this OS as a main control unit as a smart sensor in nano/composites textiles in combination with installed input devices provides an autonomous work according to program through AI. This smart AI features can help out in sensing, signal stimulating, monitoring, decision making, and giving commands through e-wearables to both internal physiological and external environmental conditions. Stimulation through internal/external factors, receiving message, coding/decoding, etc., will be channelized into a hierarchy of system which are all in connection with one AI-OS and thus will be helpful in providing alertness among connected team. This AI-OS features within the nano/composites textile can help soldiers in taking quick decisions if any kind unwanted threat is there and will provide an alertness to their backup intelligence team also and thus, for real-time management, this system can provide assistance, for example, medical, expert advisor, air and aerospace help, instant navigation to leading team, etc. From a military point of use, all these messages will be encrypted between soldier's team or their experts only. Thus, to maintain a high level of security it becomes important to have encoding of this machine's natural language to only be encrypted by their users (Suvetha et al., 2019).

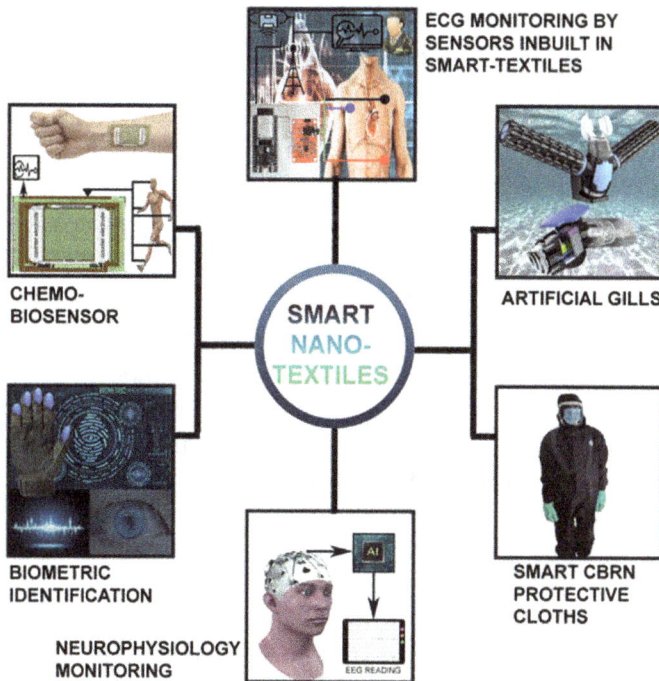

FIGURE 10.3 Research focuses on smart textiles wearable especially designed for armed forces.

With second intelligence tool, a textile can be smartly featured with 360 degree AI-smart camera, this camera vision will be fully autonomous and work like a third eye to its wearer. This AI-smart camera inbuilt in textile will provide many features in a single frame and form multiple images and match them to the successive eye so that receiver can possess information and act accordingly by filtering out the threats or any suspect movements. This AI-smart camera will be helpful when human cognitive response is decreased during high stress conditions, which impact their visual learning and viewing capabilities. The response of threat detection and suspect detection will be based on information available in the military database which is already used as a tool from many resources like spy cloud, internet sources, etc., as an image pattern recognition tool. This featured function becomes most active and gives an alert message to the wearer when a suspect resembles with AI informative data-based tools (Mehrubeoglu et al., 2011). Subsequently, this featured AI-smart camera can be merged with an X-ray vision camera which will have the capability to see through the wall (Tsuda et al., 2006) and a radiofrequency wave which will perceive the position of an object (Zhao et al., 2018).

The accomplishment of mission is important in every situation and mostly this is carried out in very unpredictable time and situation, thus environmental conditions like foggy weather and darkness may act as a bottleneck for the strike team. Power banks with a long-term battery run are required, this can be achieved by installing tiny solar panels into textile materials as well as few electrodes and battery that can be chargeable by using human energy obtained in the form of vibrational energy when soldiers are active during walking and running during mission (Taliyan et al., 2010).

Furthermore, nanofiber pipes stitched in textile material can be smartly designed with a function of AI-OS which on any battlefield fights injuries such as bullet injury, pellet injury, and stab injury. Such nanofiber pipes may release pre-stored medicines such as painkillers, antibiotics, anti-inflammatory, steroids, and other life-saving drugs, etc., for battlefield management of soldiers on the field till they can reach the hospital. Such delivery systems can immediately get activated when a projectile hits the nanofiber tubes (Choi et al., 2017) (Fig. 10.4).

10.5.7   *INTEGRATION OF INTERNET-OF-THINGS (IOT) WITH SMART TEXTILES FOR SUSTAINED PHYSIOLOGICAL MONITORING THROUGH*

Internet-of-Things (IoT) is becoming very powerful in an automated world and is a revolutionary model which can play a necessary role to remotely control and even monitor particular scenarios at the time of war (Jino Ramson et al., 2020). In the field of smart textiles, IoT can be used for performing many tasks and can act as a supporting hand for AI-OS. In the world of automation, work on several sets of data and the IoT provide synchronization between one end to another. Smart textiles can be AI programmable in nature with the capability to auto respond, monitor, adopting new things, etc., automatically. This makes control of electronic devices like sensors, displays, wireless device, energy harvesters, data storage device, etc., modifiable as per user's requirements. With the help of these devices, IoT gathers information and exercises need-based control on all devices as per requirements (Shi et al., 2020).

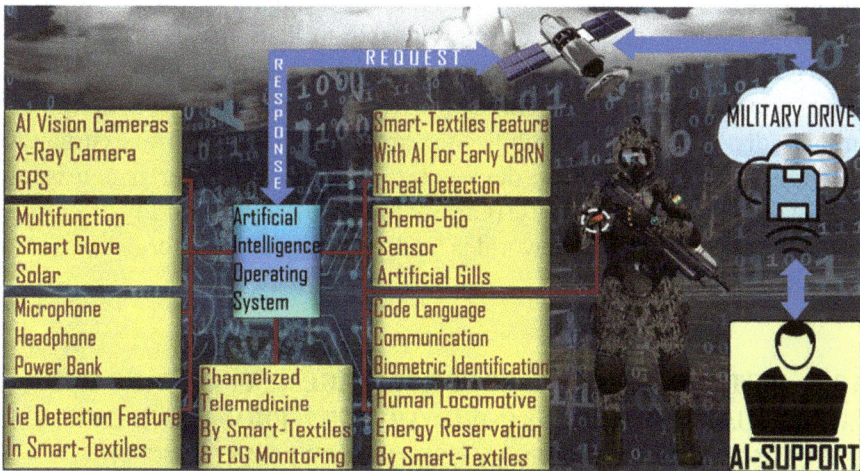

FIGURE 10.4 Featuring artificial-intelligence-based technologies in textile material as a smart solution for battlefield as well as smart inbuilt tools for earlier detection, diagnosis, and management of health physiology in extreme environment.

For physiological monitoring and pre-diagnosis via smart textiles, IoT can facilitate improved communication between the patient and physician utilizing intelligent biosensors, low-voltage-integrated circuits that are able to sense the entire human body for purposes of health monitoring. The main use of these wearable and sensing devices is to monitor all physical activities like performance, endurance, metabolic rate, etc., and physiological conditions such as cardiac fitness, pulmonary performance, cognitive functions for diagnosis and monitoring. Continuous physiological monitoring increases the data processing technique for real-time military medical assistance whereas; with the help of IoT information gathering with the help of biosensors incorporated inside textile materials become an advanced tool. IoT presents updated information for health awareness by using smart sensors in textile create protection level according to real-time requirements (Shabnam et al., 2019). IoT assistance system plays an important role and it provides an assist while a person is in movement and loss in communication take place than military drive will use for relating data and the AI-OS will work as assistance (Do Nascimento et al., 2020).

For long term of data acquisitions by health assistance connection of military drive (i.e., cloud storage) and individual people can be used for overall physiological monitoring for long period of time. The collection of various physiological indicators or biosensors on human body creates records of health; each physiological monitoring makes a huge amount of data which can be easily accessible by using military drives. With the use of IoT technology, it is possible to enhance the speed of solving problems and making data available to military drive in a fast manner. IoT, along with military drives, connected with health experts can act also as an assistant to diagnose diseases and as these devices get incorporated in smart nanotextiles, there would be better ease of use (Chen et al., 2016). With rapid advancements in the technology of IoT, the world is likely to witness a transformation in the way smart nanotextiles work.

10.6 FUTURE INNOVATIONS FOR MANAGEMENT OF PANDEMIC-LIKE SITUATIONS SUCH AS COVID-19, H1N1, EBOLA, AND OTHER COMMUNICABLE DISEASES BY USING SMART NANO/TEXTILE APPROACHES

Future innovations for infections and communicable disease control such as hospital born infections, H1N1 (swine flu), Ebola, Bird flu, and a very recent pandemic COVID-19 are required to controls such kind of unknown infections by using smart nanotextile approach with an inbuild function of detection and management through AI-based integrated systems.

Communicable diseases and other infections mostly spread due to cross-infections by human-to-human transmission or by using garments and other wearable provided in hospitals. To manage such situation, many hierarchy of research has been proposed such as antimicrobial-coated nanoparticles into smart fabrics with laundry durability of more than 50 washes (Bshena et al., 2011); chelating agent, monocarboxylic acid and carbamic acid derivative-coated fabric to impose a smart textile material to control many bacterial infections due to Gram-positive and Gram-negative bacteria, fungi, and mildew (Rajendran and Anand, 2001), insect repellant textile materials (Achee et al., 2019), etc. A representation of smart textile mask which changes color upon pathogen attacks reveals the level of contamination and manages further spread of diseases has depicted in Figure 10.3.

Antistatic textile materials also possess a sustainable and safe textile material used in PPEs kits, mask, gloves, etc. Their anti-adherent properties provide a good material for disabling pathogen and toxins attachments (Ibrahim et al., 2013a, 2013b).

Swiss innovation textile industry (Stylus, 2020) reported an antimicrobial and antiviral textile treatment (Viroblock NPJ03 from HeiQ) which uses combined vesicles of silver technologies which specifically targets lipid-enriched enveloped viruses such (SARS-CoV2). This innovative smart textile material has been tested and found effective in reducing virus infectivity over 99.99%. This innovative technology can be applied on a wide range of wearables such as PPE kits, hospital gowns, surface disinfectant wipes, gloves, towels (Fig. 10.5).

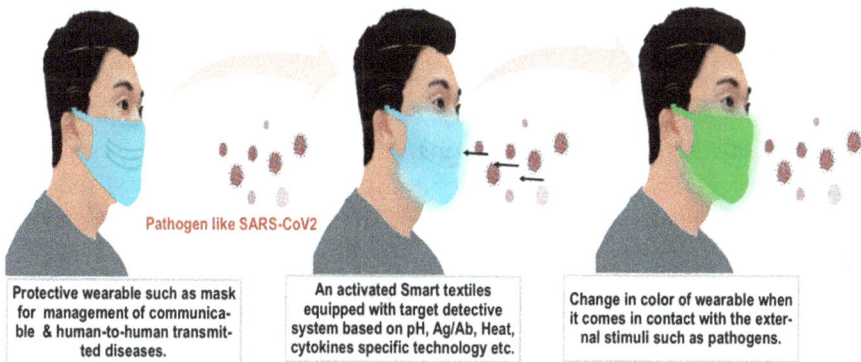

Pathogen like SARS-CoV2

| Protective wearable such as mask for management of communicable & human-to-human transmitted diseases. | An activated Smart textiles equipped with target detective system based on pH, Ag/Ab, Heat, cytokines specific technology etc. | Change in color of wearable when it comes in contact with the external stimuli such as pathogens. |

FIGURE 10.5 A schematic representation of color changing smart textile designed mask for the prevention and management of transferable diseases like COVID-19, Ebola, H1N1, and other flu.

10.7 INTELLIGENT AND SMART TEXTILES FOR FUTURISTIC APPLICATIONS IN ARMED FORCES

The use of AI in military is increasing day-by-day due to the need of faster operational responses in combat scenarios and thus it becomes imperative to develop tools for swift decision making, battlefield management, real-time soldier health and physiology monitoring and medical management, unique friend-foe identification, espionage management, artificial data recognition, etc. In future, smart textiles with inbuilt features employing AI can better serve as a platform for empirical data recognition to structured user-centered instant activity gathering virtually, which may be in hidden form of smart textile wearables (Johnson, 2019). The operational capabilities of the Armed forces can be improved by the use of smart gadgets connected to intelligence-based operating system with electronic devices and/or chips, including e-wearables that can be fixed to soldier-wearables like goggles, helmet, gloves, boot, mask, belt, uniform, etc., to make them from an ordinary textile to smart textile. In helmet, on upper position a high-resolution camera with day and night vision capability can be installed that can be monitored through an AI-based tiny chip system which can be fixed to the uniform by making channels which will be also protected from any external damages. Smartly installed camera helps in capturing images from long range which can sense a suspect or hidden movement from a long range and immediately will send message(s) to the operating system, which can further receive warnings that can be decoded by the connected team (Jalui et al., 2019). Through-the-wall imaging sensors can be installed in the helmet that can help to see behind the wall or behind non-transparent coverings. By using this technology of image capturing with intelligence system, it is now possible to create a virtual image form behind the wall which can be helpful in various kinds of operational missions (Bane and Hollerer, 2004). Use of smart goggles as a part of integrated uniform, will be helpful in displaying messages to the connected team so that a silent operation can be successfully operated and this will be also being helpful to view alert notifications by gathering AI-based information from the enemy (Taya et al., 2008). A smart textile mask can be used by armed forces to cover their face and to control and filtered air quality (Her-mou, 2012) as well as fulfill the purpose of AI detection of any harmful/poisonous/chemical warfare agent like nerve gases an alert signal to smart textile further activated an extra covering tool upon the face mask. With more features of smart textiles for ballistic protection, an elbow pad and knee pad can be smartly designed which can provide proper cushion as well as also be able to release the requisite medication if any injury takes place, like a self-healing armor that has the capability of repairing by breaching within few minutes. The sensor(s) present in the smart textile will update the health capability of armor (Showalter et al., 2016). Furthermore, a smart global positioning system watch can be integrated in smart textile to find the actual direction and it will also able to update the real-time location (RT-Location) to their headquarters, so that safeguard or quick response team can be sent to the operational team based upon requirement (Baijal and Arora, 2018). Subsequently, in shoes/boots, a metal detector sensor can be used to detect metal mines and other ballistic material(s). This sensor used in smart shoes/boot should be able to recognize improvised explosive devices from a distance of 10 m while running and 20–25 m while walking. An electric shock weapon can be fixed in hand gloves which delivers an electric shock and temporarily is able to enhance muscular functions through ion-gated channels. This will work according to need basis, firstly when the hand is punching someone and makes contact with skin on hitting and secondly when it controls through command control. All of the above devices can be equipped in textile/nanotextile materials which can be connected through an operating system based on their functions, that is, IoT. This will help to get connected with the team and support system so that a real-time update can be gathered for the benefit of the operational team as well as for managing the intelligence team. These devices can be further enabled with AI system, which would aid in making decisions by using intelligence decision support system so that the armed forces can make quick responses and decisions on the battlefield in real-time. The need for future is to harness AI to develop safe, skin compatible/mimicking, eco-friendly, programmable smart fabrics that can see, sense, hear, perform and if needed even respond as per military operational requirements.

We are living in the time of digital solutions for every need. Since AI has application in many fields but when we think about some futuristic approaches for soldiers, we need to think for a resistant solution for many things. Likewise, for a solution for extreme environmental conditions like extreme cold/heat, stress, and hypoxia, an intelligent and smart textile can be designed to sense and control external changes and should be able to maintain human hemostasis. Secondly, a futuristic approach of smart textiles for pilots can be AI-built anti-G suit as well as AI-built a quick escape pilot protective suit (AI-featured flame retardant dungarees) in case of engine failure or fire. Another smart textile can be an overall light material suit made of 3D printed smart textile materials, conductive-textile fibers (graphene, nanocarbon, etc.) and other genetically engineered textile materials with intelligent features of AI like sensing, actuating, decision making, message receiving, responding, and protecting features. This can be a futuristic novel designing in material and textile engineering (Fig. 10.6).

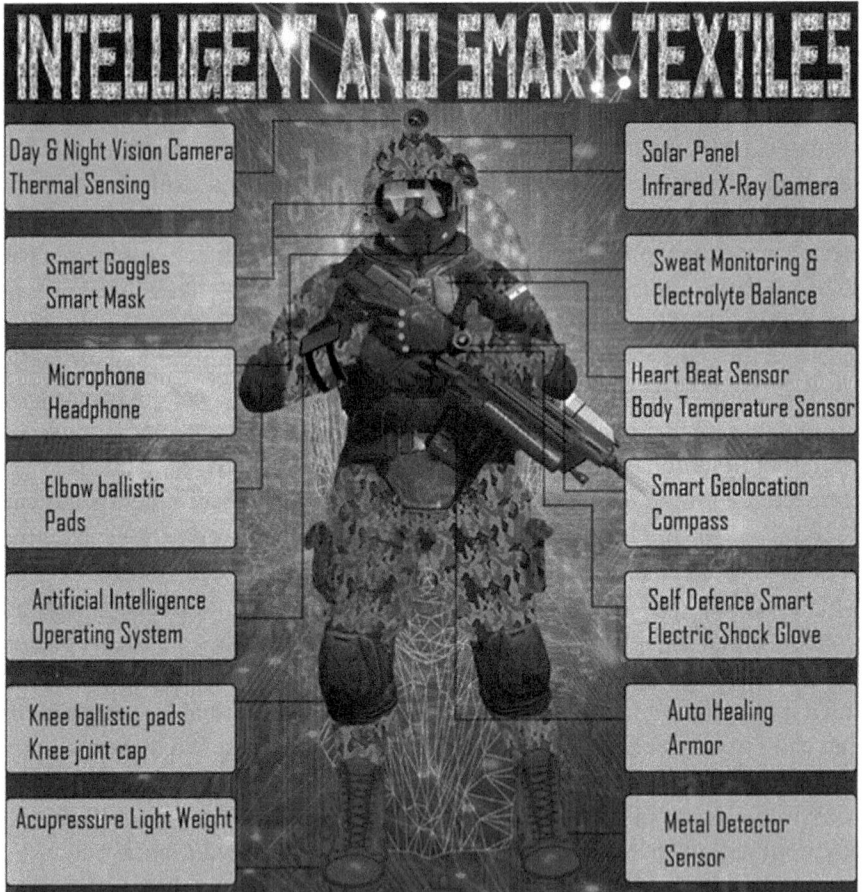

FIGURE 10.6 Intelligent and smart textiles for armed forces as an advanced system with smart wearables to combat critical situations with use of high-end technologies

10.8 CONCLUSION

Smart nanotextiles will be useful for the military and general public if the advanced approaches used in smart nanotextiles are unequivocally accepted by the military and society. Pre-diagnosis and continuous monitoring of any disease condition can reduce the burden of use of excessive medicine and wrong therapeutics. Life-threatening problems such as cardiovascular and respiratory diseases can be prevented if real-time system and data processing units are available in the form of wearables. Smart clothing and smart wearable textiles are designed to protect from any immediate hazards which may include chemical, biological, radiological, and nuclear hazards. The protective clothing also acts as a first line of defence against many substances which can be directly absorbed from skin, oropharyngeal track, respiratory tract, etc. Thus, it becomes imperative to provide overall protection to our soldiers from the deleterious effects of extreme environments, hazards and even bio-terror attacks or bio-threats, etc. The introduction of smart wearable(s) into the armed forces can be beneficial for the entire security system as it can improve activeness and alertness via monitoring for operational preparedness and checking rapid response. In future, these smart textile approaches will most likely be personalized similar to the personalized medicine system because nowadays work is channelized on a real-time monitoring basis. In case of health and diseases, security and privacy would be paramount and, therefore, smart nanotextiles may be one of the customized smart fabric systems. However, the extensive use of smart nanotextiles would eventually depends on the advancements and the acceptance will largely be based on the benefits accrued and most importantly the ease of use perceived by the user, whether in a civil or military scenario. It is anticipated that in view of the tremendous benefits offered, in future it would be smart nanotextiles that would dominate the textile arena mainly on the occupational front.

ACKNOWLEDGMENT

The authors are thankful to Defence R&D Organization for support. They would like to thank Director, Defence Institute of Physiology and Allied Sciences, Delhi, India for support and encouragement in multifarious ways. RBY is also thankful to DRDO for her fellowship.

DISCLAIMER

The views and opinions expressed in this article, being purely of academic/ scientific nature, are entirely those of the authors and do not necessarily reflect the official policy or position of DIPAS, Delhi/DRDO/Government of India or the Delhi Institute of Pharmaceutical Sciences and Research (DIPSAR), Delhi or any other organization(s) whatsoever. This article contains only unclassified information and the pictures have been used for illustrative purposes only. No endorsement of any commercial product(s)/ technology is made. The authors declare no conflict of interest.

KEYWORDS

- **smart nanotextile material**
- **nanofabrication**
- **antimicrobial**
- **artificial intelligence**
- **smart textiles for armed forces**

REFERENCES

Abdallah, S.; Wilkinson, M. C.; Waskiw-Ford, M; Abdallah, I.; Lui, A.; Smith, B.; Bourbeau, J.; Jensen, D. Validation of Hexoskin Biometric Technology to Monitor Ventilator Responsed at Rest and during Exercise in COPD. *Eur. Resp. J.* **2017,** *50,* PA 1359.

Abdullaeva, Z. Nanomaterials for Clothing and Textile Products. *Nanomat. Daily Life.* **2017,** 11–132.

Achee, N. L.; Grieco, J. P.; Vatandoost, H.; Seixas, G.; Pinto, J.; Ching-Ng, L.; David, J. P. Alternative Strategies for Mosquito-Borne Arbovirus Control. *PLoS Negl. Trop. Dis.* **2019,** *13* (1), e0006822.

AiQBioman+. https://www.aiqsmartclothing.com/product-service/bioman-plus/ (accessed on Mar 9, 2020).

Andrea, R.; Roberto, R.; Maurizio, M.; Dauide, V.; Justin, S.; David, A. Real-Time Smart Textile-Based System to Monitor Pressure Offloading of Diabetic Foot Ulcers. *J. Diabetic Sci. Tech.* **2017,** 1–5.

Baijal, R.; Arora M. K. GPS: A Military Perspective. *Geospatialworld,* 2018. Available:https://www.geospatialworld.net/article/gps-a-military-perspective.

Balint, R.; Cassidy, N. J.; Carmell, S. H. Conductive Polymers: Towards a Smart Biomaterial for Tissue Engineering. *Acta Biometr.* **2014,** *10,* 2341–2353.

Bane, R.; Hollerer, T. Interactive Tools for Visual X-ray Vision in Mobile Augmeneted Reality. *ISMAR 2004 Proc. Third IEEE ACM Int. Symp. Mix. Augment Real.,* 2004; pp 231–239.

Bheekhun, N.; Talid, A. R. A.; Hassan, M. R. Aerogel in Aerospace: An Overview. *Adv. Mater. Sci. Eng.* **2013,** 1–18.

Biresaw, G.; Carriere, C. J. Compatibility and Mechanical Properties of Blends of Polystyrene with Biodegradable Polyesters. *Composites Part A: Appl. Sci. Manuf.* **2004,** *35,* 313–320.

Bshena, O.; Heunis, T. D. J.; Dicks, L. M. T.; Klumperman, B. Antimicrobial Fibers: Therapeutic Possibilities and Recent Advances. *Future Med. Chem.* **2011,** *3* (14), 1821–1847.

Cai, G.; Xu, Z.; Yang, M.; Tang, B.; Wang, X. Functionalization of Fabrics through Thermal Reduction of Graphene Oxide. *Appl. Surf. Sci.* **2017,** *393,* 44–448.

Castano, L. M.; Flatau, A. B. Smart Fabric Sensors and E-Textile Technologies: A Review. *Smart Mater. Struct.* **2010,** *23,* 053001.

Chen, M.; Ma, Y.; Song, J.; Lai, C. F.; Hu, B. Smart Clothing: Connecting Human with Clouds and Big Data for Sustainable Health Monitoring. *Mobile Net. Appl.* **2016,** *21* (5), 825–845.

Choi, S.; Kwon, S.; Kim, H.; Kim, W.; Kwon, J. H.; Lim, M. S.; Lee, H. S.; Choi, K. C. Highly Flexible and Efficient Fabric-Based Organic Light Emitting Devices for Clothing-Shaped Wearable Displays. *Sci. Report.* **2017**, *7* (1), 1–8.

Christle, R. M. *Chromic Materials for Technical Textile Applications*; Woodhead Publishing Limited: Cambridge, **2013**.

Cochrane, C.; Mordon, S. R.; Lesage, J. C.; Koncar, V. New Design of Textile Light Diffusers for Photodynamic Therapy. *Mater. Sci. Eng. C.* **2013**, *33*, 1170–1175.

Crank, M.; Patel, M.; Marcheider-Weidemann, F.; Schleich, J.; Husing, B.; Angerer, G.; Wolf, O. Techno-Economic Feasibility of Large Scale Production of Bio-Based Polymers in Europe. *Technical Report EUR 22103 EN*. European Science and Technology Pbservatory, European Communities, 2005.

Dickson, E. F. G. *Personal Protective Equipment for Chemical, Biological and Radiological Hazards: Design, Evaluation, and Selection*; John Wiley & Sons, Inc.: Hoboken, NJ, **2013**.

Ding, Y.' Invernale, M.A; Sotzig, G. A. Conductivity Trends of PEDOT-PSS Impregnated Fabric and the Effect of Conductivity on Electrochromic Textile. *ACS Appl. Mater. Interf.* **2010**, *2*, 1588–1593.

Do Nascimento, L. M. S.; Bonfati, L. V.; Freitas, M. L. B.; Mendes Junior, J. J. A.; Siqueira, H. V.; Stevan, S. L. Sensors and Systems for Physical Rehabilitation and Health Monitoring- A Review. *Sensors* **2020**, *20* (15), 1–28.

DUPONT. https://dupont.co.in/personal-protection/chemical-protection-military-police,html/ (accessed on Mar 16, 2020).

Eugene, W., Ed. *Military Textiles*. Woodhead Publishing in Textiles: CRC Press, **2008**, 73; pp. 23, 74, 191.

Friedl, K. E. Military Applications of Soldier Physiological Monitoring. *J. Sci. Med. Sport* **2018**, *1875*, 1–7.

Grancaric, A. M.; Jerkovic, I.; Koncar, V.; Cochrane, C.; Kelly, F. M.; Soulat, D.; Legrand, X. Conductive Polymers for Smart Textile Applications. *J. Indust. Text.* **2017**, 1–31.

Gugliuzza, A.; Drioli, E. A Review on Membrane Engineering for Innovation in Wearable Fabrics and Protective Textiles. *J. Memb. Sci.* **2013**, 446, 350–375.

Hadgson, E. The Chameleon Suit-a Liberated Future for Space Explorers. *Gravit. Space Biol. Bull.* **2003**, *16* (2), 107–119.

Her-mou, L. Mask with an Air Filtering Device. *Patent/US5104430A*, **2012**.

Hexaskin and Carre Technologies Inc. Hexoskin Smart Shirt-Cardiac Respiratory Sleep and Activity Metrics. Carre Technologies Inc (Hexoskin).

Hexoskin Smart Garments Specifications. https://www.hexoskin.com/ (accessed on Mar 9, 2020).

Ibrahim, N. A.; Elmaaty, T. A.; Eid, B. M.; El-Aziz, E. A. Combined Antimicrobial Finishing and Pigment Printing of Cotton/Polyester Blends. *Carbohydr. Polym.* **2013a**, *95* (1), 379–388.

Ibrahim, N. A.; El-Sayed, Z. M.; Fahmy, H. M.; Hassabo, A. G.; Abo-Shosha, M. H. Perfume Finishing of Cotton/Polyster Fabric Cross-Linked with DMDHEU in Presence of Softeners. *Res. J. Text. Appar.* **2013b**, 17 (4), 58–63.

Jalui, S.; Hait, T.; Hathi, T.; Ghosh, S. Advanced Military Helmet Aided with Wireless Live Video Transmission, Sensor Integration and Augmented Reality Headset. *Proc. 4th Int. Conf. Commun. Electron. Syst. ICCES*, **2019**; pp 123–127.

Javaid, N.; Sher, A.; Nasir, H.; Guizani, N. Intelligence in IoT-based 5G Networks: Opportunities and Challenges. *IEEE Commun. Mag.* **2018**, *56*, 94–100.

Jhala, P. B. Development of High Altitude Clothing for Soldiers Using Plasma Enhanced Angora Wool. *International Exhibition-cum- Seminar on the Future Infantry Soldier As a System (F-INSAS), CII-Indian Army, New Delhi*, 2006, Nov 23–24.

Jino Ramson, S. R.; Vishnu, S.; Shanmugam, M. Applications of Internet of Things (IoT)-An Overview. *ICDCS 2020–2020 5th International Conference on Devices, Circ. Syst.*, 2020; pp 92–95.

Johnson, J. Artificial Intelligence and Future Warfare: Implications for International Security. *Def. Secur. Anal.* **2019**, 35 (2), 147–169.

Karpagam, K. R.; Saranya, K. S.; Gopinathan, J.; Bhattacharya, A. Development of Smart Clothing for Military Applications Using Thermodynamic Colorants. *J. Textt. Inst.* **2017**, *108* (7), 1122–1127.

Kelvin, J. C.; Heather, A. C. Nanosensors and Nanomaterials for Monitoring Glucose in Diabetes. *Trend. Mol. Med.* **2010**, *16* (12), 584–593.

Keri, J. H.; Stephen, W. P. Accuracy of the Life-shirt® (Vivometrics) in the Detection of Cardiac Rhythms. *Bio. Pshychol.***2007**, *75* (3), 300–305.

Kiu, Y.; Tang, J.; Wang, R. H.; Lu, H.; Li, L.; Kong, Y.; Qi, K.; Xin, J. H. Artificial Lotus Leaf Structures from Assembling Carbon Nanotubes and Their Applications in Hydrophobic Textiles. *J. Mater. Chem.* **2007**, *17*, 1071–1078.

Lah, A. S.; Fajfar, P.; Kugler, G.; Rijavec, T. A. NiTi Alloy Weft Knitted Fabric for Smart Firefighting Clothing. *Smart Mater. Struct.* **2019**, *28*, 065014.

Li, J. H.; Chen, J. H.; Xu, F. Sensitive and Wearable Optical Microfiber Sensor for Human Health Monitoring. *Adv. Mater. Tech.* **2018**, 3, 1–8.

Liu, Y.; Wang, X.; Qi, K.; Xin, J. H. Functionalization of Cotton with Carbon Nanotubes. *J. Mater. Chem.* **2008**, *18*, 3454–3460.

Lu, H.; Kezhi, L.; Menghao, W.; Kefeng, W.; Liming, F.; Haiting, C.; Jie, Z.; Xiong, L. Mussel-Inspired Adhesive and Conductive Hydrogel with Long-Lasting Moisture and Extreme Temperature Tolerance. *Adv. Funct. Mat.* **2018**, *28*, 1704195.

Matteo, S.; Alessandro, C. Wearable Electronics and Smart Textiles: A Critical Review. *Sensors* **2014**, *14*, 11957–11992.

Mehrubeoglu, M.; Linh, M. P.; Hung, T. L.; Muddu, R.; Dongseok, R. Real-Time Eye Tracking Using a Smart Camera. *IEEE Applied Image Pattern Recognition Workshop (AIPR)*, 2011; pp 1–7.

Neuronaute, BioSerenity. *Available online: https://www.bioserenity.com/* (accessed on Mar 9, 2020).

Panhuis, Marc.; Wu, J.; Ashraf, A. A.; Wallance, G. G. Conducting Textiles from Single-Walled Carbon Nanotubes. *Synth. Met.* **2007**, *157* (8–9), 358–362.

Paradiso, R.; Loriga, G.; Taccini, N.; Gemignani, A.; Ghelarducci, B. Wealthy-a-Wearable Healthcare System: New Frontier on E-Textile. *J. Telcomm. Info. Tech.* **2005**, *4*, 105–113.

Potter, A. W.; Blanchard, L.; Friedl, K. E.; Cadarette B. S.; Hoyt, R. W. Mathematical Prediction of Core Body Temperature from Environment, Activity and Clothing: The Heat Strain Decision Aid (HSDA). *J. Thermal Bio.* **2017**, *64*, 78–85.

Potter, A. W.; Karis, A. J.; Gonzalez, J. A. *Biophysical Characterization and Predicted Human Thermal Responses to US Army Body Armor Protection Level (BAPL)*; Army Research Institute of Environmental Medicine: Natick, MA, 2013.

Quandt, B. M.; Scherer, L. J.; Boesel, L. F.; Wolf, M.; Bona, G. L.; Rossi, R. M. Body Monitoring and Health Supervision by Means of Optical Fiber-Based Sensing Systems in Medical Textiles. *Adv. Healthcare Mater.* **2015**, *4*, 330–355.

Quartinello, F.; Tallian, C.; Auer, J.; Schon, H.; Vielnasher, R.; Weinberger, S.; Wieland, K.; Weihs, A. M.; Alexandra, H. R.; Lendl, B.; Teuschl, A. H.; Pellis, A.; Guebits, G. M. Smart Textiles in Wound Care: Functionalization of Cotton/PET Blends with Antimicrobial Nanocapsules. *J. Mat. Chem. B.* **2019**, *7*, 6592–6603.

Rajendran, S.; Anand, S. C. In: *Development of a Versatile Antimicrobial Finish for Textiles Materials for Healthcare and Hygiene Applications, Proceedings, Medical Textiles'99 Conferences*; Woodhead Publishing Ltd: Cambridge, Bolton, UK, 2001, August 24–25.

Redmond, D. P.; Hegge F. W. Observations on the Design and Specification of a Wrist-Worn Human Activity Monitoring System. *Behav. Res. Meth.* **1985**, *17* (6), 659–669.

Shabnam, F.; Azmi Hoque, S. M.; Faiyad, S. A. IoT Based Health Monitoring Using Smart Devices for Medical Emergency Services. *2019 IEEE International Conference on Robotics, Automation, Artificial-Intelligence and Internet-of-Things, RAAICON 2019*, 2019; pp 69–72.

Shi, J.; Liu, S.; Zhang, L.; Yang, B.; Shu, L.; Yang, Y.; Ren, M.; Wang, Y.; Chen, J.; Chen, W.; Chai, Y.; Tao, X. Smart Textile-Integrated Microelectronic Systems for Wearable Applications. *Adv. Mater.* **2020**, *32* (5).

Shi, Q.; Sun, J.; Hou, C.; Li, Y.; Zhang, Q.; Wang, H. Advanced Functional Fiber and Smart Textile. *Adv. Fiber Mat.* **2019**, 1–29.

Showalter, D. D.; Gooch, W. A.; Burkins, M. S.; Montgomery J. S.; Squillacioti, R. J. Development and Balastic Testing of a New Class of High Hardness Armor Steel. *AMMTIAC Quart.* **2016**, *4* (4).

SKIIN Textile Computing™ Plateform. https://mynant.ca/technology/ (accessed on Mar 9, 2020).

Stylus. https://www.stylus.com/new-textile-treatment-effective-against-covid19 (accessed on Aug 23, 2020).

Suvetha, M.; Swathi, S.; Rani, M.; Vinoth, S.; Suriya, R. A Study on Artificial Intelligence. *Bonfr. Int. J. Indust. Eng. Manage. Sci.* **2019**, *9*, 06–09.

Syduzzaman, M.; Patway, A. U.; Farhana, K.; Ahmed, S. Smart Textiles and Nano-Technology: A General Overview. *J. Text. Sci. Eng.* **2015**, *5*, 1.

Taliyan, S. S.; Biswas, B. B.; Patil, R. K.; Srivastava, G. P.; Basu, T. K. Electricity from Footsteps. *BARC News Lett.* **2010**, *313*, 47–50.

Tamburri, E.; Orlanducci, S.; Toschi, F.; Tessranova, M. L.; Passeri, D. Growth Mechanisms Morphology and Electroactivity of PEDOT Layers Produced by Electrochemical Routes in Aqueous Medium. *Synt. Mater.* **2009**, 159, 406–414.

Tao, X., Ed. *Handbook of Smart Textiles*; Springer: Singapore, 2015.

Tao, X., Ed. Smart Fibers, Fabrics and Clothing; CRC Press: Boca Raton, FL, USA, 2001.

Taya, M.; Island, M.; Examiner, P.; Mai, H. K. Smart Sunglasses, Helmet, Faceshields and Goggles Based on Electrochromic Polymers. *Patent/US787466B2* **2008**.

Tingting, G.; Zhi, Y.; Chaoji, C.; Yiju, L.; Kun, F.; Jiaqui, D.; Emily, M. H.; Hua, X.; Boyang, L.; Jianwei, S.; Bao, Y.; Liangbing, H. Three Dimensional Printed Thermal Regulation Textiles. *ASC Nano* **2017**, *11*, 11513–11520.

Tsuda, T.; Yamamoto, H.; Kameda, Y.; Ohta, Y. Visualization Methods for Outdoor See-Through Vision. *IEICE Transc. Commun.* **2006**, *E89-B* (6), 1781–1789.

Turaga, U.; Kendall, R. J.; Singh, V.; Lalagiri, M.; Ramakumar, S. S. Advances in Material for Chemical, Biological, Radiological and Nuclear (CBRN) Protective Clothing. In *Advance in Military Textiles Personal Equipment*; Woodhead Publishing Series in Textiles, 2012; pp 260–287.

Vagott, J.; Parachuru, R. An Overview of Recent Developments in the Field of Wearable Smart Textiles. *J. Text. Sci. Eng.* **2018**, *8*, 4.

Zhang, G.; Liu, Y.; Gao, X.; Chen, Y. Synthesis of Silver Nanoparticles and Material Property of Silk Fabrics Treated by Silver Nanoparticles. *Nanoscale Res. Lett.* **2014,** *9* (1), 216.

Zhao, M.; Li, T.; Alsheikh, M. A.; Tian, Y.; Zhao, H.; Torralba, A.; Katabi, D. Through-wall human pose estimation using radio signals. *Proc. IEEE Comp. Socie. Confer. Comp. Vis. Patt. Recogn.*, 2018; pp 7356–7365.

Zindan, G.; Ziyang, X.; Xia, O. Y.; Jun, Z.; Newman, L.; Jie, Z.; Chi, C. C. Wearable Fiber Optic Technology Based Smart Textiles: A Review. *Material* **2019,** *12*, 3311.

CHAPTER 11

A New Dimension on Novel Application of Nanotechnology for High-Performance Clothing

D. GOPALAKRISHNAN*, M. PARTHIBAN, and P. KANDHAVADIVU

Department of Fashion Technology, PSG College of Technology, Coimbatore 641004, Tamil Nadu, India

Corresponding author. E-mail: dgk.psgtech@gmail.com

ABSTRACT

The latest revolution in functional textile is the emergence nanoengineering and nanotechnology. The potential of this technology has led to the development of new materials and products to fulfill the end user. According to the end users, the uniqueness of materials and their characteristics will be extended economically with huge potential, and also are attracting the new ventures and researchers. Globally, the research interest has been provoked on nanoengineered materials and textiles. Researchers further advanced in their effort to develop the tool to enhance the performance and feature of the textile product. Some contradictory issues on the environment and human health are also known. Massive researches reveal the developed products with their commercial reality and production viability. The potential of nanotechnology on textiles shows only the way to develop the product with multifunctionalit attributes of the textiles. The main areas of textile processing, such as nanofibers production, nanocoatings, nanofinishing and nanocomposites are producing promising results in textile revolution in the form of antibacterial textiles, self-cleaning fabrics, absorbency and wicking-controlled textiles, fire-proof textiles, conductive fabrics, and fabrics with UV protection without affecting the features and properties of fibers. The properties of fibers and nanomaterials can be used to manufacture the fabric with good chemical resistance, antimicrobial, water repellency, etc., The method of nanotechnology overcomes the conventional method of processing with their limitations and the material properties. The effect of toxicity of nanoparticles on consumer products must also be understood. This chapter focuses on the development and potential applications of nanotechnology in developing multifunctional textiles, such as sensing textiles, drug release application of textiles, photonic technologies, electronics in textiles, graphene yarns, graphene-coated textiles, and nanofinishing of textile substrates.

11.1 INTRODUCTION

Nanotechnology is one of the entry level or infant sectors with textile associated with different applications. Nanotechnology extends its application to industrial textiles, to increase the durability of clothing. (Liam Critchley, 2018). Nanotechnology has also incorporated nanomaterials to enhance the properties and features of textiles. These nanomaterials may range from the carbon nanotubes, metal oxides, and graphene to various nanoparticles also. Many manufacturers attempted to make nanofabrics and garments with nanomaterials; the new area that has been significantly benefitted in utilizing nanomaterials is wound and hygiene management in biomedical field. Nanotechnology has not only attracted the academic researchers but also made many companies across the globe to produce commercial and viable nanotextiles.

Nanotechnology is the base of interdisciplinary sector that influences different technological fields, such as aerospace, nanoelectronics, biomedicine, and material science (Joshia et al., 2008). It also produces promising results and enhances the textiles and clothing properties and various performances (Gashti et al., 2012). The studies related to nanotechnological aspects which means the application of nanomaterials on textile substrates includes immobilization, synthesis and cross-linking.

11.2 NANOMATERIALS

In the order of nanometer, materials possessing one dimension are called nanomaterials which are less than 100 nm (Gashti et al., 2012). These nanomaterials are promisingly utilized in high performance and functional textiles due to their high specific characteristics that are stemming from great surface area-to-volume ratios (Joshia et al., 2008). The nanomaterial is perceived as a kind of material that has novel design, for example, carbon black. Carbon black has been tremendously in use in automobile industry (Raab et al., 2017). Incorporation of nanomaterial with textile substrate for enhancing the functional properties as well as performance-oriented applications, such as electromagnetic shielding, electrical conductivity etc. has been carried out (Gashti et al., 2016; Zhu et al., 2006). Mostly, the application of nanomaterials requires definite dimension of the particle with narrow variation . In some way, properties and production parameters could be manipulated. The properties include crystallization, shape geometry and dimensions, and chemical composition. Also, the production parameters include temperature, pH, chemical type, and concentration. Based on manipulated synthesis of materials, diverse shapes of nanomaterials can be obtained, such as nanospheres, nanoprism, nanowires, nanorods, etc. (Gashti et al., 2016). The durability of the textile material is the biggest challenge in the development of nanomaterials. Due to van der waals forces, nanoparticles agglomerate, also the dispersion of nanoparticles is difficult due to f double-layer attraction in electrostatic parameters. From the affinity point of view, there is no affinity between nanomaterials and textile materials. Embedding nanomaterial substrates in matrices by chemical and physical techniques has been suggested to overcome the lacking of functional sites on the surface (Gashti et al., 2016). One of the novel abilities of nanoscale materials on textile application is the smartness which shows its promising use in smart textiles too. Smart textiles provide nanotechnological components, such as nanofibers, nanowires, nanogenerators, nanocomposites, and nanostructured polymers. Smart nanotextiles are developed to use especially in aerospace, military, and biomedical applications. (Joshia et al., 2008). The nanotechnological components, their properties, production techniques, and nano technical characterization methods are required to develop nano smart textiles.

11.2.1 NANOFIBERS

Apart from nanorods, nanospheres, etc., more expectation of superior characteristics get forefront for fiber form, it has higher surface area, superior performance, and flexibility also. Nanofibers are used to reinforce in aerospace industries. Nanofibers refer to the material that is linear solid-state material having the aspect ratio of 1000:1. They are flexible and the dimensions are less than or equal to 100 nm. The flexibility is million times increased due to decrease in fiber diameter based on reduction from 10 m to 10 nm. As a result of which, surface reactivity and specific surface area is increased (Wan et al., 2018). Polyaniline, polyacetylene, PNIPAAm, PEG, and polypyrrole are the polymers that are used to manufacture the nanofiber with various functional components,, such as graphene, azo benzene, montmorillonite nanoclay, and carbon nanotubes (Pu et al., 2018).

11.2.2 NANOSOLS

Metal oxides and their particles are converted to colloidal solutions in nanoscale by organic and water as a solvents. Nanosols include inorganic nanoparticles prepared via the sol–gel method (Mahltig, 2018). Nanosols present metastable property due to their high ratio of surface-to-volume. Nanoparticle aggregation is developed to form a network structure in three dimensions and coating is done by solvent evaporation technique (Berendjchi et al., 2013). The precursor materials, such as acetyl acetonate or metal or semimetal alkoxides are hydrolyzed to form nanosols. Hydroxides are unstable in high concentration conditions, and are subjected to condensation reaction. The condensation reaction results in the particle formation in the nanoscale range. For example, precursors used are tetraethoxysilane (TEOS), and titanium(IV) isopropoxide $Ti(OC_3H_7)_4$ (Berendjchi et al., 2013). Using higher specific area and dimensions of 100 nm, coating of nanosols are preparedin the range of several hundred nanometer thickness. (Mahltig et al., 2007). Modification of textile materials was carried out due to bulk and surface properties of nanosol substrates imparted on textiles (Berendjchi et al., 2013). Nanosols are categorized based on four functions as optical, chemical, biological, and functional.

Isopropyl, ethanol, and water are used as solvents for the preparation of nanosols. It is a very simple method to change the functionality of substrates. The water used in nanosol results in limited stability (Mahltig et al., 2007). by If improper postthermal treatment is performed, the presence of nanosol coating creates an amorphous structure called xerogel. Due to undesirable properties of organic solvents, water is chosen (Mahltig et al., 2018) as the solvent. The application of nanosol coating on textile substrate forms adhesion between textile substrate and coating, and also causes problems in synthetic fibers. Various methods including corona, plasma treatment, use of cross-linking agents and thermal treatments are used to activate the surfaces and improve the adhesion (Berendjchi et al., 2013).

11.2.3 BIOMEDICINE

Using of nano fibers in biomedical applications has numerous advantages based on the path of orientation that leads to mimic the biosystem (Yilmaz et al., 2016). In their natural environment, cells live in the surrounding of nano- and/or micro featured environment. The cells proliferate to attach on the medium, the cell has higher dimension than the dimension at present. The functions of alignment, adhesion, migration, and proliferation of cells were affected due to the nanoscale surface (Wan, 2018).

11.2.4 RESPONSIVE POLYMERS

An integration of miniaturized electronic devices and textile materials is called smart textiles. The environment and its conditions, such as temperature, pressure, and light will trigger the polymer to react and response within the situation (Ebara et al., 2014). These polymers are defined as environmentally sensitive, stimuli-responsive, intelligent, or smart polymers (Yilmaz, 2017). Even though the materials are nanostructured, the reversible effect causes macro level responses (Peppas, 2004). The response-triggering stimuli are categorized as physical, chemical, and biological. The remote and local activation is performed by physical stimuli, which leads to the interaction of molecules to some extent. The biological category is one of the subcategory of chemical stimuli. The biochemical process is similar to responses in human body. Biomolecules are more important in this bio chemical process, and responsive stimuli, such as pH and ion exchange are also included in this category. This category is laid under multiple stimuli, so it is called as multi responsive polymer system or dual responsive polymer system (Niiyama, 2018). According to polymer, the response mechanism will vary. The strength of hydrogen bond is changed while the charged groups are under neutral charge due to change in pH and opposite charges get added with chemical species (Kumar, 2012). The property of the polymer changed with responses that are mentioned along with the critical point. The smart functions are mechanized by switchable solubility (Peppas, 2004). PNIPAAm and its derivative are having solubility behavior of reversible responses. The local critical solution temperature (LCST) is characterized by specific temperature below which PNIPAAm is soluble and is never soluble over this point of temperature (Niiyama, 2018). HEMA, ethylene glycol, acrylic acid, and vinyl acetate are used as monomer in the preparation of environmentally responsive polymer

(Yilmaz, 2017). When insoluble materials get incorporated, the manipulation of solubility behavior is carried out. The responsive rate is determined by dimension used and also based on higher surface area. Precursor materials are used to fabricate the different responsive stimuli. The responsive rate and response mechanisms are obtained by good design of components and its structures (Ebara et al., 2014). Engineered polymers are synthesized by species of chemical precursors. There also exist environmentally responsive polymers based on natural polymers beside synthetic polymers. The hybrid polymers contain together both natural and synthetic components. The natural component – proteins, such as gelatin, collagen, chitosan and alginate are used in environmentally responsive polymers (Yilmaz, 2017). The production of environmentally responsive polymers is also based on the bonding of hydrogen, heating and cooling, coacervation and maturation, chemical and radiation grafting, double networks, self-assembly, and slide cross-linking with genetically engineered copolymers especially block copolymers (Kopeček, 2007, Yilmaz, 2017, Sandeep, 2012).

11.2.5   MOISTURE MANAGEMENT

The topographical properties are considered very important than the chemical properties (Periolatto et al., 2013). Nanofibers are used in managing moisture functions, such as switchable hydrophilicity–hydrophobicity and superhydrophobicity. . The lotus effect is formed by the property of superhydrophobicity. Both hydrophobic surface and microstructure in rough nature together form the superhydrophobicity (Yamamoto at al., 2015). Membranes are formed by nanofibrous superhydrophobic structures with polystyrene, polyvinylidene fluoride, and polyurethane. Tthis structure enhances both hydrophobic as well as hydrophilic properties. Incorporation of pores, dents, microgrooves, and rods has improved the rough microstructure. This process is carried out by means of electrospinning. The bead-on-string format is followed and obtained in dope factors with electrospinning variables. By sonicating process, the nanoparticles are separated away and incorporation of particles on fibers in nanoform, at same time, the effect of superhydrophobicity is attained. Fluorinated polymers with low surface energy can be used on fibrous materials, such as nanomembrane to get improved hydrophobicity. A similar study shows that roughness has affected both oleophobic and hydrophobic properties at the same time. Nanoparticle of SiO_2 is used to enhance the surface roughness (Mahltig, 2018).

11.2.6 NANOWIRES

The nanoscale of the wire is 5100 nm in diameter and length is 100 nm (Ranzoni, 2017), when compared with nanoparticles that appear in other shapes, such as rods and spheres, nanowires exhibit high aspect ratios (Song, 2018). A final product transparency is decided by the diameter of the nanowires. The signal-to-noise ratios and ultra-high sensitivities of Si nanowires are high compared with conventional nanomaterials (Singh, 2016). Lowering resistance in electrical transmission by their shape and easy orientation to other shapes are provided by nanowires. Metal nanowires are superior when compared with the carbon nanotubes, within the junctures, they show high resistance (Song, 2018). Silver, copper, and gold are used in making nanowires, and addition of metal oxides, conductive polymers as polythiophene, polypyrrole, semiconductive (ZnO, SnO_2, CuO), sulfides (Cu_7S_4, CoS_2) and Si, CdTe, ZnSe are also utilized in making nanowires. The biosensors are produced by silicon nanowires of semiconductive type. ZnO is an oxide semiconductor and the control of conductivity of which is done by insulating dopants to highly conductive level.

Semiconductive silicon nanowires have been used for preparing biosensors (Clark,2016). Chemical composition is manipulated to fine-tune the conductive properties. Nano wires with flexible and rigid substrates are used for making gates, resistors, diodes, and transistors (Ranzoni et al., 2017). The resultant properties in precursor materials are used in making electromechanical devices, nanoelectronics, etc. (Song, 2018). Nanowires have also been used in nanodrug delivery systems (Sharma, 2009), personal thermal management, photocatalysis, strain sensors, lithium batteries, photodetectors, supercapacitors, and nanogenerators (Song, 2018 and Pu et al., 2018), biological sensors, chemical sensors (Helman et al., 2008 and Tisch, 2013). A numerous substrates are formed by using nanowires treated with different solutions. Due to high surface area, the preparation of miniature devices is limited because nanoscale structures get exploited for the final product, especially textiles (Duan, 2010).

11.2.7 NANOGENERATORS

The supply of electrical energy in smart wears activated the smart functions in wearables and integrated textile products that charge the appliances. Energy harvesting is one of the emerging research field in smart textiles. Though lithium rechargeable batteries are well advanced, durable and comfort properties pose challenge to fulfill the end use (Song, 2018). Energy from the ambient source get dissipated in the body of the wearer is the way of power generation (Yang et al., 2017). Energy activation is performed by response mechanisms and the functions from smart wears. Also, selection of energy source is the biggest task and risk indeed. Textile application needs flexibility, energy density performance, less weight, and safety (Lin et al., 2017). Many researchers have concentrated to develop the power devices, such as supercapacitors and batteries to be included in textile materials. The novel energy-harvesting devices are designed based on the failures of batteries. These devices collect energy from the environment (mechanical energy, body regulation, and sunlight) and convert into electricity (Hyland M, 2016). Solar cells are one of the developments to source the electricity from solar energy, integrated in textile also. The supply of solar energy is limited and is based on season and weather conditions. Hence it is not a sustainable source for the supply of power (Pu, 2018 and Chen et al., 2016). In the presence of temperature gradient, thermal energy can be produced by thermoelectric generators, solid-state p- and n- semiconductors. Nanogenerators' output and efficiency are not enough to use them in smart textiles (Pu, 2018 and Lin et al., 2017). Mechanical sources have numerous advantages to source electricity, in the same power source, triboelectric and piezoelectric generators are the best examples. Li et al. constructed the triboelectric nanogenerators, they made from membranes with nanofibrous matrix using poly (vinylidene fluoride) coating of polydimethylsiloxane (PDMS) and PAN nanofibers coated on the surface of polyamides. Triboelectric nanogenerators work based on triboelectrification and electrostatic induction. Silk, PTFE and polyamides are stretchable and flexible materials are considered for this use (Wan, 2018 and Wang, 2017). Triboelectric nanogenerators having optimized design and materials with area power density of $500W/m^2$ and 85% of efficiency have been achieved (Zhu et al., 2014 and Xie et al., 2014). By mechanical energy, electromagnetic generators produce electricity, but, they are not used in textiles due to that they require heavy rigid magnet, movement of low frequency, and flexibility. On the other hand, in nanogenerators, different materials, flexibility, and low frequency performance can be used. One more fact that is important is the impact of nanotextiles on human health and environment. In future, new classes of materials and contact impact of human skin are factors to be considered in vivo testing. After the post-use and reuse of the product, production line wastes, wastewater management will be the biggest task to protect on behalf of environment and human society. Disposal of the material should be considered in priority. Structural bonding and nano/Micro encapsulation process will prevent the unwanted releases.

DNA damage and skin damages (tissue membrane impairs, digestive, skin reactions, nerve system impairment and respiratory disorders) are reliable effects by nanomaterials (Gashti, 2014). Direct mixing in the blood also is a negative point in the effect of nanomaterials (Yilmaz, 2016). In the case of nanogenerators, smart elements that harvest and store the energy are possible. But, not possible in conventional batteries. Nanogenerators utilized in textiles will render the fashionable effect, comfort, and versatility (Pu et al., 2018). Biomedicine, IOT (Internet of Things) are more hopeful trends in smart nanotextiles and also smart nanotextile has the combination or mixed effect of different disciplines.

11.3 NANOFIBERS FOR SMART TEXTILES

In the developed era, the society evolves. People are caring more about their life quality. Clothing as one of the necessities of life is drawing more and more attention. Besides the aesthetic requirement, functionality and adaptiveness become the most important selling points that most brands are competing on. From the manufacturer point of view, functional and adaptive textiles help them gain loyal customers and earn greater interests for the added value. From the selection point of view, garment or textile has involved subjective and objective factors to deal. Subjective functions focus on satisfying a wearer's aesthetic and comfort properties. Adaptive functions relate more to transitional conditions. An adaptive garment is required to satisfy a wearer's aesthetic and comfort requirements regardless of his/her activities and the varying environment. In the field of textiles, comfort is described mainly in terms of softness, smoothness, warmth/coolness, and dryness. Bulk properties are renewed and stiffness and roughness of a fabric are considered, which can be achieved by selection of a proper material. In fact, the textile materials having lower friction coefficient and get low Modulus (Young's)have shown good softness as well as smoothness also. On the other hand, warmth/coolness and dryness are more challenging to be realized, as they are always associated with activity and the environment. An adaptive garment is often required to be able to keep body temperature constant and be independent of the wearer's activity and the ambient environment, provide comfortness (Even with perspirations) prevention of rain wind, be easy caring, etc. The functionality factors are thermoregulations, moisture regulation, waterproof, wind proof, and self-cleaning. Special functions, such as biological and chemical protection are also needed in specific applications. Nowadays, the boundaries of smart textiles have been pushed wider as intelligent wearables have become an emerging technology. The future of wearables is believed to reside in smart fabrics with condition of where that fabric is used as a variety of contexts, from sports and fitness, protection, and health monitoring to chronic disease management (Ellison et al., 2007).

11.3.1   *PERSONAL PROTECTION*

Electrospinning has boomed as a nanofiber fabrication technology after the nanofiber research for soldier protection, which evoked people's enormous enthusiasm in exploring the potential applications of electrospun nanofibers. The concept of nanofiber-based fabrics that can provide ballistic protection against extreme cold, hot, and chemical environment. For commercial textiles, the research on nanofiber is more focused on body protection from environmental threats. Fire protection is one of the major functions of protective clothing. Currently, high-performance flame-retardant fibers, such as Nomex® and polybenzimidazole (PBI) have been used for firefighter and racer suits demanding excellent fire-proof. However, application of these fibers for mass-market consumer products is expensive. The potential of composite nanofibers with low-cost flame-retardant components has been explored considering the excellent thermal insulating property of porous nanofiber mats. Flame-retardant polyamide(PA) 6(nylon 6) nanocomposite nanofibers have been prepared by introducing montmorillonite. Their microscale combustion calorimeter testing results showed that FR particles played a major role in reducing flammability of the material in both solution- and melt-compounded samples and exhibited higher effectiveness as it is in melt-compounded samples. Through a typical same method, a multi component FR-carbon nanotube (CNT)-PA6 nanocomposite nanofibers have been produced (Yin et al., 2014). The combustion properties of the fibers including both heat release rate and total heat release were significantly improved while increasing the thermal stability. With proper FR additive concentrations, synergism between multiwalled carbon nanotubes (MWCNTs) and nanoclay was observed. Metal oxides are photocatalytic materials that absorb light, degraded by chemicals and destroy bacteria in addition to catalysts. ZnO is one of the most commonly used catalysts. By simply mixing ZnO nanoparticles with a polymer solution and electrospinning the mixture into fibers, UV-protective fiber mat was obtained. This ZnO-embedded nanofiber mat was layered onto fabric. The layered fabric systems obtained UV-protective property from the ZnO nanofiber mat. The protection effect increased with increasing zinc oxide concentrations of the nanocomposite fiber mat and the area density of the mat. Aiming to solve the aggregating problem, Vitchuli et al. combined electrospray process with electrospinning. Zinc nanoparticles that are dispersed over the Nylon 6 nanofiber surface. The prepared ZnO/ Nylon 6 nanofiber had excellent antibacterial efficiency (99.99%) against both Gram-negative *Escherichia coli* and *Gram-positive Bacillus cereus* bacteria, and it also showed 95% detoxifying efficiency against chemicals with high toxicity. Depositing this ZnO/Nylon 6 nanofiber mat onto nylon/ cotton woven fabrics did not impair the water vapor and air permeability of the fabrics significantly. Therefore, ZnO/Nylon 6 nano fibermat can be a promising material for protective applications in textiles. Besides ZnO, other antibacterial and chemical warfare agents, such as silver, activated charcoal, zirconium metalorganic frameworks (UiO-66-NH2), silica-encapsulated peppermint oil (PO), octenidine 2HCl (OCT), and chlorhexidinedigluconate were also investigated for electrospinning technology-based antibacterial and chemical decontamination materials. The electromagnetic interference (EMI) shielding effect of electrospun nanofibers can be gained in two ways: embodiment of reflection mechanism. Poyaniline-based (PANI) fiber were also prepared using conductive MWCNTs and polyethylene oxide (PEO) x. To improve the electromagnetic interference shielding efficiency, carbon nanotubes (CNTs) were fluorinated, then adhesion and dispersion is enhanced in polyaniline electrospun filaments. The fluorination treatment enhanced carbon nanotubes with electrical conductivity to reach the value of 4.8×10^3 siemens/m with an EMI shielding effectiveness (SE) of 42 dB. For better EMISE, materials having electrical conductivity and magnetic properties are combined. Kim et al., prepared EMI shielding conductive nanowebs by metallizing electrospun nanowebs with Cu, Ni, Ag in addition to PVA nanofibers. They found that the EMI SE was dominated by the absorption mechanism as the metal thickness strongly affected the SE value of each metal-deposited nanofiber mat. They believed that the unique nanofibrous porous structure also played an important role in enhancing the absorption effect. In another research, they prepared Ag-coated PVA/Fe_2O_3 composite nanofibers, and with the resultant fiber mat, they obtained an EMI SE as high as~45.2 dB. Since carbon, as a conductive fiber matrix, is more attractive for EMI shielding application, magnetic nanoparticles-incorporated nanofibers. By electrospinning ferrous acetate ($FeAc_2$)/PVA solution and subsequent calcination of the resultant fibers at high temperature, hybrid Fe_3O_4/carbon nanofibers were produced by Zhu et al. Zhu et al. (2006) directly dispersed Fe_3O_4 nanoparticles in PAN solution and fabricated Fe_3O_4/carbon nanofiber. The maximum EMI SE of 67.9 dB was obtained for a 0.7-mm-thick carbon nanofiber composite mat weighing 5% weight of Fe_3O_4. The shielding efficiency was also found to be dominated by absorption of electromagnetic radiatives. With respect to screening against telltale infrared heat radiation (IR radiation), Al-doped zirconia nanofibrous membranes were believed to be a promising one for defense applications, aircraft, satellites, etc. This material exhibited low infrared emissivity levels of 0.589 and 0.703 in the 35 and 814 m wavebands, respectively. Shielding efficiency of nanoparticles in air or liquid mainly relies on the geometry of the electrospun fiber mat. Fiber mat thickness was found to be the key parameter that strongly affects nanoparticle blocking/filtering results based on the Nylon 66 nonwoven substrates. In these studies, it was shown that the filtration efficiency could be increased to over 99% when the fiber mats were thicker in nature. By directly electrospinning PVDF/PTFE NP polymer solution into nanofiber, electret membrane by nanofiber with charges were successfully fabricated with PVDF as the matrix polymer and PTFE nanoparticles (PTFE NPs) as the charge enhancer. The resultant fibrous membrane exhibited a high filtration efficiency of 99.972% with a pressure drop of 57 Pa, as well as a satisfactory quality factor of 0.14 Pa1 and superior long-term service performance.

11.3.2   WEARABLE AND SENSORS

Wearables are smart electronic devices, such as electronics and sensors that can be worn on the body. Wearables are usually associated with fabrics for the sake of lightweight, long-lasting, flexible, conformable, and wearable devices. Commercially available wearables made with flexible textiles are still rare. In research, reports on wearables made with nanofibers mainly fall as energy-harvesting devices and storage devices, and sensors. A high-quality power source is essential for sustainable wearables. Besides, harvesting of energy by batteries from multiple sources that are available in our personal and daily environments, such as body movement, pressure, and friction is highly important. Various energy-harvesting systems have been successfully assembled. Piezoelectric materials are one of the most important multifunctional materials that have been extensively used for energy harvesting and sensor applications.

Many piezoelectric materials, such as lead zirconatetitanate (a piezoelectric ceramic, known as PZT) (Wu, 2012), PVDF (Fang, 2011) and P(VDF-TrFE) (Mandal, 2011) have been used to produce electrospun nanofiber web- for nanogenerator construction. The source is the selective factor on the harvester's functionality and performance (Anton, S.R. et al., 2007). PZT is widely used as a power-harvesting material with a high piezoelectric coefficient. However, ceramic which was used is brittle, so it is difficult to manipulate during the fabrication process and fatigue crack growth and cyclic loading occur with high frequency. Wu et al. (2012) successfully developed a suspending sintering technique for the fabrication of flexible, dense, and tough PZT textile composed of aligned parallel nanowires from electrospun nanofibers. This PZT textile was then transferred onto a PDMS-coated PET film and made into a flexible and wearable nanogenerator. The maximum output voltage and current of the nanogenerator were reported to be 6 V and 45 nA, respectively. This nanogenerator was demonstrated to light a commercial LCD and power a ZnO nanowire UV sensor to detect UV light. Another common piezoelectric material is PVDF. Thanks to the polymeric nature, PVDF exhibits considerable flexibility compared with PZT and other piezoelectric ceramics. PVDF is made by PVDF nanofibers. Single PVDF fibers get deposited across a pair of electrodes using a near-field electrospinning process (direct writing) to harvest small mechanical vibrations (Chang, 2010). The resulting nanogenerators exhibited a maximum voltage of 30 mV and current of 3 nA. The peak output voltage of randomly oriented PVDF nanofiber-based nanogenerator was reported to be 1 V (Gheibi et al., 2014). By massively depositing aligned PVDF fibers, Fuh (2013) successfully increased the peak output voltage and current of PVDF nanofiber-based nanogenerator to 1.7 V and 300 nA, respectively. The conversion of efficiency (mechanical to electrical energy) shows improvement in the increase in the crystal phase of PVDF with high electric field involving needleless electrospinning. The voltage output of randomly oriented PADF nanofiber-based nanogenerator increased from just less than 1 V to about 2.6 V, as the current boosted from around 1.4–4.5 A (Fang, 2013). Another type of electrospun nanofiber-based power generator worth mentioning is triboelectric nanogenerator (TENG). Triboelectrification which is usually taken as a negative effect. Triboelectric nanogenerators are the combination of electrostatic induction and electrification and they can utilize the most common materials available in our daily life, such as PTFE, PDMS, PVC, etc. (Wang, 2013; Wang, 2015 and Li, 2016). Using PDMS-coated electrospun PVDF and PA6-coated PAN nanofiber membranes as the power-generating materials, Li et al. fabricated a nanofibrous membrane-constructed wearable triboelectric nanogenerator. With an area of 16 cm^2 gentle hand tapping, this nanogenerator demonstrated a current and voltage output of up to 110 A and 540 V, respectively, and a capability of sustainably powering an electronic thermometer, an electronic watch, and 560 LEDs. A considerable amount of work on electrospun nanofiber-based sensors, such as chemical sensors, optical sensors, and biosensors, has been done (Macagnano et al., 2015; Ding et al., 2010; Ding B et al., 2009). Mechanical sensors are mainly based on the principle of change in the resistance/conductance of the incorporated conductive sensing component responding to external stimuli, such as pressure and strength/strain (Wang et al., 2011). Strain sensors are fabricated with conductive fibers (Sharma et al., 2013; Tiwari et al., 2008). When the fiber or fiber network is stretched or elongated, the resistance/electric conductance will increase/decrease. Elongation and resistivity are correlated and strain deformation is measured by measuring the resistivity of the sensor. Strain sensor, on the other hand, can be regarded/used as a pressure sensor as the sensor is elongated perpendicularly to the compressed direction. Pressure sensor can also be made with piezoelectric materials (Wang et al., 2011 and Ren et al., 2013). When pressure is applied to the pressure sensor, an electric signal is measured to quantify the pressure. In most cases, pressure sensors are sensitive to mechanical deformation (bending). Two soft objects are pressed one over the other and the mechanical stress is measured independently. To reduce the bending sensitivity, Lee et al. used carbon nanotubes and grapheme composite nanofibers to fabricate their pressure sensor. When compressed, these fibers are changing their alignments to withstand beding deformation . It reduces the individual fiber strain. The sensor has shown area, pressure, bending conditions, and time variables.

In order to broaden the utilization of displays in wearable electronics, transparent and flexible LEDs that can be integrated with fabrics suggest another field of wearables that requires development. Using keratin that is extracted from human hair, Park et al. fabricated a transparent nanofiber layer for the cover of flexible consolidated polymer light-emitting diodes (PLEDs). PLEDs maintained the device efficiencies similar to PLEDs without the cover layer. This research extends the application of electrospun nanofibers to transparent protection for wearable electronics.

11.4 NANOFIBER-BASED YARN AND FABRIC FORMATION

As electrospun nanofiber takes the form of a continuous linear shape, convert the nanoscale material into macroscale structures by textile processes. In electrospinning system with lower productivity, production of nanofiber yarns is mainly demonstrated in lab studies. Continuous nanofiber yarns have been fabricated by coating a core yarn/wire (Dabirian et al., 2012; He et al., 2014; Zhou et al., 2017 and Liu et al., 2016) with nanofibers or directly twisting nanofiber bundles (Fennessey et al, 2004; Teo et al., 2005; Ali et al., 2012; Ali et al., 2011; He et al., 2014 and Mokhtari et al., 2016). Fennessey and Farris twisted unidirectional aligned, structurally oriented PAN bundles into continuous yarns. The twisted yarn has higher angle of twist. Wang et al. combined self-bundling electrospinning technique with posttreatments such as stretching and annealing under conditions similar to those used for conventional fibers to fabricate nanofiber yarns. The strength of the obtained nanofiber yarns approached to values equivalent to conventional fibers as the crystallinity and molecular orientation of PAN nanofibers were improved. With regard to commercialization of electrospun nanofiber products, direct lamination of nanofiber possess the characteristics of high air permeability, programmable water proof, excellent sensibility, and high efficient filtration capability-enabled protective performance. A few attempts for fabrication of nanofiber mat-laminated fabrics have been conducted (Lee et al., 2006 & 2008; Bagherzadeh et al., 2012). The durability of nanofiber mat laminates, such as adhesion strength, tear and tensile strength and resistance against abrasion was found to be the greatest challenge due to the weakness of electrospun nanofibers (Knizek, 2017 & Heinisch et al., 2013). Solvent-laminated samples were found to have lower adhesive force, which stand for 10 washes, when hot-melting laminates have higher adhesive force, the adhesive force decreases quickly after wash. Rombaldoni et al. explored the potential of low-temperature oxygen plasma treatment for enhancement of various mats with PP nonwoven. The mean adhesion energy was found to have significantly increased from 0.58 and 0.39 J/m^2 to 4.80 and 0.89 J/m^2 for PEO and PA6 nanofibers, respectively. Chemical treatments on conventional fabrics for enhanced adhesion between the fabrics and PA6 nanofibers were also studied (Varesano, 2014). The treatments improved bonding of nanofibers toward fabrics, in particular, adhesion energy, for 60% and 51% on alkali-treated cotton fabrics and ethanol-treated nylon-66 fabrics, respectively. Adhesion is one-order-of-magnitude greater than these reported values (Varesano, 2014).

11.4.1 FABRICATION TECHNOLOGY

There are various nanofiber fabrication technologies that have been developed, among which only limited technologies can produce continuous nanofibers. Although electrospinning is regarded as one of the most promising techniques and is also considered a competitive technology. These technologies include but are not limited to composite spinning, melt blowing, CVD, and 3D printing. Composite spinning has been applied more and more often in textile industry for mass production of microfibers. Though the fiber diameter is still beyond 1000 nm, composite spinning has demonstrated its potential for creating nanofibers. One of the challenges might be the limitation created by the harsh textile fabrication processes. As the filament becomes finer, its tensile strength gets weaker. If the filament is too fine, then it will be too weak to withstand texturizing, weaving, or knitting on conventional textile equipment. Eastman has created an innovative way to generate finer fibers (Avra®) by protecting the ultrathin filament with a proprietary removable polymer: the filaments are extruded and held together by the proprietary polymer, weaving and knitting was washed away in hot water after the fabric is made. This innovative work demonstrated that synthetic filament with a diameter below 1 m may emerged. Melt blowing has proven to be capable of spinning nanofibers (Ellison, 2007). Though the fibers are currently restricted to the nonwoven format, the productivity of melt blowing is much higher than electrospinning. Besides, there is still a possibility for proper yarn-collecting systems to be invented if needed. CVD is one of the methods mainly involved in the production of discrete inorganic nanomaterials, such as nanoparticles, nanowires, nanotubes, and nano platelets, among which, CNT is a quite special product that has attracted many efforts in seeking of continuous yarn fabrication manners. It has been demonstrated that The CNT yarns could be made without an apparent limit to the length (Singh et al., 2003 and Li, 2004) by mechanical drawing of CNTs from the CVD synthesis dump zone of a furnace.

Different orientation of fibers, fiber density can be achieved through control of the process parameters, including carrier gas, rate of gas flow, and rate of winding. This direct spinning process allows one-step production of CNT yarns with potentially excellent properties (Koziol, 2007). It might not be applicable to organic materials, but should still be regarded as a potential approach for continuous nanofiber yarns. 3D printing, also known as additive manufacturing (AM), has gained immense popularity as a fast prototyping technology. The reported high-resolution 3D printing techniques are mainly based on two-photo polymerization (TPP) micro fabrication (TPPMF) and soft nanolithography (SL). TPPMF is one of the promising latest micro/nanoscale 3D manufacturing tool is speedy and flexible fabrication of arbitrary and ultra-precise 3D structures with sub-100-nm resolution (Mao et al., 2017). TPPMF is realized through continuous TPP by moving the focused beam in 3D route with the resolution beyond the optical diffraction limit (Kawata et al., 2001; Farsari et al., 2009 & Xing et al., 2015). Many typical and useful microstructures, such as photonic crystals (Sakellari, I; Kabouraki, E; Gray et al., 2017), mechanical devices, and 3D hydrogels have been fabricated by TPPMF. SL provides low-cost and convenient access to micro- and nanoscale 3-D structures with well-defined and controllable surface structures (Xia et al., 1998; Qin et al., 2010). A typical SL can be divided into four major steps: (1) pattern design, (2) fabrication of the mask and the master, (3) fabrication of the soft stamp, and (4) Fabrication of structures. SL has been widely applied for the generation of 3D surfaces, microfluidic devices, protein and cell patterning, microoptics, microsensors, microactuators, etc.

11.4.2 *CONDUCTIVE NANOWIRES*

Conductive textiles (fabric, thread, or yarn) are an important element in the realization of smart textiles. They can act as an interconnect between components to transfer signals or power, such as health-monitoring sensors (Rai et al., 2014) and solar cells (Chen et al., 2013). Meanwhile, it can also be used in the applications of electromagnetic interference (EMI) shielding, resistive fiber sensors, and wearable antennas, because of their own features. During the practical applications, these smart textiles still have to undergo great deformation, so the electrical conductivity against flexibility and stretch ability are critical for electrically conductive textiles. Commonly, solid metal wires, such as Cu and stainless steel, or a nonconductive thread coated with an ~1-m-thick metal film (e.g., Ag). These options for conductive textiles are less in stiff and brittle nature. Furthermore, they can easily breakdown (resistance greatly increases) after repeated bends (Lam Po Tang et al., 2006). With the rise of composite materials, a variety of conductive materials, including conductive polymers, graphene flakes, and carbon NTs are used as the conductive layers when commercial polymer fibers, such as nylon 6,6, Lycra, PU, and polyoliefins, polypyrrole (PPy) and poly (3,4 ethylene dioxythiophene): poly (styrene sulfonate) (PEDOT: PSS) are the two mostly explored conducting polymers because of their acceptable electrical conductivity and piezoresistive properties. They can be coated onto commercial polymer fibers to form a conductive layer through various techniques, such as solution dipping, in situ polymerization, and chemical vapor deposition (Xue et al., 2014). However, the conductive layers are easily degenerated during stretching because of the smooth surfaces of polymer fibers and they are also shows unstable nature due to the absorption of moisture and oxygen. Additionally, the conductivity of polymer-coated textiles is low and therefore cannot be used in many applications. One-dimensional nanostructure, such as wires, rods, belts, and tubes are produced on the surface having excellent electrical conductivity and mechanical flexibility. Unlike other nanostructure materials (e.g., NPs and nanosheets (NSs)), the elongated shape of NWs has lower particle density and fewer junctions. Then, the NWs connect with each other and form a conducting network that makes the conductive layer with high degree of bending and tensile properties. Meanwhile, the empty space between neighboring NWs provides transmittance for visible light (Doganay et al., 2016). Among various electrical conductive NWs, metal NWs and polymer NWs are widely used in smart textiles and are also the focus of this section.

11.4.3   METAL NANOWIRES

Since the successful syntheses of various metal NWs were reported, metal nanowires have become one of the promising candidates to replace the metal oxide (e.g., indium tin oxide) as transparent conductor. Similarly, carbon NTs with the same one-dimensional nanostructures are also competitive due to their metallicity and extremely low resistance. However, the junction resistance between two overlapping NTs in the film is very high, which leads to resistances per unit length of thread on the order of 1 kΩ cm^{-1} (Xue, 2007). NW coatings are in network state rather than continuous films, much fewer materials will be used to assemble films with the considerable optical and electrical properties. The NW coatings can be simply deposited without vacuum or complex processes. Currently, various metal NWs can be used as electrically conductive thread, including gold (Au), Ag, and Cu NWs.

11.4.4   POLYMER NANOWIRES

Conductive polymers, such as PPy, polyaniline (PANI), polythiophene (PTh), and poly-(p-phenylenevinylene) (PPV) are particularly appealing because of their flexibility, easy process ability, and potential for low-cost fabrication combined with unique electrical, electronic, and optical properties similar to metals or semiconductors. Polymer nanowires exhibit many excellent properties, such as increased electrical conductivity, size-dependent excitation or emission, easier band gap turn-ability, wide range of technological areas, such as sensors, energy storage, photo detectors, solar cells, electrolyte-gated transistors, and conversion devices. Besides, the flexible NW structure is quite suitable to be assembled on common textiles, such as cotton, polyester, flax, PET. Carbon fibers and Nylon are usually used for energy storage and solar cells. The natural fiber nonwoven is the flexible substrate, and conductive cloth was acquired by dipping the cloth into SWCNT ink (SWCNT/cloth). PANI NW arrays were then deposited on the substrates (pure cloth and SWCNT/cloth) by an in-suit dilute polymerization method in order to improve the capacitance of the electrodes. The sandwich-structured flexible supercapacitors with the electrodes of SWCNT/cloth, PANI/cloth, and PANI/SWCNT/cloth were produced. High capacitance of 410 Fg1 was obtained in PANI/SWCNT/cloth electrode-based supercapacitors, which was much higher than that of SWCNT/cloth (60 Fg1) and PANI/cloth (290 Fg1)-based supercapacitors. Two coated yarns were twisted together to form a two-ply yarn supercapacitor. Then, hierarchical PANI-coated $NiCo_2S_4$ NWs grown on carbon fibers ($NiCo_2S_4$ @PANI/CF) were fabricated through hydrothermal and electrodeposition method (Liu, 2017).

11.4.5 SEMICONDUCTING NANOWIRES

Semiconductor NWs have been studied extensively for over 20 years due to their novel electronic, photonic, thermal, electrochemical, and mechanical properties. They can enable the read of diagnostic assays with consumer electronic devices, such as cell phones, smart phones, wearable electronics etc. Their properties of novel subwavelength optical phenomena, large tolerance for mechanical deformations, compatibility with other micro/nano scopic systems, length-scale decoupling associated with different physical phenomena in radial and axial directions, high surface-to-volume ratio, and so on, have brought a wide range of applications. In order to achieve the multifunction of smart textiles, various semiconductor NWs have been also assembled on appropriate fabrics by microfabrication and dip-coating techniques. Furthermore, concerning various active materials, current smart textiles can be of nanowires of oxide, sulfide, and others.

11.4.6 OXIDE NANOWIRES

Among the inorganic semiconductor nanomaterials, metal oxide NW is the focus of research efforts in nanotechnology since they exhibit excellent chemical and physical properties. They have been widely used in many areas, such as chromic devices such as optical and electrical, transducers from piezo electric, sensors, ceramics (Kolmakov et al., 2004 and Hul et al., 2011). Integration of oxide NWs into cotton/textiles to form wearable devices is most significant. Meanwhile, metal oxide NWs, such as ZnO, SnO_2, In_2O_3, etc. are synthesized by facile electrodeposition method (Hul et al., 2011) or by in situ hydrothermal synthesis (Yuan et al., 2011 and Zhai et al., 2009). ZnO NWs are perhaps the most widely studied materials among highly conductive materials with the use of external dopants. Meanwhile, intrinsic defects, such as oxygen vacancies and zinc interstitials act as donors, which make the electrical conductivity of ZnO easy to control (Janotti et al., 2007).

11.4.7   SULFIDE NANOWIRES

Similar to the oxides mentioned above, semiconducting metal sulfide NWs show electrical and optical properties, photovoltaic properties, such as dye-sensitized cells, all inorganic NP solar cells, and hybrid nano crystal–polymer composite solar cells. Some of the sulfide nanowires are also combined with textiles including Ni 2.3% CoS_2 (Fang et al., 2016), Ni3-xCoxS4 (Ding et al., 2015), Cu_7S_4 (Javed et al., 2015), and CoS_2 (Huang et al., 2015) for electrochemical devices,. In the early stage of sulfide NW smart textiles, Li et al. synthesized well-aligned NiCo sulfide NW arrays with a NiCo molar ratio of 1:1 on three-dimensional nickel foam by a facile two-step hydrothermal method (Li et al., 2015). Owing to the low electro negativity of sulfur, NiCo sulfide nanowires show more flexible structure and higher conductivity when compared with NiCo oxide NW arrays when used as active materials in supercapacitors. More importantly, the asymmetric supercapacitor is composed of NiCo sulfide NW arrays as positive electrode. NW arrays are promising for high-performance supercapacitor applications.

11.4.8   OTHER NANOWIRES

In addition to oxides and sulfides, there are still other semiconductor NWs, such as ZnSe, CdSe, CdTe, InP, GaAs, PbS, PbSe, PbSexS1-x, Si, and potassium vanadate (KVO). KVO NWs have been assembled on textiles or fibers, and few works have been done about NW-based smart textiles or fibers. In 2007, the Si NWs were first grown on CC via the vapor–liquid–solid reaction using silane gas as the Si source and Au has been used as a catalyst decomposition of hydrogen in the form of Au tetrachloride. A low operating electric field has been achieved with the emission current density of 1 mA/cm^2 at an operating electric field of 0.7 V/m. The small operating electric field has resulted from a high field factor with high aspect ratio and geometry.

11.5 TEXTILE-BASED PENGS

Bai et al. developed a two-dimensional woven nanogenerator, the fabric PENG was woven from two kinds of fibers. The working principle of the fabric PENG is similar to the fiber PENG, only that the two contacting fibers are perpendicular to each other. The tiny mechanical energies, such as sound and wind are influenced to generate electrical energy by fabric PENG. The open-circuit voltage and short-circuit current of the woven PENG were 3 mV and 17 pA. Zhang et al. also reported a woven fabric PENG. Cu wires were woven into a configuration of interdigitated electrodes.

PVC fibers grown with $BaTiO_3$ nanowires were laid by weaving with Cu wires but were perpendicular to the Cu finger electrodes. The fabric PENG when wornaround the elbow pad, can give output 1.9 V voltage and 24 nA current, and can power an LCD. Khan et al. reported a textile-PENG with ZnO nanowire arrays grown on a commercial conductive fabric (Argen Mesh), which was made by weaving 55% silver and 45% nylon fibers with a final resistivity of <1 /sq. A chemical aqueous bath in low temperature is enough to form ZnO nanowires on textile substrates. The growth density of ZnO nanowires was measured to be 240±50 nanowires/m^2 with a dominant c-axis direction. A nanoindentation technique was adopted for measuring mechanical and piezoelectric behaviors of ZnO nanowires. The gener-ated output potential increased from 0 to 0.0048 V when the applied force increases from 0 to 300 mN. Other than chemical growing methods, screen printing had also been considered. Almusallam et al. reported a fabric PENG that used piezoelectric nanocomposite film on flexible textile substrates. The particles of Ag, PZT, and polymer composite ink have been printed by screen printing on a conductive electrode-coated textile. Another Coating of top electrode was applied to cover the piezoelectric composite film. The maximum energy density of the PENG under 800 N compressive forces was reported to be of 34 J/m^3 on kermel textile substrate, while for bending forces, the maximum energy density was found to be 14.3 J/m^3 on a cotton textile. Polymeric PVDF is an ideal piezoelectric material for textile-based PENG. Soin et al. designed three-dimensional piezoelectric fabrics based on 3D spacer textile technology, Piezoelectric PVDF fibers functioned as spacer yarns, which joined together and separated two knitted substrate fabrics as two electrodes. Yarns were used for fabricating the top and bottom conductive fabric electrodes. The PVDF spacer yarns were manufactured by a continuous melt-spinning extrusion method. At an optimized drawing ratio, a high content of phase (80%) was observed. The polyester yarns have two functions: providing mechanical reinforcement to the 3D structure, and insulating the top and bottom conductive electrodes. The 3D spacer fabric PENG exhibited a power density in the range of 1.10 to 5.10 W/cm^2 at applied impact pressures of 0.02–0.10 MPa. Zeng et al. reported a fabric PENG with simple stacked configuration. $NaNbO_3$ and PVDF composite nonwoven fabrics were synthesized by an electrospinning process. The mass ratio of $NaNO_3$/PVDF in the precursor mixture was 5:100. The piezoelectrical composite fabric was then sandwiched between two elastic conductive electrode fabric and woven together with segmented polyurethane multifilament yarns and Ag-coated polyamide multifilament yarns. The fabric PENG was then tested under a compressive pressure of 0.2 MPa. StableVoc and Isc are measured to be 3.4 V and 4.4 mA, respectively. Fiber/textile-based PENGs are viable with various designs. Ceramic wires are better flexibility than bulk counterparts. Piezoelectric polymers, such as PVDF, are ideal for fiber/textile-PENGs due to their better flexibility and facile synthesis into yarns/textiles. Piezoelectric composite materials combining polymers and ceramics have also been used in textile-based PENG.

11.5.1 TRIBOELECTRIC (NANO) GENERATORS

Due to the universality of triboelectrification effect, it is feasible to integrate a TENG directly into a fabric or cloth. Commonly used materials in textiles can be good candidates for triboelectrification with strong triboelectric effects, including the natural leather, silk, cotton, and synthetic polymers (polyester, nylon, etc.). Furthermore, yarns and textiles are composed of numerous microsized fibers, so the effective surface area for generating tribo-charges is large. A rich source of mechanical energy is human body and clothing is a medium for the TENG, so that the human motion energies can be harvested for sustainable power supply to various electronic devices.

11.5.2 TEXTILE 1D YARNS/FIBERS

A textile-based TENG can be generally designed by two approaches: (1) bottom up approach, (2) direct approach of 2D-treated fabrics. Strips of commercial silver fiber fabric were sandwiched in nylon and polyester strips, respectively. Subsequently, the two strips were woven into a fabric as wefts and warps. All weft silver fabric strips serve as one electrode and warp in silver. With contact separation motion relative to another fabric, an alternating current can be generated through external circuit between the two electrodes in the textile TENG. A short-circuit current (Isc) of 0.5 A and open-circuit voltage (Voc) of 27 V were observed on polyester fabric. Electricity generation has also been demonstrated when attaching the textile TENG to the moving parts of a human body, such as feet, legs, and arms. This TENG textile was composed of common textile materials; therefore, excellent flexibility and comfort can be expected. Pu et al. used electroless deposition method to apply a layer of conformal Ni coating on a wavy polyester textile so as to convert insulating polyester textile into conductive material. Then, a thin layer of parylene was subsequently coated on the Ni-coated textile by a chemical vapor deposition (CVD) process. The woven Ni-coated strips and parylene Ni-coated strips acted as wefts and warps, respectively. The weft and warp Ni coatings were connected through an external circuit, the TENG cloth was made of strips, all the processes were applicable to yarns, fibers, or threads. Comparing with typical conductive textiles dip-coated with CNTs or graphene, the conductivity of Ni cloth is found to be higher. Meanwhile, the whole process is of low cost and suitable for scale-up synthesis. Furthermore, the woven TENG cloth retained the mechanical flexibility, air breathability, water washability, and comfortability of the original polyester cloth. Zhao et al. attempted Cu-coated on the PET yarns and polyimide (PI)Cu-coated PET yarns and used them as the wefts and warps, respectively. They also conducted machine wash experiments to demonstrate the washability of the textile TENG. The textile TNG withstands 20 washes and linear resistance of the Cu–PET yarn was maintained as <0.6/cm. Output current of TENG is reduced from 13.78 to 3.52 mA/m^2 after the first three washes, but maintained 92% after the following wash cycles. Lai et al. reported a highly stretchable single-thread textile TENG. Stainless steel coated with rubber was sewn in an elastic textile substrate into a serpentine shape, so the whole final textile can be stretched to a tensile strain up to ~100%. The radius of the fabricated thread is about 750 m. The working mode was similar to a contact separation mode, but the human skin served both as the positive material in the tribo-series and a reference electrode. The generating electric output reached peak values up to 200 V and 200 A. The capability to harvest different kinds of mechanical energy from human body, including the movement of joints, walking, tapping, etc., was demonstrated. The above textile TENGs are of 2D weaving structures. Dong et al. have reported a 3DOW structure for higher power output. They utilized 3-ply-twisted stainless steel (SS)/ polyester fiber-blended yarn (warp yarn), polydimethylsiloxane (PDMS)-coated energy-harvesting yarn (weft yarn), and binding yarns in thickness direction (Z-yarn) for obtaining the 3DOW structure.

One layer of SS/polyester fibers was sandwiched in two layers of parallel PDMS-coated yarns. Z-yarns were woven to combine the top/bottom weft yarns and middle warp yarns. The 3D configuration of the design was also observed. Single-electrode modes, including d1 (2D woven, one dielectric layer), d2 (2D woven, two dielectric layers), d31 (3DOW, weft-connection single-electrode mode), and d33 (3DOW, warp-connection single-electrode mode) showed much smaller Isc and Voc than the double-electrode mode d32. Maximum peak power density can reach 263.36 mW/m^2 under a tapping frequency of 3 Hz. The 3DOW structures are used for lighting up a warning indicator charging a commercial capacitor, driving a digital watch, and tracking human motion signals. This study demonstrated that proper design of the weaving method was also an effective approach to improve the performances of textile TENG.

11.5.3 TEXTILE 2D FABRICS

Two-dimensional (2D) fabrics are treated with conductive coatings and dielectric films for TENGs. 2D-treated fabrics can be directly used as electrodes without any extra weaving or sewing processes. Seung et al. successfully demonstrated the excellent durability mechanical textile based nano-patterned wearable TENG. Aligned ZnO nanorods were grown on Ag-coated textile by a hydrothermal process, followed by dip-coating a layer of PDMS on the top. The nanopatterned morphology of the ZnO nanorods was maintained for the final PDMS coatings. Therefore, when having contact separation motion relative to another Ag textile, the effective contacting area for triboelectrification was enhanced. According to the comparing experiment, a high output voltage and output current of about 120 V and 65 A, respectively, were observed for a nanopattern. TENG has output current of 30V and 20 A, respectively, obtained flat PDMS coating under the same mechanical compressive force. Finally, they demonstrated the self-powered operation of LEDs, an LCD, and vehicle scanning system with output power of the textile TENG. In some textile TENGs, the structure should be stacked by attaching conductive textiles with common nonconductive textiles, but the polymer films or sealing tapes will make the whole device not air-breathable. Zhang et al. adopted a simple approach to sew conductive and dielectric textiles together with a stacked structure and sewed at the backside of cotton/nylon fabric. A process of salinization was further conducted to introduce functional groups on the surface of the cotton fabric. Trichloro (1H, 1H, 2H,2H-perfluorooctyl) silane was used to transform normally neutral cotton fabrics into a strong negative triboelectric, due to the grafted polymeric fluo-roalkylated siloxane film on cotton. Considering nylon is a strong positive triboelectric material, the final device can have larger triboelectrification.

11.6   FUNCTIONS OF NANOCOMPOSITES

11.6.1   SENSORS

Sensors are smart component for devices whichever is applicable to measuring a physical quantity and converting it into a signal that can be detected by a human observer or an electronic device. Electrode takes a critical part in the sensor determining the sensitivity and efficiency. Mediators facilitate ion transfer between the analyte and the electrode. Composition of mediator materials may includedifferent materials, such as conducting polymers, metallocenes, and conductive nanocellulose composites. Electrode surface is covered by mediators taking place between the analytic and the electrode. The properties of the mediators heavily influence the response rate of the sensor. Low response rate is a very critical issue in terms of sensor applica-tions. In bulk polymers, the response rate is low with the time necessary for target molecules to penetrate into the polymer. However, in the case where nanostructuring is achieved via nanofibers or nanotubes, response rate can be highly increased. These render nanocomposites ideal materials for use in sensors (Abdi, 2013). Sensors, as key elements of smart devices, can function based on electrical, mechanical, or optical mechanisms. They can be subgrouped as electrochemical, piezoelectric, acoustic, fluorescent, and colorimetric sensors, etc. Biosensors constitute an important class of sensors capable of sensing biochemicals including glucose, estrogen, and urea. The electrochemical sensor can produce an electrical signal from a concentra-tion of enzyme used as chemical species. Nanocellulose has found use in nanocomposites for biosensor applications (Abdi, 2013). Glucose biosensors show new hope in diabetes treatment research. Here, glucose oxidase enzyme is used (Yilmaz et al., 2015). An active layer of this enzyme is used to react with glucose and transmits current into the electrode. Here, the strength of the electric signal produced at the electrode is proportional to the concentration of glucose. Another biosensor application of nanocellulose is in wound care via detecting elastase, a biomarker for inflammatory diseases (Edwards, 2013). Nag et al. developed a flexible wearable sensor for monitoring respiration and other body motions. The multilayered sensor comprises a polydimeth-ylsiloxane layer and a membrane of nanocomposite including functionalized multiwalled carbon nanotubes. The polydimethylsiloxane and the carbon nanotubes acted as electrodes. strain sensitivity is the main source for the function of sensor. The thickness was selected to give the best strain and conductivity behavior. Polydimethylsiloxane was selected based on its cost efficiency, hydrophobicity, inert structure, and nontoxic nature. Carbon nanotube was preferred based on thermal stability and tensile strength.

11.6.2   DEFENSE APPLICATIONS

For body armor and their improvement, nanocomposites are considered. Shear thickening fluids are used in body armors. When under shear, the speedy stiffening response occurred in shear thickening fluids. Recent work has been reported to improve the shear stiffening effect by using Silica with PEG fluid matrix. PEG fluid is applied on Kevlar fabric and also has led to higher energy absorption under impact (Talreja, 2017). Nanocomposites also have found use in microwave protection. In a related study, a two-ply nanocomposite thickness under 1 mm is prepared. One layer is formed of polyaniline (absorbing layer) and the other is formed of polyanilinemagnetite (Fe_3O_4; matching layer). The nanocomposite was reported to have better microwave absorption in comparison to single-layer absorbers (Xu, F, 2015). In this nanocomposite structure, polyvinylpyrrolidone acted as the matrix. The nanocomposite exhibits double functionality, such as electrical conductivity and magnetic effect. The developed system may find use in electromagnetic interference shielding and microwave absorption (Yang et al., 2016).

11.6.3   FIRE PROTECTION

Polymers generally exhibit poor resistance to fire. During ignition, they burn quickly and emit toxic fumes and heat. Incorporation of nanoclay leads to substantial increase in flame-retardancy. Nanoclay addition results in substantial reduction in the rate of burning and interferes with volatile diffusion in air.

11.6.4   ACTUATORS

Nanocomposites enabled the development of actuators that minimize the material and energy consumption. Lower particle dimensions, higher surface areas, and lower nanofiller loading rates allow the production of cost-efficient actuators. Actuators have the capability to function in hostile environments that humans cannot withstand. An actuator of nanocomposites including cellulose, polypyrrole, and ionic liquid was developed. Here, wet cellulose films were produced from cotton pulp via spin coating. Then, pyrrole was polymerized and adsorbed on cellulose; further, activation was carried out in ionic liquid of 1-butyl-3-methylimidazolium chloride solution. Higher mobility and conductivity of nano composites compared with pristine cellulose, conventional electro active paper, and conducting polymer (Mahadeva et al., 2011).

11.6.5 SELF-CLEANING

Another function that nanocomposites can serve is self-cleaning. Self-cleaning mechanism can be induced by both UV and visible light via use of nanocomposites. TiO_2 presents promising effectivity for photocatalytic stain removal. Exposure to ultraviolet radiation for longer period is necessary whereas success under visible light is very limited. Impartation of titania with nanoparticles, dyes, and compounds including SiO_2 (Xu et al., 2013) is performed. New development of nanocomposite system is TiO_2 and Porphyrin to achieve photocatalytic self-cleaning. Self-cleaning effect is induced by visible light due to the presence of porphyrin. Porphyrin dye molecules are excited under visible light and electrons are injected to the conduction band of titania. This leads to the formation of superoxide anions, which take part in stain decomposing. For the improvement of porphyrin durability against light, coating of metals as a layer is done. Superhydrophobicity also contributes to self-cleaning. As the water contact angle on a surface exceeds 150°, the surface is considered as superhydrophobic. Micro/nano surface roughness will result superhydrophobicity to mimicking lotus leaves as well as using components of hydrophobic chemistry such as fluorosilane (Xu et al., 2013).

11.7 CONCLUSION

Textile properties that are interconnected to electronics and intertwined by their function of applications. In this chapter, we reviewed critically a wide range of nano elements that are focused with high-performance application. Globally, many of the researchers have attempted to develop nanomaterials to obtain miniature or microelectronics. This new enhancements followed under every step of development will cause value-added novel products. It will lead the world with healthier and make everyday life with harmless and also to provide satisfaction. Many of the interfaces will form the bridge between human and machine elements. In future, technology should come much closer to human life, such as IOT, AI, etc., from the economic point of view, normal clothing is much less cost than the smart cloths. Totality of the smart textiles gets flourished and it is hopeful for further achievements in textile field.

KEYWORDS

- **nanotechnology**
- **nanoelectronics**
- **power textiles**
- **graphene yarns**
- **wearable textiles**
- **energy scavenging fabrics**
- **nanofinishing**
- **technical textiles**
- **nanomaterials**

REFERENCES

Abdi, M. M.; Abdullah, L. C.; Tahir, P. M.; Zaini, L. H. Cellulosic Nanomaterials for Sensing Applications. In *Handbook of Green Materials. 3. Self and Direct Assembling of Bionanomaterials*; Oksman, K., Mathew, A. P., Bismarck, A., Rojas, O., Sain, M., Eds.; World Scientific Publishing, 2014; pp 197–212.

Ali, U.; Zhou, Y.; Wang, X.; et al. Electrospinning of Continuous Nanofiber Bundles and Twisted Nanofiber Yarns. In *Nanofibers—Production, Properties and Functional Applications*; Lin, T., Ed.; InTech, 2011.

Ali, U.; Zhou, Y.; Wang, X.; et al. Direct Electrospinning of Highly Twisted, Continuous Nanofiber Yarns. *J. Text. Ind.* **2012,** *103*, 80–88.

Anton, S. R.; Sodano, H. A. A Review of Power Harvesting Using Piezoelectric Materials (2003–2006). *Smart. Mater. Struct.* **2007,** *16*, R1.

Bagherzadeh, R.; Latifi, M.; Najar, S. S.; et al. Transport Properties of Multilayer Fabric Based on Electrospun Nanofiber Mats as a Breathable Barrier Textile Material. *Text. Res. J.* **2012,** *82*, 70–76.

Berendjchi, A.; Khajavi, R.; Yazdanshenas, M. E. Application of Nanosols in Textile Industry. *Int. J. Green. Nanotechnol.* **2013**, *1*, 1–7.

Buguin, A.; Li, M.-H.; Silberzan, P.; et al. Micro-actuators: When Artificial Muscles Made of Nematic Liquid Crystal Elastomers Meet Soft Lithography. *J. Am. Chem. Soc.* **2006**, *128*, 1088–1089.

Burroughes, J.; Bradley, D.; Brown, A.; Marks, R.; Mackay, K.; Friend, R.; Burns, P.; Holmes, A. Light-emitting Diodes Based on Conjugated Polymers. *Nature* **1990**, *347*, 539–541.

Chang, Y.; Lye, M. L.; Zeng, H. C. Large-scale Synthesis of High-Quality Ultralong Copper Nanowires. *Langmuir* **2005**, *21*, 3746–3748.

Chang, C.; Tran, V. H.; Wang, J.; Fuh, Y.; Lin, L. Direct-Write Piezoelectric Polymeric Nanogenerator with High Energy Conversion Efficiency. *Nano Lett.* **2010**, *10*, 726–731.

Chen, T.; Qiu, L.; Yang, Z.; Peng, H. Novel Solar Cells in a Wire Format. *Chem. Soc. Rev.* **2013**, *42*, 5031–5041.

Cheng, H.; Dong, Z.; Hu, C.; Zhao, Y.; Hu, Y.; Qu, L.; Chen, N.; Dai, L. Textile Electrodes Woven by Carbon Nanotube–Graphene Hybrid Fibers for Flexible Electrochemical Capacitors. *Nanoscale* **2013**, *5*, 3428–3434.

Clark, D. P.; Pazdernik, N. J. Nanobiotechnology. In *Biotechnology*, 2nd ed.; Clark, D. P., Pazdernik, N., Eds.; Elsevier, 2016.

Curtis, A.; Wilkinson, C. Nanotechniques and Approaches in Biotechnology. *Mater. Today* **2001**, *4*, 22–28.

Dabirian, F.; Ravandi, S. A. H.; Hinestroza, J. P.; et al. Conformal Coating of Yarns and Wires with Electrospun Nanofibers. *Polym. Eng. Sci.* **2012**, *52*, 1724–1732.

Das, A.; Krishnasamy, J.; Alagirusamy, R.; Basu, A. Electromagnetic Interference Shielding Effectiveness of SS/PET Hybrid Yarn Incorporated Woven Fabrics. *Fiber Polym.* **2014**, *15*, 169–174.

Desai, T. Micro- and Nanoscale Structures for Tissue Engineering Constructs. *Med. Eng. Phys.* **2000**, *22*, 595–606.

Ding, B.; Wang, M.; Yu, J.; Sun, G. Gas Sensors Based on Electrospun Nanofibers. *Sensors* **2009**, *9* (3), 1609–1624.

Ding, B.; Wang, M.; Wang, X.; Yu, J.; Sun, G. Electrospun Nanomaterials for Ultrasensitive Sensors. *Mater. Today* **2010**, *13* (11), 16–27.

Ding, R.; Gao, H.; Zhang, M.; Zhang, J.; Zhang, X. Controllable Synthesis of $Ni_{3-x}Co_xS_4$ Nanotube Arrays with Different Aspect Ratios Grown on Carbon Cloth for High-Capacity Supercapacitors. *RSC Adv.* **2015**, *5*, 48631–48637.

Doganay, D.; Coskun, S.; Genlik, S. P.; Unalan, H. E. Silver Nanowire Decorated Heatable Textiles. *Nanotechnology* **2016**, *27*, 435201.

Duan, X. Nanowire Thin Films for Flexible Macro Electronics. In *Encyclopedia of Materials: Science and Technology*, 2nd ed.; Buschow, K. H. J., Cahn, R. W., Flemings, M. C., Ilschner, B., Kramer, E. J., Mahajan, S., Veyssiere, P., Eds.; Elsevier, 2010; pp 1–10.

Avra™. *Eastman Launches Avra™ Performance Fibers, Introducing a Polyester Fiber with a Unique Combination of Size, Shape, and Performance to the Active Apparel Market*, 2017. http://www.eastman.com/Company/News_Center/2016/Pages/Eastman-launches-Avra-Performance-Fibers.aspx.

Ebara, M. *et al. Smart Biomaterials*; Springer: Japan, 2014.

Edwards, J. V.; Prevost, N.; French, A.; Concha, M.; DeLucca, A.; Wu, Q. Nanocellulose-Based Biosensors: Design, Preparation, and Activity of Peptide-linked Cotton Cellulose Nanocrystals Having Fluorimetric and Colorimetric Elastase Detection Sensitivity. *Engineering* **2013**, *5*, 20–28.

Ellison, C.; Phatak, A.; Giles, D.; Macosko, C.; Bates, F. Melt Blown Nanofibers: Fiber Diameter Distributions and Onset of Fiber Breakup. *Polymer* **2007,** *48,* 3306–3316.

Fan, F. R.; Tian, Z. Q.; Wang, Z. L. Flexible Triboelectric Generator. *Nano Energy* **2012,** *1,* 328–334.

Fang, J.; Wang, X.; Lin, T. Electrical Power Generator from Randomly Oriented Electrospun Poly (Vinylidene Fluoride) Nanofibre Membranes. *J. Mater. Chem.* **2011,** *21,* 11088–11091.

Fang, J.; Niu, H.; Wang, H.; Wang, X.; Lin, T. Enhanced Mechanical Energy Harvesting Using Needleless Electrospun Poly(Vinylidene Fluoride) Nanofibre Webs. *Energy Environ. Sci.* **2013,** *6* (7), 2196–2202.

Fang, W.; Liu, D.; Lu, Q.; Sun, X.; Asiri, A. M. Nickel Promoted Cobalt Disulfide Nanowire Array Supported on Carbon Cloth: An Efficient and Stable Bifunctional Electrocatalyst for Full Water Splitting. *Electrochem. Commun.* **2016,** *63,* 60–64.

Farsari, M.; Chichkov, B. N. Materials Processing: Two-Photon Fabrication. *Nat. Photonics* **2009,** *3,* 450.

Fennessey, S. F.; Farris, R. J. Fabrication of Aligned and Molecularly Oriented Electrospun Polyacrylonitrile Nanofibers and the Mechanical Behavior of Their Twisted Yarns. *Polymer* **2004,** *45,* 4217–4225.

Flemming, R.; Murphy, C.; Abrams, G.; et al. Effects of Synthetic Micro-and Nano-structured Surfaces on Cell Behavior. *Biomaterials* **1999,** *20,* 573–588.

Fuh, Y. K.; Chen, S. Y.; Ye, J. C. Massively Parallel Aligned Microfibers Based Harvester Deposited via *in Situ,* Oriented Poled Near-Field Electrospinning. *Appl. Phys. Lett.* **2013,** *103,* 033114.

Gashti, M. P.; Alimohammadi, F.; Song, G.; Kiumarsi, A. Characterization of Nanocomposite Coatings on Textiles: A Brief Review on Microscopic Technology. *Current Microscopy Contributions to Advances in Science and Technology*; Formatex Research Centre: Badajoz, Spain, 2012; pp 1424–1437.

Gashti, M. P.; Pakdel, E.; Alimohammadi, F. Nanotechnology-based Coating Techniques for Smart Textiles in Active Coatings for Smart Textiles. Elsevier Ltd., 2016; pp 243–268.

Garnett, E. C.; Cai, W.; Cha, J. J.; Mahmood, F.; Connor, S. T.; Christoforo, M. G.; Cui, Y.; McGehee, M. D.; Brongersma, M. L. Self-Limited Plasmonic Welding of Silver Nanowire Junctions. *Nat. Mater.* **2012,** *11,* 241–249.

Gheibi, A.; Latifi, M.; Merati, A. A.; Bagherzadeh, R. Piezoelectric Electrospun Nanofibrous Materials for Self-Powering Wearable Electronic Textiles Applications. *J. Polym. Res.* **2014,** *21,* 469.

Guo, F.; Xu, X. X.; Sun, Z. Z.; Zhang, J. X.; Meng, Z. X.; Zheng, W.; Zhou, H. M.; Wang, B. L.; Zheng, Y. F. A Novel Amperometric Hydrogen Peroxide Biosensor Based on Electrospun Hb-Collagen Composite. *Colloids Surf., B: Biointerfaces* **2011,** *86* (1), 140–145.

He, J. X.; Qi, K.; Zhou, Y. M.; et al. Fabrication of Continuous Nanofiber Yarn Using Novel Multi-nozzle Bubble Electrospinning. *Polym. Int.* **2014a,** *63,* 1288–1294.

He, J. X.; Zhou, Y. M.; Wu, Y. C.; et al. Nanofiber Coated Hybrid Yarn Fabricated by Novel Electrospinning-airflow Twisting Method. *Surf. Coat. Technol.* **2014b,** *258,* 398–404.

He, W.; Wang, C.; Zhuge, F.; Deng, X.; Xu, X.; Zhai, T. Flexible and High Energy Density Asymmetrical Supercapacitors Based on Core/Shell Conducting Polymer Nanowires/ Manganese Dioxide Nanoflakes. *Nano Energy* **2017,** *35,* 242–250.

Heinisch, T.; Bajzík, V.; Knížek, R.; et al. Effect of the Process of Lamination Microporous Nanofiber Membrane on the Evaporative Resistance of the Two-Layer Laminate. In *Advanced Materials Research*; Trans Tech Publ., 2013.

Helman, A.; Borgström, M. T.; Van Weert, M.; Verheijen, M. A.; Bakkers, E. P. A. M. Synthesis and Electronic Devices of III–V Nanowires. In *Encyclopedia of Materials: Science and Technology*, 2nd ed.; Buschow, K. H. J., Cahn, R. W., Flemings, M. C., Ilschner, B., Kramer, E. J., Mahajan, S., Veyssiere, P., Eds.; Elsevier Ltd., 2008; pp. 1–6.

Hu, L.; Chen, W.; Xie, X.; Liu, N.; Yang, Y.; Wu, H.; Yao, Y.; Pasta, M.; Alshareef, H. N.; Cui, Y. Symmetrical MnO_2–Carbon Nanotube-Textile Nanostructures for Wearable Pseudocapacitors with High Mass Loading. *ACS Nano* **2011**, *5*, 8904–8913.

Huang, Y.; Paloczi, G. T.; Yariv, A.; et al. Fabrication and Replication of Polymer Integrated Optical Devices Using Electron-Beam Lithography and Soft Lithography. *J. Phys. Chem. B* **2004**, *108*, 8606–8613.

Huang, J.; Hou, D.; Zhou, Y.; Zhou, W.; Li, G.; Tang, Z.; Li, L.; Chen, S. MoS_2 Nanosheet-coated CoS_2 Nanowire Arrays on Carbon Cloth as Three Dimensional Electrodes for Efficient Electrocatalytic Hydrogen Evolution. *J. Mater. Chem. A* **2015**, *3*, 22886–22891.

Hsu, P. C.; Wang, S.; Wu, H.; Narasimhan, V. K.; Kong, D.; Lee, H. R.; Cui, Y. Performance Enhancement of Metal Nanowire Transparent Conducting Electrodes by Mesoscale Metal Wires. *Nat. Commun.* **2013**, *4*, 2522.

Hyland, M.; Hunter, H.; Liu, J.; Veety, E.; Vashaee, D. Wearable Thermoelectric Generators for Human Body Heat Harvesting. *Appl. Energy* **2016**, *182*, 518–524.

Janotti, A.; Van de Walle, C. G. Native Point Defects in ZnO. *Phys. Rev. B* **2007**, *76*, 165202.

Javed, M. S.; Dai, S.; Wang, M.; Xi, Y.; Lang, Q.; Guo, D.; Hu, C. Faradic redox Active Material of Cu_7S_4 Nanowires with a High Conductance for Flexible Solid State Supercapacitors. *Nanoscale* **2015**, *7*, 13610–13618.

Jin, L. N.; Shao, F.; Jin, C.; Zhang, J. N.; Liu, P.; Guo, M. X.; Bian, S. W. High-Performance Textile Supercapacitor Electrode Materials Enhanced with Three-dimensional Carbon Nanotubes/Graphene Conductive Network and *in Situ* Polymerized Polyaniline. *Electrochim. Acta* **2017**, *249*, 387–394.

Joshia, M.; Bhattacharyya, A.; Ali, S. W. Characterization Techniques for Nanotechnology Applications in Textiles. *Indian J. Fibre Text. Res.* **2008**, *33*, 304–317.

Jung, Y. S.; Jung, W.; Tuller, H. L.; Ross, C. Nanowire Conductive Polymer Gas Sensor Patterned Using Self-assembled Block Copolymer Lithography. *Nano Lett.* **2008**, *8*, 3776–3780.

Kaiser, A. B. Electronic Transport Properties of Conducting Polymers and Carbon Nanotubes. *Rep. Progress Phys.* **2001**, *64*, 1.

Kamei, K. I.; Mashimo, Y.; Koyama, Y.; et al. 3D Printing of Soft Lithography Mold for Rapid Production of Polydimethylsiloxane-based Microfluidic Devices for Cell Stimulation with Concentration Gradients. *Biomed. Microdev.* **2015**, *17*, 36.

Kawano, K.; Pacios, R.; Poplavskyy, D.; Nelson, J.; Bradley, D. D.; Durrant, J. R. Degradation of Organic Solar Cells Due to Air Exposure. *Sol. Energy Mater. Sol. C* **2006**, *90*, 3520–3530.

Kawata, S.; Sun, H. B.; Tanaka, T.; et al. Finer Features for Functional Microdevices. *Nature* **2001**, *412*, 697.

Kim, A.; Won, Y.; Woo, K.; Kim, C.; Moon, J. *ACS Nano* **2013**, *7*, 1081–1091.

Kim, I; Lee, E. G; Eunji, J; Cho, G. Characteristics of Polyurethane Nanowebs Treated with Silver Nanowire Solutions as Strain Sensors. *Text. Res. J.* **2017**, *88* (11).

Knittel, D.; Schollmeyer, E. Electrically High-Conductive Textiles. *Synth. Met.* **2009**, *159*, 1433–1437.

Knizek, R.; Karhankova, D.; Fridrichova, L. Two and Three-layer Lamination of Nanofiber. *World. Acad. Sci. Eng. Technol. Int. J. Chem. Mol. Nucl. Mater. Metall. Eng.* **2017**, *9*, 1266–1269.

Ko, F. K.; Yang, H. J. Functional Nanofibre: Enabling Materials for the Next Generation SMART Textiles. *Text., Bioeng., Inform. Symp. Proc.* **2008**, *1* (2), 1217–1228.

Kolmakov, A.; Moskovits, M. Chemical Sensing and Catalysis by One-dimensional Metal-oxide Nanostructures. *Annu. Rev. Mater. Res.* **2004,** *34,* 151–180.

Kopeček, J. Hydrogel Biomaterials: A Smart Future. *Biomaterials* **2007,** *28* (34), 5185–5192.

Kumar, A. *Smart Polymeric Biomaterials: Where Chemistry & Biology can merge*; 2012.

Lam Po Tang, S.; Stylios, G. An Overview of Smart Technologies for Clothing Design and Engineering. *Int. J. Cloth. Sci. Technol.* **2006,** *18,* 108–128.

Lee, S.; Kay Obendorf, S. Developing Protective Textile Materials as Barriers to Liquid Penetration Using Melt-electrospinning. *J. Appl. Polym. Sci.* **2006,** *102,* 3430–3437.

Lee, S.; Kay Obendorf, S. Transport Properties of Layered Fabric Systems Based on Electrospun Nanofibers. *Fiber Polym.* **2007,** *8,* 501–506.

Lee, S. C.; Kwon, I. K.; Park, K. Hydrogels for Delivery of Bioactive Agents: A Historical Perspective. *Adv. Drug. Deliv. Rev.* **2013,** *65* (1), 17–20.

Lee, S.; Reuveny, A.; Reeder, J.; Lee, S.; Jin, H.; Liu, Q.; Yokota, T.; Isoyama, T.; Abe, Y.; Suo, Z.; Someya, T. A Transparent Bending-insensitive Pressure Sensor. *Nat. Nanotechnol.* **2016,** *11* (5), 472–478.

Li, Y.; Cao, L.; Qiao, L.; Zhou, M.; Yang, Y.; Xiao, P.; Zhang, Y. Ni–Co Sulfide Nanowires on Nickel Foam with Ultrahigh Capacitance for Asymmetric Supercapacitors. *J. Mater. Chem. A* **2014,** *2,* 6540–6548.

Li, T.; Xu, Y.; Willander, M.; Xing, F.; Cao, X.; Wang, N.; Wang, Z. L. Lightweight Triboelectric Nanogenerator for Energy Harvesting and Sensing Tiny Mechanical Motion. *Adv. Funct. Mater.* **2016,** *26* (24), 4370–4376.

Li, Z.; Shen, J.; Abdalla, I.; Yu, J.; Ding, B. Nanofibrous Membrane Constructed Wearable Triboelectric Nanogenerator for High Performance Biomechanical Energy Harvesting. *Nano Energy* **2017,** *36,* 341–348.

Liu, H.; Kameoka, J.; Czaplewski, D. A.; Craighead, H. Polymeric Nanowire Chemical Sensor. *Nano Lett.* **2004,** *4,* 671–675.

Liu, N.; Fang, G.; Wan, J.; et al. Electrospun PEDOT: PSS–PVA Nanofiber Based Ultrahigh-strain Sensors with Controllable Electrical Conductivity. *J. Mater. Chem.* **2011,** *21,* 18962–18966.

Liu, C. K.; He, H. J.; Sun, R. J.; et al. Preparation of Continuous Nanofiber Corespun Yarn by a Novel Covering Method. *Mater. Des.* **2016,** *112,* 456–461.

Macagnano, A.; Zampetti, E.; Kny, E. *Electrospinning for High-performance Sensors*; Springer, 2015.

MacDiarmid, A. G. "Synthetic Metals": A Novel Role for Organic Polymers (Nobel Lecture). *Angew. Chem. Int. Ed.* **2001,** *40,* 2581–2590.

Mahadeva, S. K.; Yang, S. Y.; Kim, J. Electrical and Electromechanical Properties of Cellulose-polypyrrole-ionic Liquid Nanocomposite: Effect of Polymerization Time. *IEEE Trans. Nanotechnol.* **2011,** *10* (3), 445–450.

Mahltig, B. Nanosols for Smart Textiles. In *Smart Textiles: Wearable Nanotechnology*; Yılmaz, N. D., Ed.; Wiley-Scrivener, 2018.

Mahltig, B.; Textor, G. *Nanosols and Textiles*; World Scientific Publishing, 2007.

Mandal, D.; Yoon, S.; Kim, K. J. Origin of Piezoelectricity in an Electrospun Poly(Vinylidene Fluoride-Trifluoroethylene) Nanofiber Web-based Nanogenerator and Nano-pressure Sensor. *Macromol. Rapid Commun.* **2011,** *32,* 831–837.

Mao, M.; He, J.; Li, X.; et al. The Emerging Frontiers and Applications of High-resolution 3D Printing. *Micromachines* **2017,** *8,* 113.

Marx, S.; Jose, M. V.; Andersen, J. D.; Russell, A. J. Electrospun Gold Nanofiber Electrodes for Biosensors. *Biosens. Bioelectron.* **2011,** *26* (6), 2981–2986.

Mokhtari, F.; Salehi, M.; Zamani, F.; et al. Advances in Electrospinning: The Production and Application of Nanofibres and Nanofibrous Structures. *Text. Prog.* **2016,** *48,* 119–219.

Nag, A.; Mukhopadhyay, S. C.; Kosel, J. Flexible Carbon Nanotube Nanocomposites Sensor for Multiple Physiological Parameter Monitoring, *Sens. Actuators A: Phys.* **2016,** *251,* 148–155.

Niiyama, E.; Fulati, A.; Ebara, M. Responsive Polymers for Smart Textiles. In *Smart Textiles: Wearable Nanotechnology*; Yılmaz, N. D. Ed.; Wiley-Scrivener, 2018.

O'Brien, G. A.; Quinn, A. J.; Tanner, D. A.; Redmond, G. A Single Polymer Nanowire Photodetector. *Adv. Mater.* **2006,** *18,* 2379–2383.

Paleos, G. A. *What Are Hydrogels*?, 2012. https://www.scribd.com/document/366622248/What-Are-Hydrogels.

Park, S. E.; Kim, S.; Lee, D. Y.; Kim, E.; Hwang, J. Fabrication of Silver Nanowire Transparent Electrodes Using Electrohydrodynamic Spray Deposition for Flexible Organic Solar Cells. *J. Mater. Chem. A* **2013,** *1,* 14286–14293.

Park, M.; Lee, K. S.; Shim, J.; et al. Environment Friendly, Transparent Nanofiber Textiles Consolidated with High Efficiency PLEDs for Wearable Electronics. *Org. Electron.* **2016,** *36,* 89–96.

Peppas, N. A. Hydrogels. In *Biomaterials Science: An Introduction to Materials in Medicine*; Ratner, B. D., Ed.; Academic Press: California USA, 2004; pp 100–107.

Periolatto, M.; Ferrero, F.; Montarsolo, A.; Mossotti, R. Hydrorepellent Finishing of Cotton Fabrics by Chemically Modified TEOS-based Nanosol. *Cellulose* **2013,** *20* (1), 355–364.

Pu, X.; Hu, W.; Wang, Z. L. Nanogenerators for Smart Textiles. In *Smart Textiles: Wearable Nanotechnology*; Yılmaz, N. D., Ed.; Wiley-Scrivener, 2018.

Qin, D.; Xia, Y.; Whitesides, G. M. Soft Lithography for Micro- and Nanoscale Patterning. *Nat. Protoc.* **2010,** *5,* 491.

Raab, C.; Simkó, M.; Fiedeler, U.; Nentwich, M.; Gazsó, A. Production of Nanoparticles and Nanomaterials. *Nano Trust. Dossiers* **2017,** *6,* 1–4.

Rai, P.; Oh, S.; Shyamkumar, P.; Ramasamy, M.; Harbaugh, R. E.; Varadan, V. K. Nano-bio-textile Sensors with Mobile Wireless Platform for Wearable Health Monitoring of Neurological and Cardiovascular Disorders. *J. Electrochem. Soc.* **2014,** *161,* B3116–B3150.

Ranzoni, A.; Cooper, M. A. The Growing Influence of Nanotechnology in Our Lives. In *Micro and Nanotechnology in Vaccine Development*; Skwarczynski, M., Toth, I., Eds.; Elsevier Inc., 2017; pp 1–10.

Rathmell, A. R.; Nguyen, M.; Chi, M.; Wiley, B. J. Synthesis of Oxidation Resistant Cupronickel Nanowires for Transparent Conducting Nanowire Networks. *Nano Lett.* **2012,** *12,* 3193–3199.

Ren, G.; Cai, F.; Li, B.; et al. Flexible Pressure Sensor Based on a Poly(VDF-TrFE) Nanofiber Web. *Macromol. Mater. Eng.* **2013,** *298,* 541–546.

Sandeep, C.; Harikumar, S. L.; Kanupriya. Hydrogels: A Smart Drug Delivery System. *Int. J. Res. Pharm. Chem.* **2012,** *2* (3), 603–614.

Sakellari, I.; Kabouraki, E.; Gray, D.; et al. Quantum Dot Based 3D Photonic Devices. In *Advanced Fabrication Technologies for Micro/Nano Optics and Photonics X*; International Society for Optics and Photonics, 2017.

Salvado, R.; Loss, C.; Gonçalves, R.; Pinho, P. Textile Materials for the Design of Wearable Antennas: A Survey. *Sensors* **2012,** *12,* 15841–15857.

Samad, Y. A.; Li, Y.; Alhassan, S. M.; Liao, K. Non-destroyable Grapheme Cladding on a Range of Textile and other Fibers and Fiber Mats. *RSC Adv.* **2014,** *4,* 16935–16938.

Schoen, D. T.; Schoen, A. P.; Hu, L.; Kim, H. S.; Heilshorn, S. C.; Cui, Y. High Speed Water Sterilization Using One-Dimensional Nanostructures. *Nano Lett.* **2010,** *10,* 3628–3632.

Servati, P.; Soltanian, S.; Ko, F. Electrospun Nanofiber Based Strain Sensors for Structural Health Monitoring. In *Structural Health Monitoring 2013: A Roadmap to Intelligent Structures: Proceedings of the Ninth International Workshop on Structural Health Monitoring,* September 10–12, 2013, DEStech Publications, Inc., 2013.

Shakir, I.; Ali, Z.; Bae, J.; et al. Conformal Coating of Ultrathin Ni(OH)$_2$ on ZnO Nanowires Grown on Textile Fiber for Efficient Flexible Energy Storage Devices. *RSC Adv.* **2014,** *4,* 6324–6329.

Sharma, H. S.; Muresanu, D. F.; Sharma, A.; Patnaik, R.; Lafuente, J. V. Nanoparticles Influence Pathophysiology of Spinal Cord Injury and Repair. In *Progress in Brain Research*; Howard, C. J., Ed.; Elsevier, 2009; pp. 154–180.

Sharma, T.; Langevine, J.; Naik, S.; Aroom, K.; Gill, B.; Zhang, J. J. Aligned Electrospun PVDF-TrFE Nanofibers for Flexible Pressure Sensors on Catheter. In *Solid-State Sensors, Actuators and Microsystems, 2013 Transducers & Eurosensors XXVII: The 17th International Conference on,* IEEE, 2013.

Singh, A. K. The Past, Present, and the Future of Nanotechnology. In *Engineered Nanoparticles Structure, Properties and Mechanisms of Toxicity*; Singh, A. K., Ed.; Elsevier, 2016; pp 515–525.

Soltanian, S.; Rahmanian, R.; Gholamkhass, B.; et al. Highly Stretchable, Sparse, Metallized Nanofiber Webs as Thin, Transferrable Transparent Conductors. *Adv. Energy Mater.* **2013a,** *3,* 1332–1337.

Soltanian, S.; Rahmanian, R.; Gholamkhass, B.; Kiasari, N. M.; Ko, F.; Servati, P. Highly Stretchable, Sparse, Metallized Nanofiber Webs as Thin, Transferrable Transparent Conductors. *Adv. Energy Mater.* **2013b,** *3,* 1332–1337.

Soltanian, S.; Servati, A.; Rahmanian, R.; et al. Highly Piezoresistive Compliant Nanofibrous Sensors for Tactile and Epidermal Electronic Applications. *J. Mater. Res.* **2015,** *30,* 121.

Stegmaier, T. Recent Advances in Textile Manufacturing Technology. In *The Global Textile and Clothing Industry*; Shishoo, R., Ed.; Woodhead Publishing, 2012.

Sun, Y.; Gates, B.; Mayers, B.; Xia, Y. Crystalline Silver Nanowires by Soft Solution Processing. *Nano Lett.* **2002,** *2,* 165–168.

Sun, B.; Long, Y. Z.; Liu, S. L.; et al. Fabrication of Curled Conducting Polymer Microfibrous Arrays Via a Novel Electrospinning Method for Stretchable Strain Sensors. *Nanoscale* **2013,** *5,* 7041–7045.

Syduzzaman, M.; Patwary, S. U.; Farhana, K.; Ahmed, S. Textile Science & Engineering Smart Textiles and Nano-technology: A General Overview. *Text. Sci. Eng.* **2015,** *5* (1), 1–7.

Talreja, K.; Chauhan, I.; Ghosh, A.; Majumdar, A.; Butola, B. S. Functionalization of Silica Particles to Tune the Impact Resistance of Shear Thickening Fluid Treated Aramid Fabrics. *RSC Adv.* **2017,** *7* (78), 49787–49794.

Teo, W.; Ramakrishna, S. Electrospun Fibre Bundle Made of Aligned Nanofibres over Two Fixed Points. *Nanotechnology* **2005,** *16,* 1878.

Tisch, U.; Haick, H. Arrays of Nanomaterial-Based Sensors for Breath Testing. In *Volatile Biomarkers*; Amann, A., Smith, D., Eds.; Elsevier BV, 2013.

Tiwari, M. K.; Yarin, A. L.; Megaridis, C. M. Electrospun Fibrous Nanocomposites as Permeable, Flexible Strain Sensors. *J. Appl. Phys.* **2008,** *103,* 044305.

Tokonami, S.; Shiigi, H.; Nagaoka, T. Micro- and Nanosized Molecularly Imprinted Polymers for High-throughput Analytical Applications. *Anal. Chim. Acta* **2009,** *641,* 7–13.

Torgersen, J.; Qin, X. H.; Li, Z.; et al. Hydrogels for Two-photon Polymerization: A Toolbox for Mimicking the Extracellular Matrix. *Adv. Funct. Mater.* **2013,** *23,* 4542–4554.

Tsukada, S.; Nakashima, H.; Torimitsu, K. Conductive Polymer Combined Silk Fiber Bundle for Bioelectrical Signal Recording. *PLoS ONE* **2012,** *7,* e33689.

Wan, L. Y. Nanofibers for Smart Textiles. In *Smart Textiles: Wearable Nanotechnology*; Yılmaz, N. D., Ed.; Wiley-Scrivener, 2018.

Wang, Z. L. Triboelectric Nanogenerators as New Energy Technology for Self-powered Systems and as Active Mechanical and Chemical Sensors. *ACS Nano* **2013,** *7,* 9533–9557.

Wang, Z. L. Triboelectric Nanogenerators as New Energy Technology and Self-Powered Sensors—Principles, Problems and Perspectives. *Farad. Discuss.* **2015,** *176,* 447–458.

Wang, B. Z.; Chen, Y. The Effect of 3D Printing Technology on the Future Fashion Design and Manufacturing. *Appl. Mech. Mater.* **2014,** *496* (500), 2687–2691.

Wang, X.; Zhang, K.; Zhu, M.; et al. Enhanced Mechanical Performance of Self-bundled Electrospun Fiber Yarns Via Post-treatments. *Macromol. Rapid Commun.* **2008,** *29,* 826–831.

Wang, W. K.; Sun, Z. B.; Zheng, M. L.; et al. Magnetic Nickel–Phosphorus/Polymer Composite and Remotely Driven Three-dimensional Micromachine Fabricated by Nanoplating and Two-photon Polymerization. *J. Phys. Chem. C* **2011a,** *115,* 11275–11281.

Wang, K.; Zhao, P.; Zhou, X.; Wu, H.; Wei, Z. Flexible Supercapacitors Based on Cloth-Supported Electrodes of Conducting Polymer Nanowire Array/SWCNT Composites. *J. Mater. Chem.* **2011b,** *21,* 16373–16378.

Wang, Y.; Zheng, J.; Ren, G.; et al. A Flexible Piezoelectric Force Sensor Based on PVDF Fabrics. *Smart. Mater. Struct.* **2011c,** *20,* 045009.

Wang, K.; Meng, Q.; Zhang, Y.; Wei, Z.; Miao, M. High-performance Two-ply Yarn Supercapacitors Based on Carbon Nanotubes and Polyaniline Nanowire Arrays. *Adv. Mater.* **2013,** *25,* 1494–1498.

Wang, F.; Li, S.; Wang, L. Fabrication of Artificial Super-hydrophobic Lotus-leaf-like Bamboo Surfaces through Soft Lithography. *Colloids Surf. Physicochem. Eng. Aspects* **2017a,** *513,* 389–395.

Wang, Y.; Yang, Y.; Wang, Z. L. Triboelectric Nanogenerators as Flexible Power Sources. *NPJ Flex. Electron.* **2017b,** *1.*

Whitesides, G. M.; Ostuni, E.; Takayama, S.; et al. Soft Lithography in Biology and Biochemistry. *Annu. Rev. Biomed. Eng.* **2001,** *3,* 335–373.

Wu, J.; Zhou, D.; Too, C. O.; Wallace, G. G. Conducting Polymer Coated Lycra. *Synth. Met.* **2005,** *155,* 698–701.

Wu, W.; Bai, S.; Yuan, M.; Qin, Y.; Wang, Z. L; Jing, T. Lead Zirconate Titanate Nanowire Textile Nanogenerator for Wearable Energy-harvesting and Self-powered Devices. *ACS Nano* **2012,** *6,* 6231–6235.

Xia, Y.; Whitesides, G. M. Soft Lithography. *Annu. Rev. Mater. Sci.* **1998,** *28,* 153.

Xia, X.; Tu, J.; Zhang, Y.; Wang, X.; Gu, C.; Zhao, X. B.; Fan, H. J. High-quality Metal Oxide Core/Shell Nanowire Arrays on Conductive Substrates for Electrochemical Energy Storage. *ACS Nano* **2012,** *6,* 5531–5538.

Xie, Y.; Wang, S.; Niu, S.; Lin, L.; Jin, Y.; Wu, Z.; Wang, Z. L. Grating-structured Freestanding Triboelectric-layer Nanogenerator for Harvesting Mechanical Energy at 85% Total Conversion Efficiency. *Adv. Mater.* **2014,** *26* (38), 6599–6607.

Xing, J. F.; Zheng, M. L.; Duan, X. M. Two-Photon Polymerization Microfabrication of Hydrogels: An Advanced 3D Printing Technology for Tissue Engineering and Drug Delivery. *Chem. Soc. Rev.* **2015**, *44*, 5031–5039.

Xu, J.; Wang, K.; Zu, S. Z.; Han, B. H.; Wei, Z. Hierarchical Nanocomposites of Polyaniline Nanowire Arrays on Graphene Oxide Sheets with Synergistic Effect for Energy Storage. *ACS Nano* **2010**, *4*, 5019–5026.

Xu, Q. F.; Liu, Y.; Lin, F. J.; Mondal, B.; Lyons, A. M. Superhydrophobic TiO_2–Polymer Nanocomposite Surface with UV Induced Reversible Wettability and Self-cleaning Properties. *ACS Appl. Mater. Interfaces* **2014**, *5* (18), 8915–8924.

Xu, F.; Ma, L.; Huo, Q.; Gan, M.; Tang, J. Microwave Absorbing Properties and Structural Design of Microwave Absorbers Based on Polyaniline and Polyaniline/Magnetite Nanocomposite. *J. Magn. Magn. Mater.* **2015**, *374*, 311–316.

Xue, P.; Wang, J.; Tao, X. Flexible Textile Strain Sensors from Polypyrrole Coated XLA Elastic Fibers. *High Perform. Polym.* **2014**, *26*, 364–370.

Xue, P.; Park, K.; Tao, X.; Chen, W.; Cheng, X. Electrically Conductive Yarns Based on PVA/Carbon Nanotubes. *Compos. Struct.* **2007a**, *78*, 271–277.

Xue, P.; Tao, X; Tsang, H. *In Situ* SEM Studies on Strain Sensing Mechanisms of PPy-coated Electrically Conducting Fabrics. *Appl. Surf. Sci.* **2007b**, *253*, 3387–3392.

Yamamoto, M.; Nishikawa, N.; Mayama, H.; Nonomura, Y.; Yokojima, S.; Nakamura, S.; Uchida, K. Theoretical Explanation of the Lotus Effect: Superhydrophobic Property Changes by Removal of Nanostructures from the Surface of a Lotus Leaf. *Langmuir* **2015**, *31* (26), 7355–7363.

Yang, M.; et al. Single Flexible Nanofiber to Simultaneously Realize Electricity—Magnetism Bifunctionality. *Mater. Res.* **2016a**, *19*, 2.

Yang, M.; Sheng, S.; Ma, Q.; Lv, N.; Yu, W.; Jinxian, W. Single Flexible Nanofiber to Simultaneously Realize Electricity Magnetism Bifunctionality. *Mater. Res.* **2016b**, *19*, 2.

Yilmaz, N. D. Multi-Component, Semi-Interpenetrating-Polymer Network and Interpenetrating-Polymer-Network Hydrogels: Smart Materials for Biomedical Applications Abstract. In *Functional Biopolymers*; Thakur, V. K., Thakur, M. K., Eds.; Springer, 2017.

Yilmaz, N. D.; Koyundereli Cilgi, G.; Yilmaz, K. Natural Polysaccharides as Pharmaceutical Excipients. In *Handbook of Polymers for Pharmaceutical Technologies, Volume 3, Biodegradable Polymers*; V. K. Thakur., M. K. Thakur., Eds.; Wiley Scrivener, 2015; pp. 483–516.

Yilmaz, N. D.; Konak, S.; Yilmaz, K.; Kartal, A. A.; Kayahan, E. Characterization, Modification and Use of Biomass: Okra Fibers. *Bioinspired, Biomim. Nanobiomater.* **2016**, *5* (3), 85–95.

Yilmaz, N. D.; Khan, G. M. A.; Yilmaz, K. Biofiber Reinforced Acrylated Epoxidized Soybean Oil (AESO) Composites. In *Handbook of Composites from Renewable Materials, Physico-chemical and Mechanical Characterization*; Thakur, V. K., Thakur, M. K., Eds.; Wiley-Scrivener, 2017; pp. 211–251.

Yuan, C.; Hou, L.; Li, D.; Shen, L.; Zhang, F.; Zhang, X. Synthesis of Flexible and Porous Cobalt Hydroxide/Conductive Cotton Textile Sheet and its Application in Electrochemical Capacitors. *Electrochim. Acta* **2011**, *56*, 6683–6687.

Zhai, S.; Karahan, H. E.; Wei, L.; Qian, Q.; Harris, A. T.; Minett, A. I.; Ramakrishna, S.; Ng, A. K.; Chen, Y. Textile Energy Storage: Structural Design Concepts, Material Selection and Future Perspectives. *Energy Storage Mater.* **2016**, *3*, 123–139.

Zhai, T.; Fang, X.; Liao, M.; Xu, X.; Zeng, H.; Yoshio, B.; Golberg, D. A. Comprehensive Review of One-Dimensional Metal-Oxide Nanostructure Photodetectors. *Sensors* **2009**, *9*, 6504–6529.

Zhang, L.; Shi, E.; Ji, C.; Li, Z.; Li, P.; Shang, Y.; Li, Y.; Wei, J.; Wang, K.; Zhu, H. Fiber and Fabric Solar Cells by Directly Weaving Carbon Nanotubes Yarns with CdSe Nanowire-based Electrodes. *Nanoscale* **2012**, *4*, 4954–4959.

Zhao, F.; Shi, Y.; Pan, L.; Yu, G. Multifunctional Nanostructured Conductive Polymer Gels: Synthesis, Properties, and Applications. *Acc. Chem. Res.* **2017**, *50*, 1734–1743.

Zhou, Y.; He, J.; Wang, H.; et al. Continuous Nanofiber Coated Hybrid Yarn Produced by Multi-nozzle Air Jet Electrospinning. *J. Text. Ind.* **2017**, *108*, 783–787.

Zhu, G.; Zhou, Y. S.; Bai, P.; Meng, X. S.; Jing, Q.; Chen, J.; Wang, Z. L. A Shape-adaptive Thin-film-based Approach for 50% High Efficiency Energy Generation through Micrograting Sliding Electrification. *Adv. Mater.* **2014**, *26* (23), 3788–3796.

Zhu, Y.; Zhang, J. C.; Zhai, J.; Al, E. Multifunctional Carbon Nanofibers with Conductive, Magnetic and Superhydrophobic Properties. *ChemPhysChem* **2006**, *7*, 336–341.

PART III

Nanotechnology: An Innovative Way for Removal of Textile Waste and Nano-Textiles Used for Healthcare Perspective

CHAPTER 12

Nanotechnology Applications for the Management of Textile Effluent

A.S.M. RAJA*, A. ARPUTHARAJ, G. KRISHNAPRASAD,
T. SENTHIL KUMAR, and P.G. PATIL

*ICAR- Central Institute for Research on Cotton Technology
Mumbai-400019, India*

Corresponding author. E-mail: raja.asm@icar.gov.in

ABSTRACT

Effluent management is one of the important aspects of textile processing due to the stringent norms introduced by the authorities. Untreated effluent causes various ill effects to the environment and human beings. The current methods of effluent management are not cost-effective for the adoption by the industries. Another limitation of current effluent treatment methods is the formation of huge amount of solid wastes due to the large-scale addition of flocculation and precipitating agents. The disposal of solid wastes from the effluent treatment methods is still a researchable issue. Nanotechnology has a wide range of applications in the textile effluent management. It offers effective solution to the management of effluent compared with the conventional treatment processes. Nano-based materials can be employed for the effective treatment of effluent for adsorption, disinfection, filtration, photocatalysis, and color and removal of toxic substance. One of the limitations of nanoparticles application in the effluent management is that it is difficult to separate the nanoparticles after the effluent treatment. This problem is addressed by immobilizing the nanoparticles in the polymer substrates, that is, formation of polymer nanocomposites. The present chapter reviews various nanotechnology-based applications in the effluent management along with the

Fundamentals of Nano-Textile Science. Prashansa Sharma, Devsuni Singh & Vivek Dave (Eds.)
© 2022 Apple Academic Press, Inc. Co-published with CRC Press (Taylor & Francis)

advantages and disadvantages. The chapter also envisages the scope of nano-based technologies in the conventional-based effluent treatment methods for improvement.

12.1   INTRODUCTION

Water is the most important element for the survival of human being. It is used for different human needs and for agricultural purposes. Water is also used by different industries as a medium for their treatment for producing different products. Once the process is completed, the water is released as effluent. The effluent contains harmful substances and waste materials. The pollution problem due to the effluent water is the major cause of concern. The release of effluent directly into the land and sea leads to irreversible damage to soil, groundwater, and sea. In the textile industries, water is used for different textile processes, such as desizing, scouring, bleaching, dyeing, printing, and finishing. Different processes require huge quantity of water. On the average, 50–100 L of water is required for processing 1 kg of textile material (Saxena et al., 2017).

Wastewater treatment is a process of cleansing the effluent of different industrial processes to meet the different standards set by the regulating authorities or end users before letting it out. Due to stringent effluent discharge norms framed by the central and state governments, it is mandatory to adhere to certain minimum requirements by the industries for effluent management. Treatment of the effluent to satisfy the standard norms is a cumbersome process. The following table depicts the various norms framed by the Central Pollution Control Board, Government of India for the effluent which can be discharged from the cotton industries (composite and processing) (Table 12.1).

12.2   ZERO-LIQUID DISCHARGE

In view of the severe pollution problem, recently, the Central Pollution Control Board of India also introduced zero-liquid discharge (ZLD) norms. ZLD refers to complete recycling of the effluent for the separation of water and dissolved chemicals and concentrating the dissolved chemicals into solid form through multiple effect evaporator or solar dryer. Other important components of ZLD include (1) the requirement of reuse of recycled water to the maximum extent by the industry and (2) it reduces the water requirement.

12.3 POLLUTION AND DRINKING WATER

One of the important problems associated with pollution is the contamination of water bodies used for drinking water purpose. Both groundwater and surface water are affected by pollution. The microparticles present in the polluted water are also entering into the water bodies used for drinking purpose. Apart from pollution, the soil erosion due to heavy rainfall also leads to the contamination of drinking water due to entry of pathogens along with soil components. The decontamination processes which are currently available, such as chlorination and ozonation require high amount of chemical agents and energy inputs and may also produce harmful and toxic by-products (Gehrke et al., 2015).

TABLE 12.1 Norms for Discharging Effluent in Textile Composite Mills (https://cpcb.nic.in/displaypdf.php?id=SW5kdXN0cnktU3BlY2lmaWMtU3RhbmRhcmRzL0VmZmx1ZW50LzQxMi5wZGY= accessed on 01.01.2020)

Parameter	Concentration not to exceed, milligram per liter (mg/L), except pH and SAR
pH	6.5–8.5
Total suspended solids	100
Biochemical oxygen demand (BOD)	30
Oil and Grease	10
Chemical oxygen demand (COD)	250
TDS	2100 (not applicable for marine disposal)
Total chromium	2
Sulfide (as S)	2
Color in platinum Cobalt Units	150
Sodium absorption ratio (SAR)	26 (not applicable for marine disposal
Ammoniacal nitrogen (as N)	50

12.4 DIFFERENT PARAMETERS USED IN EFFLUENT MANAGEMENT

12.4.1 *BIOLOGICAL OXYGEN DEMAND OF WATER AND WASTEWATER*

Biological oxygen demand (BOD) directly indicates the organic load present in the effluent. The microorganism present in water utilizes the dissolved oxygen and decomposes the organic materials. More oxygen demand will be created if greater amount of organic compounds are present in the water. Total BOD will provide only the information about the biodegradable organic impurities. Natural organic impurities, such as starch, cellulose, etc., are majorly contributing for higher BOD value. This value is useful in determining the extent of biological treatment of the effluent. For good quality water, the BOD value will be lesser than 1 ppm. Different methods can be followed to determine the BOD of water. However, BOD5 which determines the demand of dissolved oxygen after 5 days of incubation is mostly followed. Nowadays, online analyzers are also available commercially for determining the total BOD of the effluent. .

12.4.2 *CHEMICAL OXYGEN DEMAND OF WATER AND WASTEWATER*

Chemical oxygen demand (COD) measures the total organic load present in the effluent. The major difference between the BOD and COD is that even nonbiodegradable substances are also chemically oxidized using strong oxidizing agents such as potassium dichromate under reflux conditions. COD value will always be higher than the total organic carbon (TOC) and BOD values as it includes biodegradable and nonbiodegradable substances. COD values may be in the range of 2–3 times the BOD values. Determination of COD is carried out using titration method which is comparatively less time-consuming than the BOD5 determination.

12.4.3 *TOTAL DISSOLVED SOLIDS OF WATER AND WASTEWATER*

Total dissolved solid (TDS) is the very important characteristics of the effluent after BOD and COD. It provides information about the total dissolved organic and inorganic substances in the effluent. There is a direct relationship between conductivity and TDS value of the effluent.

12.5 EFFLUENT TREATMENT METHODS

12.5.1 *PRIMARY TREATMENT*

After physical separation of impurities in the preliminary treatment, primary treatment is carried out to remove the suspended solids and water-soluble organic molecules from the effluent. This is achieved by using the precipitation technique. The basic principle of this process is to reduce the water solubility of the constituents by electrostatic attractive forces. This is accomplished by coagulation followed by the flocculation process in a clarifier by maintaining optimum pH conditions. Typical coagulants possessing trivalent cations, such as Al^{3+}, Fe^{3+} are used for this action. The limitation of this treatment is that water-soluble inorganic constituents, such as NaCl, Na_2SO_2 cannot be removed.

12.5.2   SECONDARY TREATMENT

This is also called biological treatment or activated sludge process. The BOD of the effluent is reduced by the carbonization of organic matters using microorganisms. Aeration of the effluent and maintaining optimum pH play a vital role in reducing the biological load. There is no other better method available other than biological process to reduce the organic load of the effluent. Aerated lagoons, trickling filters are working on this principle. Maintaining optimum quantity of the colony forming units (CFU) and nutrient concentration is the critical factor in obtaining better efficiency using this treatment. However, the presence of heavy metals reduces the efficiency of this process.

12.5.3   TERTIARY TREATMENT

This treatment is also called as polishing treatment since it is being carried out after primary and secondary treatment. The output water from the tertiary treatment can be reused, if it is done effectively. However, better infrastructure is required to carry out the tertiary treatment process. Activated carbon treatments, reverse osmosis (RO), electro dialysis, ultra membrane filtration (UF), come under the category of tertiary treatment process. Mostly, two or more tertiary treatments are combined together to reduce the TDS and color of the effluent. Tertiary treatment of the effluent is very important in order to achieve the zero-liquid discharge (ZLD).

12.6   EFFLUENT MANAGEMENT USING NANOTECHNOLOGY

Nanotechnology has a wide range of applications in the effluent management. It offers an effective solution to the management of effluent compared with the conventional treatment processes. The nanotechnology application for the effluent management can be divided into the following categories (Table 12.2).

TABLE 12.2 Nanotechnology Applications of Different Effluent Treatment Processes.

Treatment processes	Nanomaterials used
Adsorption	CNTs/polymer nanocomposites/ zeolites/ nanoscale metal oxide and nanofibers
Antimicrobial/disinfection	Nanosilver/titanium dioxide (Ag/TiO$_2$) and CNTs
Degradation of pollutants/photo catalysis	Nano-TiO$_2$, Nano-ZnO and Fullerene
Nanofiltration and prevention of membrane-fouling	Nano-Ag/TiO$_2$/zeolites/magnetite and CNTs

12.6.1 ADSORPTION

Adsorption refers to the attracting of surface molecules by both external and internal surface of gases and solutions when they are contact to the solid substrate. The solid materials used for this process are called as adsorbents and the adsorbed liquids or gases are stated as adsorbates. The commonly known adsorbent is activated carbon which has high surface area compared with the corresponding solid substances. Nowadays, nano-based adsorbents which have very high surface area compared with activated carbon are being explored as high-efficiency adsorbents for the removal of contaminants from the effluent. The nanomaterials, such as CNT, fullerene, zeolites, etc. are the most promising substances for the adsorption process. CNT-based adsorbents have been used to remove several organic contaminants, such as pesticide residues, bisphenol, nonylphenol, phthalates, etc. which are considered as endocrine disturbing substances (Sulekha, 2016; Jame & Zhou, 2016; Bernd, 2010). The researchers used nanoparticles (NPs), nanocomposites, nanofibers, and carbon nanotubes (CNTs) for the removal of different contaminants from the effluent.

12.6.2 REMOVAL OF HEAVY METAL

A number of heavy metals are discharged from many industries, such as arsenic, cadmium, chromium, copper, nickel, lead, and mercury. The presence of heavy metal in the water resources poses serious health problem to the users. Heavy metals enter through volcanic eruptions and weathering of soils and rocks when the metals react with water bodies Also, a variety of human activities causing contamination, such as mining, processing of metals, use of metals containing metal pollutants for manufacturing various products (http://www.lenntech.com/aquatic/metals.htm#ixzz4K18F3Q6j) are responsible for the prevalence of metals in water resources. Nanotechnology-based adsorbents effectively separate pollutants from contaminated water as well as fresh water. The Al_2O_3 NPs exhibit very good adsorption capacity, reactivity toward the pollutant, and large internal surface area. Due to this extraordinary property, they are effectively used in wastewater engineering. Arsenic is released into the environment due to different human activities, such as mining, burning of fuels, production of metal, pesticides, etc. Arsenic pollution can be alleviated using magnetite NPs (MNPs) as adsorbents.

12.6.3   REMOVAL OF POLLUTANTS BY NANOPARTICLES

There are a lot of scopes and challenges need to be addressed for the removal of dyes from wastewater by NPs. The NPs-based adsorbents, in one way, are used for the removal of dye, metal ions, and other pollutants through adsorption in recent days due to their excellent advantages, such as porosity, active sites, and pollutant removal efficiency. The separation of toxic substance by adsorption process using iron NPs is very promising. The unique properties of iron NPs include their varied structures, stabilities of chemical and thermal atmosphere, sensitivity, improved adsorption kinetics, very less quantity is required for adsorption, easy functionalization with other materials, economical, and recovery and recycle are simple, separation of adsorbent from the liquid medium is very easy (Gomez et al., 2014; Mehta et al., 2015; Reddy and Yun, 2016; Kefeni et al., 2017). Many researchers reported the separation efficiency, kinetics, isotherm, and recyclability of iron NPs. Kim and Choi (2017) synthesized amorphous iron nanoparticle via a facile borohydride reduction method under controlled conditions and observed that the maximum adsorption capacity of Congo red dye for the amorphous iron nanoparticle was found to be 1735 mg/g. The spinel ferrite-based magnetic NPs CoFe1.9Mo0.1O4 were produced by the sol–gel method and the adsorption capacity of the adsorbent was found by using basic thiazine dye from aqueous solutions. From this study, at optimum condition, around 95% of methylene blue dye was removed (Amar et al., 2018). The magnetic NPs are developed from green sources also. Qiao et al. (2019) synthesized sulfobetaine-modified magnetic Fe_3O_4 NPs through one-step chemical coprecipitation method and studied the adsorption of anionic dyes, such as methyl blue and amaranth from aqueous solution. Also, they have reported that the adsorption process was extraordinarily faster and reached saturation at 15 min and the Langmuir adsorption capacity of sulfobetaine-modified magnetic Fe_3O_4 NPs was observed to be 127.06 and 57 mg/g for methyl blue and amaranth, respectively. The iron oxide-based nanomaterials can functionalize with other materials for getting better adsorption capacity. Lee et al. (2019) developed the iron oxide NPs cross-linked with CS and had immobilized this structure on dextran gel column and studied the separation efficiency of anionic dyes in aqueous solution. CS is cationic in nature and it can easily adsorb the anionic pollutants, and also helps for stable immobilization of the nanoadsorbents in the dextran gel beads. Zafar et al. (2019) synthesized zinc oxide (ZnO) NPs through coprecipitation method and studied their adsorption capacity from methyl orange and amaranth dye aqueous solution through batch adsorption studies. They found that the maximum adsorption capacity was found to be 75.9 and 65.2 mg/g for methyl orange and amaranth, respectively.

12.6.4   *REMOVAL OF POLLUTANT BY NANOCOMPOSITES*

Composite materials are materials made by combining two or more materials to form new materials which have superior properties compared with the individual constituents. Similarly, nanocomposites come under one of the composite materials classification. Nanocomposites are composites in which one of the materials or phases lies in the nanometer range (1 nm = 10^{-9} m) (Roy et al., 1986). When the material dimensions are in nanometer range, interactions between the different phases or materials during composite preparation are extensively improved. The surface area/volume ratio of the materials used in the nanocomposites preparation has played quite a lot and it is very crucial to understand their structure–property relationships (Pedro et al., 2009). From the existing literature, it becomes evident that nanocomposites are providing many advantages, such as improving compatibility between materials, enhancing material properties, and reduction of material usage. The potential application of nanocomposites in catalysis, sensors, structural materials, engineering parts, aerospace application, automotive industrial application, dental applications, batteries, medical devices, and effluent treatments has also been reported. The application of nanocomposites in the field of wastewater engineering is emerging now. Numerous researchers have reported the adsorption of dyes, metal ions, and other pollutants using nanocomposite materials. Janaki et al. (2012) produced nanocomposites using starch and polyaniline through chemical oxidative polymerization of aniline and analyzed the adsorption capacity of Reactive Black 5 and Reactive Violet 4 through batch adsorption experiment. Results showed that 99% of the Reactive Black 5 and 98% of the Reactive Violet 4 are removed by starch/polyaniline nanocomposite adsorbent. Samaneh et al. (2017) studied the magnetic nanocomposite beads made from gelatin and including CNTs which is functionalized by carboxylic acid and studied the adsorption behavior with anionic Direct Red 80 and basic dye from aqueous solution, Further, they have reported that the maximum adsorption capacity of anionic Direct Red 80 and basic dye was 465.5 and 380.7 mg/g, respectively. A new magnetic nanocomposite was prepared by polycatechol-modified iron NPs. Fe_3O_4/PCC MNPs were prepared through a facile chemical coprecipitation method, and the adsorption capacities were determined by the adsorption of dyes, such as methylene blue, cationic turquoise blue GB, malachite green, crystal violet, and Cationic Pink FG. The maximum adsorption capacities on Fe_3O_4/PCC MNPs were 60.06, 70.97, 66.84, 66.01, and 50.27 mg/g, respectively (Yani et al., 2018). Arej and Gorair (2019) investigated the removal of cationic dyes from wastewater using economically environmentally friendly nanocomposites through green strategy. They have developed Arg/Cs/GO nanocomposites by trapping gaphene oxide in arginine/CS (Arg/Cs) hydrogel and additionally introduced magnetic property to get magnetic nanocomposites Arg/Cs/GO-mg. The adsorption of capacity of Arg/Cs/GO-mg adsorbent was found using methylthioninium chloride (MB), and they found that the maximum monolayer adsorption was 1.18 mg/g, they also revealed that the reaction followed pseudo-second-order reaction. Sahar et al. (2019) prepared CS-based aluminum oxide magnetic nanocomposite (CS/Al_2O_3/Fe_3O_4) and analyzed its adsorption capacity with the removal of acid fuchsin dye from aqueous solution. Also, they have found that the maximum adsorption capacity of CS/Al_2O_3/Fe_3O_4 nanocomposite for acid fuchsin was 1666.67 mg/g and found that both film diffusion and intraparticle diffusion controlled the adsorption process.

12.6.5 *REMOVAL OF POLLUTANT BY CARBON NANOTUBES*

The CNTs have the length-to-diameter ratio of more than 1,000,000 and also they are allotropes of carbon (Saifuddin et al., 2013). Many production methods are available to produce CNTs, such as plasma-based synthesis method or Arc-discharge evaporation method, laser ablation method, thermal synthesis process, chemical vapor deposition (CVD), and plasma enhanced

CVD (PECVD) (Parijat and Mandeep, 2016). There are two types of CNTs, that is, single-walled CNTs (SWCNTs) and multiwalled CNTs (MWCNTs). The SWCNTs are comprised of single cylinder carbon layer and the diameter ranges from 0.4 to 2 nm (Klumpp et al., 2006), whereas MWCNTs consist of many coaxial cylinders and the diameter ranges from 1 to 3 nm (Rastogi et al., 2014). The major areas of application of CNTs are drug delivery, biomedical imaging, biosensors, and scaffolds in tissue engineering (Beg et al., 2011). In recent days, CNTs are used as good adsorbent for the separation of toxic substance in aqueous effluent. They are smaller in diameter with high surface area, morphology, and are easily functionalized with other materials to become a potential adsorbent for liquid adsorption (Norzilah et al., 2011). Akbar et al. (2017) studied the adsorption of Basic Red 46 by SWCNTs, carboxylate group functionalized single-walled CNTs (SWCNT-COOH), graphene (G), and graphene oxide (GO) from aqueous solution. Also, they found that the maximum adsorption capacity of SWCNTs, SWCNTs-COOH, G, GO were 38.35, 49.45, 30.52, 55.57 mg/g, respectively.

12.6.6 REMOVAL OF POLLUTANT BY ELECTROSPUN NANOFIBERS

Nanofiber is defined as stretched thread-like object with size on the nanoscale, from several hundred to several thousand nanometers. Nanofiber is an emerging, interdisciplinary area of research with important commercial applications. Nanofibers usually have the nanoscale effect of high surface energy, surface reactivity, high thermal and electrical conductivity, and high strength. The electrospun nanofibers can be used in many applications, such as drug delivery systems, reinforcing material for composite preparation, filtration, tissue engineering, catalyst, conducting polymer, wound dressing, pharmacy application, and wastewater engineering (Ji-Huan He et al. 2008). In recent years, electrospun carbon nanofibers have attracted significant attention for the removal of organic pollutant, as it is relatively inexpensive alternative to CNTs (Badr et al., 2019). The nanofiber can be produced from any chargeable polymeric materials. Table 12.3 lists the different types of electrospun fiber used for the removal of dyes and metal ions from aqueous medium.

12.6.7 NANOFILTRATION

Nanofiltration (NF) is widely used in effluent treatment process. By this process, it is possible to separate dissolved substances having the size in

TABLE 12.3 Different Source of Electrospun Nanofiber Adsorbent for Pollutant Removal.

S. No	Adsorbent	Adsorbate	Observations	Reference
1	Polyamide-6/CS nanofibrous membrane	Solophenyl Red 3BL and Polar Yellow GN	The maximum removal of Polar Yellow and Solophenyl Red 3BL were 94.827% and 96.33%, respectively.	Mozhdeh et al. (2014)
2	β-cyclodextrin-based nanofibers	Methylene Blue	The maximum Langmuir adsorption is 826.45 mg/g	Rui et al. (2015)
3	Aminated poly(ethylene terephthalate) (APET) nanofibers	Pb(II) from polluted water	The APET nanofibers adsorbed 97% of Pb(II) in continuous-flow mode adsorption	Diego et al. (2017)
4	Thermoplastic polyurethane (TPU) and poly (vinyl alcohol) (PVA)-based nanofiber and cross-linked with BTCA-PVA nanofiber	Reactive Red 141 dye	BTCA/PVA electrospun nanofiber membranes showed higher adsorption capacity of Reactive Red 141 dye. The adsorption capacity reached to 88.31 mg/g whereas TPU-based membranes showed very low adsorption capacity of 14.48 mg/g	Akduman et al. (2017)
5	Zein-based electrospun nanofibers	Reactive Black 5	Zein electrospun nanofibers used as an adsorbent for removing Reactive Black 5 dye and they found that Zein electrospun nanofibers-dye cross-linked via hydrophobic, electrostatic forces and hydrogen bond interactions	Umair et al. (2017)
6	Surface-modified amidoximated polyacrilonitrile (APAN) nanofibers	Indigo carmine dye	The maximum adsorption capacity was found to be 154.5 mg/g	Yazdi et al. (2018)
7	Gamma-Al$_2$O$_3$ nanoparticles/ethyl cellulose electrospun adsorbents	Pb(II) ions from aqueous medium	The maximum adsorption capacity was found to be 134.5 mg/g	Ehsan et al. (2018)
8	Co-electrospinning of Zein and Nylon-6 nanofiber	Reactive Blue 19	The sorption capacity Reactive Blue 19 by the nanofiber adsorbent is 70 mg/g. Also, they found that the adsorption mechanism is mainly through hydrogen bond and electrostatic means	Raheel et al. (2018)

TABLE 12.3 *(Continued)*

S. No	Adsorbent	Adsorbate	Observations	Reference
9	Polyacrylonitrile (PAN)–based nanofiber modified to contain amidino diethylenediamine chelating groups on their surface via heterogeneous reaction with diethylenetriamine (DETA) which is called APAN	Metal ions, namely Cu(II), Ag(I), Fe(II), and Pb(II) ions	The maximum adsorption capacities of APAN nanofiber mats with Cu(II), Ag(I), Fe(II), and Pb(II) ions were 150.6, 155.5, 116.5, and 60.6 mg/g, respectively	Pimolpun and Pitt (2010)
10	Electrospun PVA/CS nanofiber membrane	Cu(II) metal ion	The maximum removal of Cu(II) by PVA/CS nanofiber membrane was 90.3 mg/g.	Wu et al. (2018)
11	Electrospun carbon nanofibers (ECNFs) and oxidized electrospun carbon nanofibers (O-ECNFs)	Methylene blue (MB)	The maximum sorption of Methylene blue by O-ECNFs is 170 mg/g, whereas by ECNFs, it is 32.5 mg/g.	Badr et al. (2019)
12	Polyvinyl alcohol (PVA) nanofibrous membranes chemically cross-linked by a polycarboxylic acid (1,2,3,4 butanetetracarboxylic acid (BTCA))	Reactive Red 141	The adsorption, approximately >80% of Reactive Red 141 removal by PVA nanofibers	Akduman et al. (2019)

the range of 1–100 nm by passing the effluent through membranes having the pore size of 0.1–10 nm under pressure. Nanofiltration has the capacity to remove all the dissolved substances as well as microorganisms present in the effluent. One of the problems associated with membrane process is the fouling of membranes. This membrane-fouling problem is efficiently managed if the membrane is coated with NPs, such as nano-Ag, nano-TiO_2, iron oxide, etc. (Table 12.4).

TABLE 12.4   Different Types of Filtration and their Characteristics.

Process	Pore size in nm	Pressure(bar/psi)	Membrane
Microfiltration	50–500	0.5–2.0/15–60	Polymeric substances such as cellulose Acetate; fluoro polymers, such as teflon, polyvinylidenedifluoride (PVDF), etc.
Ultrafiltration	5–50	0.5–10/30–100	-do-
Nanofiltration	0.6–5	10–40/90–150 psi	Cellulose acetate, polyamide
Reverse osmosis	Less than 0.6	30–70	-do-

12.6.8   PHOTOCATALYSIS

Photocatalysis is the process of breaking down the organic molecules into CO_2 and H_2O using light and water with the help of strong oxidation agent and electronic holes in the presence of photocatalyst. This process finds lots of application in the effluent treatment process in order to remove the organic impurities from the water. Photocatalyst in nanoform and in composites with polymer can be used for effluent treatment. Semiconductors, such as nano-TiO_2, ZnO, etc. have been used as photocatalyst for degrading the organic contaminants present in the effluent. Fullerene derivatives are found to have photocatalytic properties in the solar spectrum.

12.6.8.1   CHEMICAL OXIDATION PROCESS

Recent technologies use chemical oxidants, such as peroxide (H_2O_2), hypochlorite (OCl^-), ozone (O_3), etc. to oxidize the organic pollutant into less toxic, harmless substances or convert them into forms which can be managed easily. However, the use of oxidizing agents, such as H_2O_2 and O_3 results in lower rates of degradation efficiency due to their inherent oxidation potential (Zheng et al., 2013). Advanced oxidation processes (AOPs) with the ability of using the high reactivity of hydroxyl (OH) radicals at room temperature and pressure conditions have become a useful technology for the wastewater treatment (Ghuge and Saroha, 2018). Following are the advantages of this process.

i)   Degradation of organic matter in aqueous phase using chemical reaction.
ii) Highly versatile process for different organic molecule.
iii) Some of the heavy metals can also be removed.

AOP can be classified into different types, such as Fenton, Photo-Fenton, ozonation, photocatalysis, etc., which depends on the oxidation chemical or source used for the production of OH· radicals. AOPs can be of two types, that is, homogeneous and heterogeneous processes, which depend upon the phase of the reactants used for the oxidation process. Peroxide, ozone, and Fenton processes are the examples for the homogeneous process. Semiconductor-based chemical oxidation is an example for the heterogeneous process.

12.6.8.2   *NANOMATERIALS FOR PHOTOCATALYSIS*

Nanotechnology can be defined as the understanding and control of materials at the nanoscale. At approximately 1 and 100 nm size dimensions, materials exhibit unique properties differing from the properties of bulk materials which can be exploited for novel applications.

Nanoform of the material is very interesting because of the size-dependent properties being exhibited by the materials in the nanoscale regime (Ede et al., 2018). Following are the important nanoscale properties of various materials: (1) enhanced surface area, (2) quantum confinement effect, (3) chemical reactivity, and (4) catalytic activity. The catalytic activity of any material depends upon the surface area of the material. Since nanomaterials have higher surface-to-volume ratio and profound exposure of catalytic sites, they have extremely higher rate of catalytic property which cannot be achieved by the bulk materials.

Photocatalysis can be defined as the process that increases the rate of a chemical reaction by lowering the activation energy of the primary reaction to occur by either direct irradiation or by the irradiation of a catalyst. A sensitized photocatalytic oxidation process works on the principle that the excitation of semiconductor materials, such as titanium dioxide (TiO_2), zinc oxide (ZnO) upon absorption of a photon of suitable energy of UV light produces highly reactive radicals which can rapidly oxidize organic compounds. The bandgap energy is the region extending from the empty conduction band

(CB) to the top of the valence band (VB). When a semiconductor material is irradiated by UV light, an electron migrates from the fully occupied valence band of the semiconductor to a higher energy level empty CB. This results in the formation of electron-hole pairs in valence bond. These holes react with water and produces highly reactive OH radicals (Saleh and Gupta, 2011; Ng et al., 2019). Oxidation of an organic compound can be easily achieved if the valence band edge of the semiconductor photocatalyst is properly positioned to the oxidation potential of the organic compound. Size reduction principle can be used for exploiting the bandgap of the metal oxides. For example, in the case of ZnO, when we reduce its size from micro to nanosize, bandgap value increases to 3.32 eV from 3.19 eV (Kamarulzaman et al., 2015). A wide range of semiconductors can be used for the photocatalytic process. Different metal oxides have different bandgap energies. Table 12.1 compares the bandgap energy of different metal oxides (Table 12.5).

TABLE 12.5 Bandgap Energy of Different Metal Oxides (Ola and Maroto-Valer, 2015).

Nature of metal oxide	Bandgap energy (eV)
ZnO	3.2
TiO_2 (Anatase)	3.2
TiO_2 (Rutile)	3.0
ZrO_2	5.0
Fe_2O_3	2.3
Cu_2O	2.2
SnO_2	3.8

NP-based photocatalyst, for example, is often easier to create with higher efficiencies, and can be used in any heterogeneous system. Much R & D works have been reported for the nanoparticle-assisted photocatalysis for the mineralization of organic compounds. Nature, size, and shape are the important parameters which influence the photocatalytic property of the nanomaterials.

a. Titanium dioxide

TiO_2 crystallizes in two main forms, anatase and rutile. Anatase form with (001) facets has higher catalytic performance than the rutile form. Titania is a much studied photocatalyst due to its cost-effectiveness, chemical stability, and highly oxidizing photogenerated holes. Fabrod and Kajbafvala, (2013) found out that more than 90% decolorization of Congo red was observed in the presence of nano-TiO_2 after 3.5 h UV radiation, otherwise, no decolorization was observed in the absence of the catalyst. Ghasemi et al. (2016) immobilized the nano-TiO_2 into the structures of Fe-ZSM-5 zeolite by sol–gel method and achieved 80% COD removal from the petroleum refinery wastewater.

Byeon and Kim (2013) fabricated nano-TiO_2 on the surface of the reduced graphene oxide for the dye degradation and found out that fabricated catalyst was showing enhanced catalytic activity than TiO_2. This increase was attributed to the aligning of nano-TiO_2 on graphene surface which enables easy electron transfer. Bennet and Keller (2011) compared the photocatalytic activity of different semiconductors for the degradation of coumarin and found that titania was superior to other metal oxides. Reports are also available for the increased combined photocatalytic activity of titania with other oxides such as bismuth trioxide (Bian et al., 2007). Pallidum-doped titania increases the degradation efficiency into manifolds due to narrowing bandgap. You et al. (2017) reported that photo-bi-catalysis using ZnO–TiO_2 core-shell nanofibers resulted high efficiency and nontoxicity degradation of methyl orange dye. This is achieved by the piezoelectric effect of ZnO.

b. Zinc oxide

ZnO crystallizes in two main forms, wurtzite and zinc blende. The wurtzite structure is the most stable hexagonal lattice at ambient conditions and commonly belongs to $P6_3mc$ space group. It has been vastly investigated for its prospective photocatalytic activity. Wide range of nanostructutres of ZnO, such as nanowires, nanorods, NPs has been synthesized for photomineralization of organic dyes. Payra et al. (2018) compared the properties of nano-ZnO derived from zinc containing zeolite imidazolate framework using aerial calcination process for the degradation of methylene blue dye. They found out that the cytotoxicity of the degraded methylene blue by nano-ZnO was less compared with the untreated dye. Doping with transition metal ions such as chromium can extend the absorbance of ZnO into visible region which can carry out the photocatalysis under visible light.

12.6.8.3   FUTURE OF PHOTOCATALYSIS

Photocatalysis faces the major limitation of activity of TiO_2 and ZnO only in the presence of high-energy UV irradiation. Though doping can shift the photo catalysis under visible light, this modification needed much more efforts to make it commercially viable alternative for organic mineralization (Ng et al., 2019). Cost-effective synthetic methods are required for the production of tailor-made NPs with optimum size distribution and shape.

12.6.8.4   NANO-BASED OZONATION

Ozonation is also type of AOP and has edge over other AOPs due to residual ozone decomposition into water and oxygen. Catalytic ozonation is gaining importance for the mineralization of organic wastewater. Catalyst enhances the decomposition of ozone and increases the rate of reaction (Faria et al., 2009). Singh et al. (2016) reported that nanometal oxides can increase the dye degradation potential during ozonation. Malik et al. (2018) compared the efficiency of catalytic ozonation process using ferrous and Fe^0 NPs. Maximum COD removal (73.5%) and color removal (87%) efficiency was obtained at 0.7 g/L of NP.

12.7   CHALLENGES IN USING NANOTECHNOLOGY FOR EFFLUENT MANAGEMENT

The challenge associated with the use of nanotechnology in wastewater treatment is the removing of NPs from the treated water after the process is completed. This problem can be solved to some extent by immobilization of the NPs on suitable materials. The other challenges, such as toxicity of the NPs to human beings, aquatic species, and environment also need to be studied in detail and different research groups are working in this direction. Apart from technical challenges, the cost of nano-based processes also plays an important factor for adaptation by the industries.

12.8   CONCLUSION

Nanotechnology-based effluent treatment technologies are promising for the removal of pollutants effectively with minimum addition of chemicals. Most of the research work done in this direction needs to be adopted at an industrial level to find their efficacy. Photocatalysis based on NPs has to be further studied for the use of visible light for activation instead of UV light currently being used. The nano-based visible light photocatalysis process would be the ultimate technology for the removal of pollutants from the effluent water. The recovery of NPs and its reuse of the same also further need to be studied for industrial adaptation.

KEYWORDS

- effluent
- effluent management
- adsorption
- nano-filtration
- removal of pollutants
- photo catalysis

REFERENCES

Abul, A.; Samad, S. A.; Huq, D.; Moniruzzaman, M.; Masum, M. Textile Dye Removal from Wastewater Effluents Using Chitosan-Zno Nanocomposite. *J. Textile Sci. Eng.* **2015,** *5* (3), 200.

Akbar, E.; Omid, M.; Ali, F.; Fahimeh, N.; Reza, A.; Vahid, H. Evaluation of the Potential Cationic Dye Removal Using Adsorption by Graphene and Carbon Nanotubes as Adsorbents Surfaces. *Arab. J. Chem,* **2017,** *10,* S2862–S2869.

Akduman, C.; Akçakoca Kumbasar, E. P.; Morsunbul, S. Electrospun Nanofiber Membranes for Adsorption of Dye Molecules from Textile Wastewater. *IOP Conf. Ser. Mater. Sci. Eng.* **2017,** *254* (10), 1–10.

Akduman, C.; Morsümbül, S. E.; Kumbasar, E. P. A. The Removal of Reactive Red 141 from Wastewater: A Study of Dye Adsorption Capability of Water-Stable Electrospun Polyvinyl Alcohol Nanofibers. *AUTEX Research Journal.* **2019.**

Amar, I. A.; Sharif, A.; Alkhayali, M.; Jabji, M.; Altohami, F.; AbdulQadir, M.; Ahwidi, M. Adsorptive Removal of Methylene Blue Dye from Aqueous Solutions Using CoFe1.9Mo0.1O4 Magnetic Nanoparticles. *Iran. J. Energy Environ.* **2018,** *9* (4), 247–254.

Arej, S.; Al-Gorair. Treatment of Wastewater from Cationic Dye Using Eco-Friendly Nano-composite: Characterization, Adsorption and Kinetic Studies. *Egypt. J. Aquat. Res.* **2019,** *45,* 25–3.

Badr, M. T.; El-Hamshary, H.; Al-Deyab, S. S.; El-Newehy, M. H. Functionalized Electrospun Carbon Nanofibers for Removal of Cationic Dye. *Arab. J. Chem,* **2019,** *12,* 747–759.

Beg, S.; Rizwan, M.; Sheikh, A. M.; Hasnain, M. S.; Anwer, K.; Kohli, K. Advancement in Carbon Nanotubes: Basics, Biomedical Applications and Toxicity. *J. Pharm. Pharmacol.* **2011,** *63,* 141–163.

Bennett, S. W.; Keller, A. A. Comparative Photoactivity of CeO_2, γ-Fe_2O_3, TiO_2 and ZnO in Various Aqueous Systems. *Appl. Catal. B: Environ.* **2011,** *102* (3–4), 600–607.

Bernd, N. *Pollution Prevention and Treatment Using Nanotechnology*; Wiley Online Library, 2010. 10.1002/9783527628155.nanotech010.

Byeon, J. H.; Kim, Y. W. Gas-Phase Self-Assembly of Highly Ordered Titania@ Graphene Nanoflakes for Enhancement in Photocatalytic Activity. *ACS Appl. Mater. Inter.* **2013**, *5* (9), 3959–3966.

Ede, S. R.; Anantharaj, S.; Sakthikumar, K.; Karthick, K.; Kundu, S. Investigation of Various Synthetic Protocols for Self-Assembled Nanomaterials and Their Role in Catalysis: Progress and Perspectives. *Mater. Today Chem.* **2018**, *1* (10), 31–78.

Ehsan Sadeghi, P.; Hooman, F.; Mohammad, F-D. Batch Removal of Pb (II) Ions from Aqueous Medium Using Gamma-Al_2O_3 Nanoparticles/Ethyl Cellulose Adsorbent Fabricated via Electrospinning Method: An Equilibrium Isotherm and Characterization Study. *Pol. J. Chem. Technol.* **2018**, *20* (2), 32–39.

Farbod, M.; Kajbafvala, M. Effect of Nanoparticle Surface Modification on the Adsorption-Enhanced Photocatalysis of Gd/TiO_2 Nanocomposite. *Powder Technol.* **2013**, *239*, 434–440.

Farghali, M. A.; El-Din, T. A. S.; Al-Enizi, A. M.; El Bahnasawy, R. M. Graphene/Magnetite Nanocomposite for Potential Environmental Application. *Int. J. Electrochem. Sci,* **2015**, *10*, 529–537.

Faria, P. C.; Monteiro, D. C.; Órfão, J. J.; Pereira, M. F. Cerium, Manganese and Cobalt Oxides as Catalysts for the Ozonation of Selected Organic Compounds. *Chemosphere* **2009**, *74* (6), 818–824.

Gehrke, I.; Geiser, A.; Somborn-Schulz, A. Innovations in Nanotechnology for Water Treatment. *Nanotechnol. Sci. Appl.* **2015**, *8*, 1–17.

Ghasemi, Z.; Younesi, H.; Zinatizadeh, A. A. Preparation, Characterization and Photocatalytic Application of TiO_2/Fe-ZSM-5 Nanocomposite for the Treatment of Petroleum Refinery Wastewater: Optimization of Process Parameters by Response Surface Methodology. *Chemosphere* **2016**, *159*, 552–564.

Ghuge, S. P.; Saroha, A. K. Catalytic Ozonation for the Treatment of Synthetic and Industrial Effluents-Application of Mesoporous Materials: A Review. *J. Environ. Manage.* **2018**, *211*, 83–102.

Gomez-Pastora, J.; Bringas, E.; Ortiz, I. Recent Progress and Future Challenges on the Use of High Performance Magnetic Nano-Adsorbents in Environmental Applications. *Chem. Eng. J.* **2014**, *256*, 187–204.

Herney-Ramirez, J.; Vicente, M. A.; Madeira, L. M. Heterogeneous Photo-Fenton Oxidation with Pillared Clay-Based Catalysts for Wastewater Treatment: A Review. *Appl. Catal. B-Environ.* **2010**, *98* (1), pp 10–26.

Jame, S. A.; Zhou, Z. Electrochemical Carbon Nanotube Filters for Water and Wastewater Treatment. *Nanotechnol. Rev.* **2016**, *5* (1), 41–50.

Janaki, V. K.; Vijayaraghavan; Byung-Taek, O.; Kui-Jae, L.; Muthuchelian, K.; Ramasamy, A. K.; Seralathan, K-K. Starch/Polyaniline Nanocomposite for Enhanced Removal of Reactive Dyes from Synthetic Effluent. *Carbohydr. Polym.* **2012**, *90*, 1437–1444.

Ji-Huan He; Yong Liu; Lu-Feng Mo; Yu-Qin Wan; Lan Xu. *Electrospun Nanofibres and Their Applications*; Smithers Rapra Technology, 2008.

Kamarulzaman, N.; Kasim, M. F.; Rusdi, R. Band Gap Narrowing and Widening of ZnO Nanostructures and Doped Materials. *Nanoscale Res. Lett.* **2015**, 10.

Kefeni, K. K.; Mamba, B. B.; Msagati, T. A. M. Application of Spinel Ferrite Nanoparticles in Water and Wastewater Treatment: A Review. *Sep. Purif. Technol.* **2017**, *188*, 399–422.

Kim, S. H.; Choi, P. P. Enhanced Congo Red Dye Removal from Aqueous Solutions Using Iron Nanoparticles: Adsorption, Kinetics, and Equilibrium Studies. *Dalton T.* **2017**, 15470–15479.

Klumpp, C.; Kostarelos, K.; Prato, M.; Bianco, A. Functionalized Carbon Nanotubes as Emerging Nanovectors for the Delivery of Therapeutics. *Biochem Biophys Acta* **2006**, *1758*, 404–412.

Lee, S. Y.; Shim, H. E.; Yang, J. E.; Choi, Y. J.; Jeon, J. Continuous Flow Removal of Anionic Dyes in Water by Chitosan-Functionalized Iron Oxide Nanoparticles Incorporated in a Dextran Gel Column. *Nanomaterials* **2019**, *9* (8), 1164.

Bian, Z.; Zhu, J.; Zhang, D.; Li, G.; Huo, Y.; Li, H.; Lu, Y. Mesoporous Titania Spheres with Tunable Chamber Structure and Enhanced Photocatalytic Activity. *J. Am. Chem. Soc.* **2007**, *129* (27), 8406–8407.

Malik, S. N.; Ghosh, P. C.; Vaidya, A. N.; Mudliar, S. N. Catalytic Ozone Pretreatment of Complex Textile Effluent Using Fe^{2+} and Zero Valent Iron Nanoparticles. *J. Hazard. Mater.* **2018**, *357*, 363–375.

Mehta, D.; Mazumdar, S.; Singh, S. K. Magnetic Adsorbents for the Treatment of Water/Wastewater—A Review. *J. Water Process Eng.* **2015**, *7*, 244–265.

Mohammad, A. M.; Nima, M. Removal of Cadmium and Zinc Ions from Industrial Wastewater Using Nanocomposites of PANI/ZnO and PANI/CoHCF: A Comparative Study. *Desalin. Water Treat.* **2016**, 20817–20836.

Morillo Martín, D.; Magdi Ahmed, M.; Rodríguez, M.; García, M.; Faccini, M. Aminated Polyethylene Terephthalate (PET) Nanofibers for the Selective Removal of Pb(II) from Polluted Water. *Materials (Basel)* **2017**, *10* (12), 1352.

Mozhdeh, G.; Ali, A. G.; Mokhtar, A.; Negar, T.; Babak, R. Fabrication of Electrospun Polyamide-6/Chitosan Nanofibrous Membrane toward Anionic Dyes Removal. *J. Nanotechnol.* **2014**, 1–12.

Ng, K. H.; Yuan, L. S.; Cheng, C. K.; Chen, K.; Fang, C. TiO_2 and ZnO Photocatalytic Treatment of Palm Oil Mill Effluent (POME) and Feasibility of Renewable Energy Generation: A Short Review. *J. Clean. Prod.* **2019**, 209–225.

Ola, O.; Maroto-Valer, M. M. Review of Material Design and Reactor Engineering on TiO_2 Photocatalysis for CO_2 Reduction. *J. Photoch. Photobio. C.* **2015**, 16–42.

Parijat, P.; Mandeep, D. Carbon Nanotubes: Types, Methods of Preparation and Applications. *Int. J. Pharm. Sci. Res.* **2016**, 15–21.

Payra, S.; Challagulla, S.; Indukuru, R. R.; Chakraborty, C.; Tarafder, K.; Ghosh. B.; Roy, S. The Structural and Surface Modification of Zeolitic Imidazolate Frameworks towards Reduction of Encapsulated CO_2. *New J. Chem.* **2018**, 19205–19218.

Pedro, H. C. C.; Kestur, G. S.; Fernando, W.; Nanocomposites: Synthesis, Structure, Properties and New Application Opportunities. *Mater. Res.* **2009**, 1–39.

Phenrat, T.; Thongboot, T.; Lowry, G. V. Electromagnetic Induction of Zerovalent Iron (ZVI) Powder and Nanoscale Zerovalent Iron (NZVI) Particles Enhances Dechlorination of Trichloroethylene in Contaminated Groundwater and Soil: Proof of Concept. *Environ. Sci. Technol.* **2015**, 872–880.

Pimolpun, K.; Pitt, S.; Preparation and Adsorption Behavior of Aminated Electrospun Polyacrylonitrile Nanofiber Mats for Heavy Metal Ion Removal. *ACS Appl. Mater. Interf.* **2010**, 3619–3627.

Qiao, J.; Gao, S.; Yao, J.; Zhang, L.; Li, N.; A Novel and Green Adsorbent Based on Sulfobetaine-Modified Magnetic Fe_3O_4 Nanoparticles (SBMNPs) was Successfully Synthesized via a Convenient One-Step Chemical Co-Precipitation Method and Applied to the Removal of the Anionic Dyes Methyl Blue (MB) and Amaranth (AM) from Aqueous Solution. *AIP Adv.* **2019**, 065308

Raheel, A. H.; Umair, A. Q.; Raja, F. Q.; Rasool, B. M.; Muzamil, K.; Farooq, A.; Zeeshan, K.; Ick, S. K.; Efficient Removal of Reactive Blue 19 Dye by Co-Electrospun Nanofibers. *Preprints* **2018**, 1–14.

Rastogi, V.; Yadav, P.; Bhattacharya, S. S.; Mishra, A. K.; Verma, N.; Verma, A. Carbon Nanotubes: An Emerging Drug Carrier for Targeting Cancer Cells. *J. Drug Deliv.* **2014**, 1–23.

Reddy, D. H. K.; Yun, Y. S. Spinel Ferrite Magnetic Adsorbents: Alternative Future Materials for Water Purification. *Coord. Chem. Rev.* **2016**, 90–111.

Roy, R.; Roy, R. A.; Roy, D. M. Alternative Perspectives on "Quasicrystallinity": Non-Uniformity and Nanocomposites. *Mater. Lett.* **1986**, 323–328.

Rui, Z. Y.; Wang, X. L.; Bolun, S. W. Synthesis of β-Cyclodextrin-Based Electrospun Nanofiber Membranes for Highly Efficient Adsorption and Separation of Methylene Blue. *ACS Appl. Mater. Interf.* **2015**, 26649–26657.

Sahar, A.; Ali, A. A.; Shahram, G. High Effective Adsorption of Acid Fuchsin Dye Using Magnetic Biodegradable Polymer-Based Nanocomposite from Aqueous Solutions. *Microchem. J.* **2019**, 103966

Saifuddin, N.; Raziah, A. Z.; Junizah, A. R. Carbon Nanotubes: A Review on Structure and Their Interaction with Proteins. *J. Chem.* **2013**, Article ID 676815.

Saleh, T. A.; Gupta, V. K. Functionalization of Tungsten Oxide into MWCNT and Its Application for Sunlight-Induced Degradation of Rhodamine B. *J. Colloid. Interf. Sci.* **2011**, 337–344.

Samaneh, S. S.; Saeed, S. S.; Hamed, J. Y.; Mojdeh, M. Adsorption of Anionic and Cationic Dyes from Aqueous Solution Using Gelatin-Based Magnetic Nanocomposite Beads Comprising Carboxylic Acid Functionalized Carbon Nanotube. *Chem. Eng. J.* **2017**, 1133–1144.

Saxena, S.; Raja, A. S. M.; Arputharaj, A. Challenges in Sustainable Wet Processing of Textiles. *Text. Cloth. Sustain.: Sustain. Text. Chem. Process.* **2017**, 43–79. doi: 10.1007/978-981-10-2185-5_2 ISSN: 2197-9863.

Singh, S.; Srivastava, V. C.; Mandal, T. K.; Mall, I. D.; Lo, S. L. Synthesis and Application of Green Mixed-Metal Oxide Nano-Composite Materials from Solid Waste for Dye Degradation. *J. Environ. Manage.* **2016**, 146–156.

Sulekha, M. Nanotechnology for Waste Water Treatment. *Int. J. Chem. Stud.* **2016**, 22–24.

Umair, A. Q.; Zeeshan, K.; Farooq, A.; Muzamil, K.; Ick-Soo, K. Electrospun Zein Nanofiber as a Green and Recyclable Adsorbent for the Removal of Reactive Black 5 from the Aqueous Phase. *ACS Sustain. Chem. Eng.* **2017**, 4340–4351.

Wang, L.; Wang, A. Adsorption Characteristics of Congo Red onto the Chitosan/Montmorillonite Nanocomposite. *J. Hazard. Mater.* **2007**, 979–985.

Wu, R.; Zheng, G. F.; Li, W. W.; Zhong, L. B.; Zheng, Y. M. Electrospun Chitosan Nanofiber Membrane for Adsorption of Cu(II) from Aqueous Solution: Fabrication, Characterization and Performance. *J. Nanosci. Nanotechnol.* **2018**, 5624–5635.

Yani, H.; Juan, X.; Qinqin, Z.; Chang, Cui; Chuan, Wang. 'Facile Synthesis of Surface-Functionalized Magnetic Nanocomposites for Effectively Selective Adsorption of Cationic Dyes. *Nanosc. Res. Lett.* **2018**, 1–9.

Yazdi, M. G.; Ivanic, M.; Alaa, M.; Uheida, A. Surface Modified Composite Nanofibers for the Removal of Indigo Carmine Dye from Polluted Water. *RSC Adv. R. Soc. Chem.* **2018**, 24588–24598.

You, H.; Wu, Z.; Jia, Y.; Xu, X.; Xia, Y.; Han, Z.; Wang, Y. High-Efficiency and Mechano-/Photo-Bi-Catalysis of Piezoelectric-ZnO@ Photoelectric-TiO$_2$ Core-Shell Nanofibers for Dye Decomposition. *Chemosphere* **2017**, 528–535.

Zafara, M. N.; Dara, Q.; Nawazb, F.; Zafarc, M. N.; Iqbald, M.; Nazar, M. F. Effective Adsorptive Removal of Azo Dyes Over Spherical ZnO Nanoparticles. *J. Mater. Res. Technol.* **2019**, *8* (1), 713–725.

Zheng, C.; Zhao, L.; Zhou, X.; Fu, Z.; Li, A. Treatment Technologies for Organic Wastewater. *Water Treat.* **2013**, 249–286.

CHAPTER 13

Recent Advances of Nanoparticles in the Removal of Textile Dyes

REKHA SHARMA[1*], ANKITA DHILLON[1], and DINESH KUMAR[2]

[1]*Department of Chemistry, Banasthali Vidyapith, Rajasthan, India*

[2]*School of Chemical Sciences, Central University of Gujarat, Gandhinagar, India*

Corresponding author. E-mail: dinesh.kumar@cug.ac.in

ABSTRACT

Nowadays, air, soil, and water pollution are the different pollutions that are contaminating the environment. Water pollution is the major pollution, which is a severe concern for mankind. Through the production of poisonous waste matters, such as toxic metals, dyes, harmful chemicals, and various other contaminants in pharmaceutical industries, the increasing industrialization and urbanization play a main part in water pollution. Eventually, the water pollution occurs because of the animals and human contact over the earth with soil and water system. Without any pretreatment and further removal in the form of degradation by various methods, these toxicants, several textile waste matters are directly disposed into water bodies. In the manufacturing of carpets and clothes, about 80% of dyes form strong covalent bonds with the textile fibers, and the residual dye turns into a waste product and is disposed of into water bodies. The auxochrome and chromophore are the two molecular components which are accountable for the preferred color production in a dye; therefore, synthetic dyes are used in the textile industries. It has been demonstrated that approximately 7×10^5 tons per year of dyes are produced, which are increasing continually. Dyes can be categorized into diverse types, such as dispersive, reactive, azo, basic, and acidic dyes. Reactive

Fundamentals of Nano-Textile Science. Prashansa Sharma, Devsuni Singh & Vivek Dave (Eds.)

dye demonstrates extra steadiness and is available in abundance among all categories of dyes. After breaking of azo bonds, the azo dyes become more carcinogenic owing to the development of aromatic amines. These dyes are extensively used in wool, in textile to color cotton, polyamide, and silk. For the removal of textile dyes, various techniques have been utilized, for example, biological, membrane separation, ion exchange, precipitation, flocculation, adsorption, chemical, and photocatalysis. However, most of the techniques have limitations, such as a small surface area, less efficiency, and cost. To overcome these inadequacies for the removal of textile dyes, recent research is centered over eco-friendly, commercial, and display elevated surface area NPs.

13.1   INTRODUCTION

Due to lack of handling of manufacturing wastes, widespread industrialization has caused widespread water pollution of various water reservoirs. Several pollutants, for example, toxic metals, fertilizers, pesticides, suspended solids, and dyes, etc. contaminate the water (Gurses et al., 2006; Karim et al., 2014). Prior to their utilization for household uses, severe regulations have been completed for the handling of these contaminants. Dyes are usually utilized in pharmaceutical, textile, tanning, paper, and food manufacturing (Ashiq et al., 2012; Yagub et al., 2014; Deka et al., 2014). Owing to their nonbiodegradable, toxic, and recalcitrant character, the removal of dyes is essential for water resources (Jain et al., 2014; Li et al., 2013). The presence of dye has a harmful result in aquatic life and affects the chemical oxygen demand (COD) of the sunlight infiltration and waste matter (Zhou et al., 2014). Basic dyes impart color to water because they have such high color intensity, which makes it adverse and unfit for utilization even at trace levels. The acute exposure to basic dyes, for example, Rhodamine B (RhB) and methylene blue (MB) results in serious health issues, though these are not as harmful as reactive or azo dyes. On inhalation, MB can cause an augmented heart rate, on ingestion through mouth, it causes nausea, vomiting, tissue necrosis, quadriplegia, and jaundice (Ashiq et al., 2012; Gürses et al., 2014). In the biotechnology field, the RhB dye is usually utilized as a staining dye. It is known to cause irritation of the respiratory tract, eyes, and skin as well as carcinogenicity and neurotoxicity (Mittal and Mishra, 2014; Bhattacharyya et al., 2014). Previously, for the removal of these dyes, various methods have been utilized. But, adsorption is the extensively adopted and simplest method as shown in Figure 13.1 (Gupta et al., 2000; Vinod and Anirudhan,

2003). Other methods include membrane filtration, coagulation/flocculation, ozonation, photodegradation, and biological and chemical degradation. If the utilized adsorbents for the removal of dyes are nontoxic, cheaper, and eco-friendly, then the adsorption method becomes further inexpensive and enviable. To bring out improved substitutes for the existing alternatives, progressively, researchers are experimenting on this method. Due to the existence of functional phenolic and carboxylic groups, humic acid (HA) can combine with various natures of materials and chemical compounds (Sedláček et al., 2014). To fabricate trihalomethanes, HA is competent in reacting with chlorine in the water that is potentially a carcinogenic agent and imparts color and odor to water (Deng and Bai, 2003). The presence of these groups creating it toxic because of the chemical and structural changes in HA. The removal of dye can be done through a binding capacity of -OH and -COOH groups, which also causes toxicity to the HA. Previously, for the efficient removal of contaminants, HA demonstrated an outstanding efficiency (Liu et al., 2008; Tan et al., 2008; Abdul et al., 1990; Li et al., 2003). In HA, when the -COOH and -OH groups are deprotonated in faintly basic or acidic media, the negatively charged HA easily absorbs the positively charged basic dyes. On the water-insoluble matrix, the HA requires its immobilization because of its water-soluble properties. The biopolymeric membrane of HEC and SA, which is cross-linked with GA, was used for immobilization. HEC and SA can form blend film because of their compatible blending through their water-soluble nature (Naidu et al., 2005).

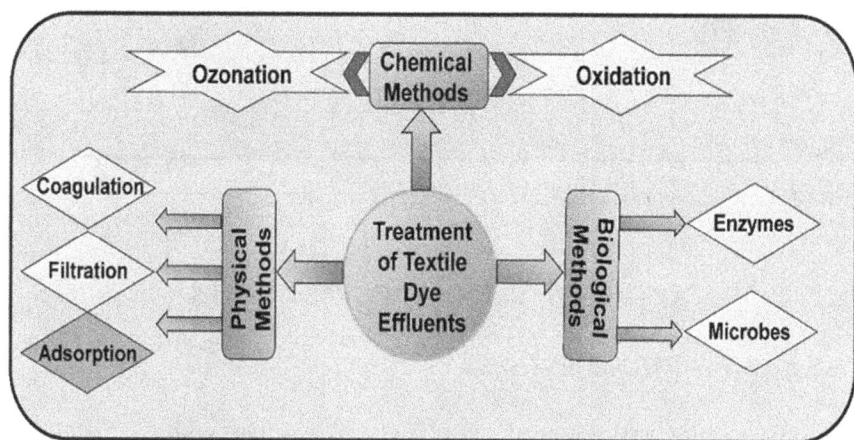

FIGURE 13.1 Various techniques for the removal of textile dye.

For the heavy metal adsorption process, Chen et al. reported the immobilization of HA on sodium alginate (Chen et al., 2011; Chen et al., 2012). Though, on this adsorbent membrane, no previous research has been done for the adsorption of dyes. This work primarily focuses on the removal of MB and RhB through adsorption by using SA/HEC/HA composite membrane which is cross-linked with GA. The adsorption of dye has been investigated by optimizing various parameters that affect the adsorption, for example, temperature, initial concentration, pH, and adsorbent dosage. The adsorption data were tested by using different isotherm models, that is, Langmuir, Freundlich, D–R, and Temkin isotherms. The adsorption mechanism was explained by the optimization of adsorption data using PFO and PSO kinetic models. The thermodynamic parameters were known to find the nature of the adsorption process and the feasibility of reaction process. Finally, the reusability studies of the membrane were performed after the adsorption of both methylene blue and Rhodamine B dyes. The mechanism of adsorption of dye using nanoparticles has been shown in Figure 13.2.

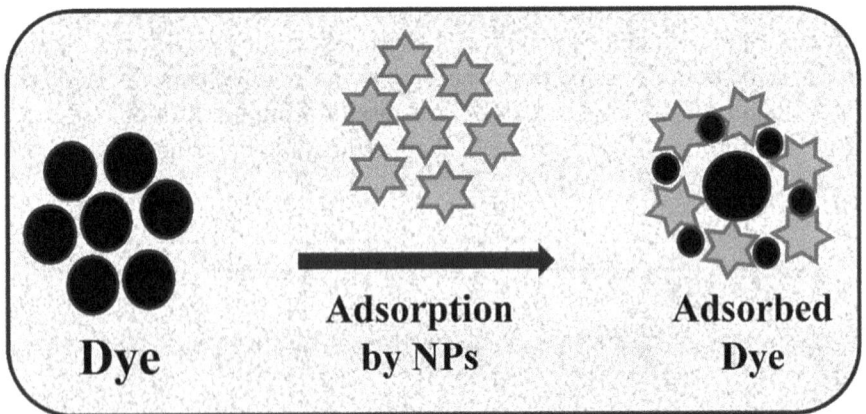

FIGURE 13.2   Schematic presentation of the removal of dye using NPs.

13.2   NANOPARTICLES FOR THE REMOVAL OF DYES

13.2.1   NANOPARTICLES (NPS)

The NPs exhibit various properties, such as they show slight dispersion resistance, benefits of enormous precise surface area, higher adsorption steadiness, and advanced adsorption ability. Because of the effect of an external magnetic field, the Fe_3O_4 NPs likewise permit the easy removal of dyes (Kong et al., 2012; Xu et al., 2013). For the adsorption of textile dyes from wastewater, these properties of NPs attracted various researcher's attention to study their potential. The research results specified that to adsorb textile dyes from water, Fe_3O_4 NPs can be efficiently utilized (Giri et al., 2011). To re-utilize surplus iron ore tailings, the synthesized Fe_3O_4 NPs could be considered a viable input. To harvest Fe_3O_4 NPs, various methods have been employed, such as the coprecipitation method and acid leaching-precipitation method. The developed Fe_3O_4 NPs exhibited rapid adsorption of CR dyes and MB dyes. The adsorbent exhibits over 85% desorption after the adsorption of both dyes, which showed the reusability of the adsorbent. Currently, for the adsorption of dyes from the binary system, that is, Direct green 6, Direct red 31, and Acid red 18, Mahmoodi (2014) developed manganese ferrite NPs using iron nitrate and manganese nitrate. The adsorption of dyes utilizing the magnetic adsorbents has resulted that in the binary system, there was no perceived selectivity. The adsorption of Crystal Violet, Malachite Green, and MB dye was reported by Debrassi et al. (2014) by the utilization of chitosan products in their magnetic NPs with 144.79, 248.42, and 223.58 mg/g of adsorption capacities, respectively. The adsorbent showed the increased adsorption potential of the nanoparticulate system through the reusability test with minor influence on adsorption performance after three repetitive cycles. In their adsorbent, Fan et al. (2013) assembled many components comprising chitosan. The simple removal, advanced adsorption capacity, and developed robustness are the properties of individual components of the resultant adsorbent. To adsorb the MB dye, the synthesized adsorbent was found to be potential. For the adsorption of textile dyes from aqueous systems, some nonmagnetic NPs have been described likewise along with the magnetic NPs. The Acid Red 14 dye removal was reported by Ahmed et al. (2016) using nanopolyaniline as adsorbent material. In the experiment with the addition of biosorbent baker's yeast, the removal efficiency of the adsorbent for Acid Red 14 was improved to 430 mg/g from 323 mg/g. These results show the potential of polyaniline nanoparticulate system for choice as an adsorbent and further reveal that this method produces promising results.. The adsorption of Malachite Green dye material has been reported by Lee et al. (2011) using nanosized AMP clay. For the adsorption process, various parameters have been utilized. However, the maximum adsorption capacities for dye removal were exhibited to be 81.72% at the value of 334.8 mg/g, and an adsorbent dosage of AMP clay was found to be 0.1 mg/mL; the adsorption process has resulted in the complete removal of dye when AMP clay dosage was 0.2 mg/mL. The adsorption of EY, a toxic dye from the aqueous solution, was reported by Assefi et al. (2014) using Co_2O_3-NP-AC as an outstanding sorbent. From the aqueous solutions, the adsorption of dyes on Co_2O_3-NP-AC can be a low-cost, green, and an effective adsorbent showing increased adsorption capacity. Various adsorbent materials are listed in Table 13.1 with their adsorption capacities and optimizing conditions.

For effective adsorption of azo dyes, spherical ZnO-NPs have been described by Zafar et al. (2018). For the efficient sustainable supply of drinking water, technology-based smart handling methods are essential to reduce the harmful contaminants of the polluted water. For the elimination of water toxins, the preferred innovative adsorbents are nanosized metal oxides, for example, the materials which possess the versatility, simplicity, high surface reactivity, and efficacy for the adsorption process. The nanostructured ZnO adsorbent synthesized by coprecipitation shows remarkable efficiency toward the adsorption of extensively used amaranth (AM), methyl orange (MO), and azo dyes from water solutions. Various analytical techniques, such as SEM, XRD, BET, and FTIR characterized the synthesized NPs. The synthesized ZnO-NPs were additionally used as adsorbent for the adsorption of toxic azo dyes (AM and MO) from water after balancing characterization. The results showed that a quantity of 0.3 g ZnO-NPs exhibited maximum adsorption competence at pH 6 of the respective dye at 40 ppm concentration. The kinetics studies exposed that the adsorption data were well-fitted to the PSO kinetics model, and the Langmuir model was fitted for the thermodynamic data for both dyes on ZnO-NPs. The results proposed that the $R\text{-}SO_3$–groups were the active sites in the prime adsorption mechanism of selected dye removal system, which may be because of the electrostatic attraction among ZnO-NPs and the dyes.

In conclusion, in future water treatment, for the removal of anionic dyes, the ZnO-NPs could be utilized under suitable circumstances and observed as a profitable adsorbent material (Zafar et al., 2019).

For deprivation and hydrogenation of water pollutants, for example, Orange G (OG), 4-nitrophenol (4-NP), and Methyl Orange (MO), Kulkarni et al. (2019) described a biodegradable green sustainable nanocatalyst Au@NiAg showing increased catalytic action. The developed Au@NiAg NPs are smaller in size and rounded in morphology, which were equipped from triangular-shaped Au@Ni NPs, ascribed to the digestive ripening of the Au@Ni, where Ag ion was placed on the surface of Au@Ni NPs acting as a host material. The reduction of organic dyes and hydrogenation of 4-nitrophenol Au@NiAg NPs

TABLE 13.1 The Various NPs for the Removal of Different Textile Dye Effluents.

S. No.	Dyes	Material	Conditions	Qm (mg g-1)	Reference(s)
1.	Congo red	Fe_3O_4 NPs	10.45–55.72 mg/L of initial dye concentration; 6.2 of pH range; 0.3–0.9 g/L of adsorbent dose; 95% of desorption capacity	172.4	Giri et al. (2011)
2.	Solvent green 7	Organo-functionalized magnetite microsphere	70–90 mg/L of initial dye concentration and acidic pH conditions	81.82–100.52	Lian et al. (2013)
3.	Rhodamine B	Fe_2O_3 NPs incorporated with Palm Kernel Shell	100–500 mg/L of initial dye concentration; 5–8 of pH range; 0.1 g of adsorbent dose	625	Panneerselvam et al. (2012)
4.	MB	Magnetic nanoparticles coated in activated maize cob powder	250 mg/L of initial dye concentration; 6.0 pH; 0.4 g/100 mL of adsorbent dose; and 45 min of contact time	70.29	Tan et al. (2012)
5.	Methylene blue	Magnetite nanoparticles	1.6–32 mg/L of initial dye concentration; 9.2 of pH range; 0.3–0.9 g/L of adsorbent dose; 85% of desorption	70.4	Giri et al. (2011)
6.	Methylene blue	Magnetic B cyclodextrin-chitosan/ graphene oxide (MCCG)	Alkaline pH; 0.01 g/25 mL of adsorbent dose	84.32	Fan et al. (2013)
7.	Basic red 18 Basic blue 41 Basic blue 9	Magnetic ferrite nanoparticles-alginate composite	50 mg/L of initial dye concentration; 8 pH	56 25 106	Mahmoodi (2013)
8.	Safranin O	SDS-modified magnetite nanoparticles	pH: 3. Desorption: >95%	769.23	Shariati et al. (2011)
9.	MB CV MG	N-benzyl-Ocarboxymethylchitosan magnetic nanoparticles	100–300 mg/L of initial dye concentration; 3–5 of pH range	223.58 248.42 144.79	Debrassi et al. (2012)

TABLE 13.1 *(Continued)*

S. No.	Dyes	Material	Conditions	Qm (mg g-1)	Reference(s)
10.	Methylene blue and Congo red	SDS-modified magnetite nanoparticles	30 mg/L of initial dye concentration; 6.2 pH; 2 min of contact time; 0.15 g/L of adsorbent dose; 95% desorption	70.4 172.4	Giri et al. (2011)
11.	Orange II	Graphene Nanosheets incorporated $Fe_2Co_3O_4$	Increases on pH (4–10), temperature (25°C–45°C), and PMS dosage (0.04–0.60 g/L), on concentration (30–120 mg/L) decreases	E_a 49.5 kJ/mol	Yao et al. (2013)
12.	Yellow and blue dye	*Solanum nigrum* AgNPs and *Cannabis sativa* AgNPs	Alkaline pH 9 and at 60°C.	—	Khatoon and Sardar (2017)

fitted to Langmuir–Hinshelwood kinetic model. The activity factors of OG, MO, and 4-NP were found to be 3810, 5476, and 3167 s^{-1}/g correspondingly on Au@NiAg NPs which revealed an excellent catalytic activity. For the complete performance improvement of Au@NiAg, the AgNPs acted as cocatalyst. With the invention of nontoxic, eco-friendly, biodegradable, effective, and sustainable core-shell nanocatalyst, the usage of *A. vasica* plant extract facilitated the formulation which will open an area with model catalytic properties.

From the results of the TEM monographs, it was observed that the produced NPs exhibit a core-shell structure with alloy formation and are identical in size. The synthesized nanocatalyst was specifically efficient when organic pollutants were decreased. Actually, for OG, MO, and 4-NP, the author has reported exceedingly increased activity properties of 3810, 5476, and 3167 s^{-1}/g, respectively. For these types of procedures, the reported standards are greater in comparison to formerly described catalysts. The accumulation of nanocatalyst is the key cause for the outstanding decoloration of dyes and a decrease of 4-nitrophenol as well, which was controlled by the shielding of Au NPs core through Ni shell. The Ag NPs increased the catalytic action of the developed core-shell NPs and act as cocatalyst. In reprocessed responses, the developed nanocatalyst has revealed an excellent catalytic activity (Kulkarni et al., 2019).

Because of the average size reduction of IO nanoclusters from 11.3 to 9.3 nm, Kaloti and Kumar (2016) described the coating of IO NPs which harvests the CIO, showing the optical characteristics and improved functionalities which were predicted through TEM analysis. Various studies, for example, FESEM, AFM, XRD, TEM, and BET confirmed the effective capping of IO NPs by chitosan. The higher saturation magnetization at 72.5–75.5 emu/g of Ag–γ-Fe$_2$O$_3$@Cs at room temperature offers a significant characteristic for their biomedical, SERS, and catalytic submissions, whereas the magnetization was found at 69.2 emu/g in the case of CIO. For the reduction of MO dye even at a low concentration of Ag up to 30 nM, the as-synthesized Ag–γ-Fe$_2$O$_3$@Cs, without any loss of efficacy, could be reused up to 10 cycles, and displayed equally increased catalytic efficacy.

The novel Ag–γ-Fe$_2$O$_3$@Cs comprises various functional groups, such as −CO, −OH, and −NH$_2$ in chitosan. In this adsorbent material, chitosan functioned as an operative additive that offers the surface and biocompatible coating for incorporating dye, bacteria, Ag, and IO. Besides antibacterial and catalytic activity, the perceived distinguishing features also designate their perspective for anticarcinogenic activities and hyperthermia. Their biomedical and ecological perspective were clearly signified through SERS activity for model dye p-ATP having a detection limit at 10 pM, and the antibacterial activity with MBC and MIC at 4.2 and 1.1 μg/mL for model bacteria *E. coli* on Ag−γ-Fe$_2$O$_3$@Cs was also demonstrated (Kaloti and Kumar, 2016).

Yang et al. (2016) reported the use of CMCT and BNCC in an innovative biopolymer-based BNCC−CMCT aerogel synthesized through a Schiff base reaction by freeze-drying a hydrogel. From the limited chlorite oxidation and consecutive periodate of cellulose, the BNCC can be acquired shadowed through the treatment of hot water. A model system covering MB (cationic) dye shows the potential of this bioaerogel in ecological handling. The dye has adsorbed due to the electrostatic complexation among acidic groups of anionic aerogel and has resulted with long-time equilibrium, dye concentration, and time-dependent dye adsorption abilities. The Langmuir isotherm model was best fitted to the adsorption data. The equilibrium constant was found at room temperature at K ~ 0.0089 and determined a removal efficacy of ~785 mg/g. An excellent removal efficiency resulted due to the negative charge of the aerogel and its extremely porous nature. Fruitful adsorption−regeneration cycles showed outstanding recyclability up to six cycles, and above an extensive pH range (for example, pH > 7), the adsorbent endured constant adsorption capacity which was 785 mg/g for MB dye. Through an eco-friendly cross-linking method, the BNCC−CMCT aerogel is prepared from recyclable starting constituents, so it is a promising "green" adsorbent (Yang et al., 2016).

For the degradation of dye, Wang et al. (2017) developed PS-b-PSAN-capped ZnO precursors. The progress and development of effective and ecologically benevolent photocatalysts driven through severe water contamination necessitates a modern method that allows the elimination of contaminants upon visible light irradiation. The ZnO/C hybrid materials have been revealed to be effective photocatalysts. The increased accessible surface area of ZnO is essential to advance the efficacy of hybrid materials of ZnO/C. Inside a porous C matrix, a good dispersal of ZnO nanocrystals was attained because the PS block inhibits the accumulation of ZnO particles during the pyrolysis process. They found the adsorption capacity of the synthesized adsorbent of dye at 125 mg/g with 0.021 min^{-1} of degradation rate constant (Wang et al., 2017).

Sharma et al. (2017) reported the adsorption of CV dye on phase-tuned quantum confined In$_2$S$_3$ nanocrystals. By using ILs as structure-guiding agents, the synthesized nanocrystals are solvothermally available. Discriminating alteration of morphology, shape, size, and above all, crystal phase of In$_2$S$_3$ is achieved through altering the aromatic π-stacking ability of the

$[C_n mim]Br$ ILs, the alkyl side chain length, and the H-bonding, here, $n = 2$, 4, 6, 8, and 10. It is perceived that the synthesis was deprived of ILs keeping the other reaction parameters the same, whereas the crystallite size is expressively less when ILs were utilized. Pure tetragonal form of β-In_2S_3 is formed once no IL is utilized at 150°C, though a pure cubic phase crystallizes at the same reaction condition in the presence of $[C_n mim]Br$ [n = 2,4]. Similarly, to obtain morphology controlled and pure phase nanocrystals, sensible selection of precursors and reaction temperature will have a deep consequence. In the presence of visible and dark light, the adsorption-determined photocatalytic and catalytic action of synthesized nanosized In_2S_3 is recognized by studying the degradation of CV dye. For the In_2S_3 nanocrystals, for the removal of CV dye, a maximum of 94.8% catalytic efficacy is attained using TMAB ionic liquid.

The results displayed that with the help of robust interface among the n-type nanomaterials and cationic dye, the cationic dye can be degraded through the adsorption method even in the dark (Sharma et al., 2017).

For the adsorption of MB dye and Cr(VI) heavy metal ion photochemically, polyaniline-capped g-C_3N_4 nanosheets have been synthesized by Wu et al. (2019). Moreover, by capping an organic acid (PA) and an inorganic acid HCl with PANI, the CNns have been improved. The results demonstrated that having an advanced precise surface area and a 3D hierarchical structure, these samples showed prolonged absorption of visible light. The reason for their increased photocatalytic presentation is due to the presence of abundant free radicals, condensed electrical resistance, and inhibited photoluminescence emission. In the Cr(VI) stalk concentration of 100 ppm, the PANI/g-C_3N_4 sample exhibited excellent photocatalytic activity having the removal capacity of 4.76 mg/min/gc. It has been the outcome that the capping of PA and HCl acid shows an important part in increasing the performance of g-C_3N_4 capped with PANI, whereas, no determination in augmented action was observed in the case of g-C_3N_4 capped with PANI monotonically doped with PA or HCl. To upsurge the electrical conductivity synergistically and generate a 3D-linked network, the enhanced photocatalytic action of PANI/g-C_3N_4 could be accredited to the intrachain and interchain doping of HCl and PA over PANI, correspondingly. For competent photocatalytic responses and the development of g-C_3N_4-based composites, a novel understanding on PANI and PA and HCl capped g-C_3N_4 may have outstanding potential (Wu et al., 2019).

For the bacterial disinfection and removal of dye, Sadasivam et al. (2017) reported PAN-templated 2D nanofibrous mats. For the removal of ecological contaminants, the polymeric nanofibers manufacturing and their probable adaptability started to substitute smart hybrid nanocomposites. For the efficient removal of dye and disinfection of bacteria, PAN-based 2D nanofibrous mats were synthesized through the electrospinning method with QA/Fe and PEI/Fe as hybrid fillers. Various analytical tools, for example, BET, FESEM, FTIR, XRD, and TEM have been utilized for the determination of fabricated nanocomposites. They examined the thermal and magnetic possessions through the capacities of TGA and superconducting quantum interference devices. To adsorb the manufacturing dye wastes, the adsorption performance of the PAN−PEI/Fe nanofibers have been confirmed by the study of isotherms and kinetic data, and additional recycling studies validated the capacity of the mats. Later, the study of bactericidal disinfection has been done on *Escherichia coli* and *Staphylococcus aureus* bacteria, which further exhibited the killing efficiency of 89.5% and 99%, respectively on the magnetic-QA-loaded PAN nanofiber mats. The bacterial disinfection experiment was done using the disk diffusion method which showed the zone of inhibitions of ~33 and 23 mm, correspondingly, which was evaluated from the microscopic studies validating the disinfection efficacy. Fascinatingly, the in vivo noxiousness researches using the zebrafish models required the real-time submission and in vitro cell culture researches in NIH 3T3, and BHK-21 cell lines showed the cytocompatibility of these nanofibrous mats. Consequently, for the removal of bacteria and manufacturing dye wastes in water and wastewater streams, the inclusive research of the synthesized 2D nanofibrous mats functionalized with PAN-template avowed to be a capable nanocomposite (Sadasivam et al., 2017).

For the degradation of dye materials, Naresh et al. (2018) developed through serially layered Aurivillius perovskites, pH-mediated selective and collective solar photocatalysis. The ecological sustainability using solar energy has fascinated significant consideration in the synthetic direction of compounds; for example, under natural sunlight, the semiconductor photocatalysis is a developing area in current materials research. For the first time, a series of $Bi_5ATi_4FeO_{18}$ (A = Pb, Sr, and Ca) are developed (five-layer Aurivillius-phase perovskites). The orthorhombic structure for the Aurivillius phases with Fe largely inhabiting the central octahedral layer which was indicated by the XRD analyses which show that the divalent cations (Pb, Sr, and Ca) are statistically dispersed over the Cubo octahedral (A-sites of the perovskite). The developed compound shows collective photocatalytic degradation of RhB−Rh 6G and RhB−MB mixture at pH 2 under visible light-absorbing ability as well as almost 100% photocatalytic discerning remediation of MB from the RhB−MB mixture at pH 11 under natural solar irradiation. Though, the compounds showed excellent COD removal efficiency and photocatalytic degradation activity under sunlight irradiation at pH 11 toward separate MB and at pH 2 for RhB, whereas, from the mixed dye solution (MB + RhB) at the neutral pH, the selective MB degradation occurs with an extensive reduction with slower MB photolysis sluggish in MB adsorption. In the alkaline and acidic media equally, the catalysts are remarkably robust and sustain decent crystallinity and are recyclable up to five cycles without any perceptible harm of activity. For the selective elimination of one or various contaminants from a mixture and communal remediation of several contaminants, the authors deliberated a significant understanding hooked on the progress of layered perovskite photocatalysts (Naresh et al., 2018).

13.3   FUTURE PERSPECTIVES

For the adsorption technique for various dye from water systems, this chapter designates the important progressions augmenting worldwide. Novel researches on elimination methods of dye are essential to display decent presentation toward progressively severe ecological guidelines and communal consciousness, obeying inside the monitoring confines executed. There are some utmost useful methods for the degradation of dye, such as biological action, adsorption, and chemical oxidation represented in Figure 13.1. Within a small time, duration, the chemical oxidation can display outstanding removal efficiency of dye. Augmented mineralization could be attained as associated through additional methods. The adsorption method delivers increased flexibility for the removal of dye from the water system. Though, on the removal of textile dyes using adsorption method, several of the research articles uninterruptedly have been publishing, whereas some facts are essential to painstaking for imminent study, for example, the conduct of the adsorbent, especially nanosorbents, pilot plant adsorption studies. To use in a real system, aggregation and separation of nanomaterials are obstacles similarly. In aqueous solution, the proclamation on the nanoadsorbent causes nanotoxicity on breathing organization. The reusability of the nanoadsorbent is vital and is still a big issue. To treat dye wastewater, biological handling is a biodegradable and easy method. However, the dye, the effluent conditions, treatment quality needed, operating conditions, flexibility, environmental impact, costs, and others are the criteria for the selection of the most appropriate method to treat dye water and wastewater. Biological handling is abandoned and a time-consuming method. Consistent with the target contaminant, biological handlings ought to be altered through these useful methods. Chemical precipitation showed augmented removal efficacy toward dye solutions having 25–300 mg/L of dye concentrations. For handling the increased amount of manufacturing surplus, the price of used chemicals, high chemical doses, and the production of large amounts of sludge are the key problems. Similarly, coagulation–flocculation is similarly the predictable technique for the removal of dye, which can remove 60–400 and 800–1500 mg/L of dye concentration, and this technique is simple to operate as well. Also, dye molecules hamper the potential of dye recovery subsequently by forming complexes with flocculants. The removal of dyes ranging from 50 to 400 mg/L and 1000 mg/L, chemical oxidation, for example, the Fenton process, photocatalytic oxidation, and ozonation were enabled. However, to treat water and wastewater streams, the degradation of dyes can give complexes, that is, extra lethal compounds. On the development of low-cost, easy to separate, showing high adsorption capacity adsorbents, many researchers are working with increased mechanical strength.

13.4   CONCLUSIONS

The customer demand in enhanced connectivity in fashion, functionality, and appearance has inspired the progress of nanotechnology-based textiles. Many nanomaterials and nanostructures counting CNTs, NPs, nanoelectronic components, and Bragg diffraction gratings have been placed or intertwined, hooked on textiles over the last two decades. The progress in the field of nanocomposites at an industrial scale similarly generates novel manufacturing approaches, including spray coating, particle impregnation, direct weaving, and multifunctional compound fiber drawing. The optics, electronics, and surface alteration proposals, functionality besides the probable enhanced performance, which are the field of application of nanomaterials. The nanotechnology applications in the removal of textiles dye are mainly using adsorption technology; they have used various nanoadsorbent materials, which also possess antibacterial properties. These nanomaterials are significant toward various dyes, such as MB, CV, RhB, and Rh6G etc., in addition to toxic heavy metal ions likewise.

This chapter deliberated various types of nanomaterials and functional-ized nanoparticles for the effective removal of toxicants using adsorption as potential nanocomposite materials.

ACKNOWLEDGMENTS

The authors gratefully acknowledge the support from the Ministry of Human Resource Development Department of Higher Education, Government of India under the scheme of Establishment of Centre of Excellence for Training and Research in Frontier Areas of Science and Technology (FAST), for providing the financial support to perform this study vide letter No, F. No. 5–5/201 4–TS. Vll. Dinesh Kumar, DST, New Delhi is also thankful for financial support to this work (sanctioned vide project Sanction Order F. No. DST/TM/WTI/WIC/2K17/124(C).

KEYWORDS

- **nanoparticles**
- **flocculatio**
- **surface area**
- **toxic effluents**
- **dye**

REFERENCES

Abdul, A. S.; Gibson, T. L.; Rai, D. N. Use of Humic Acid Solution to Remove Organic Contaminants from Hydrogeologic Systems. *Environ. Sci. Technol.* **1990**, *24*, 328–333.

Ahmed, S. M.; El-Dib, F. I.; El-Gendy, N. S.; Sayed, W. M.; El-Khodary, M. A Kinetic Study for the Removal of Anionic Sulphonated Dye from Aqueous Solution Using Nano-Polyaniline and Baker's Yeast. *Arabian J. Chem.* **2016**, *9*, S1721–S1728.

Ashiq, M. N.; Najam-Ul-Haq, M.; Amanat, T.; Saba, A.; Qureshi, A. M.; Nadeem, M. Removal of Methylene Blue from Aqueous Solution Using Acid/Base Treated Rice Husk as an Adsorbent. *Desalination Water Treat.* **2012**, *49*, 376–383.

Assefi, P.; Ghaedi, M.; Ansari, A.; Habibi, M. H.; Momeni, M. S. Artificial Neural Network Optimization for Removal of Hazardous Dye Eosin Y from Aqueous Solution Using Co_2O_3-NP-AC: Isotherm and Kinetics Study. *J. Ind. Eng. Chem.* **2014**, *20*, 2905–2913.

Bhattacharyya, K. G.; Sen Gupta, S.; Sharma, G. K. Interactions of the Dye, Rhodamine B with Kaolinite and Montmorillonite in Water. *Appl. Clay Sci.* **2014**, *99*, 7–17.

Chen, J. H.; Liu, Q. L.; Hu, S. R.; Ni, J. C.; He, Y. S. Adsorption Mechanism of Cu(II) Ions from Aqueous Solution by Glutaraldehyde Crosslinked Humic Acid-Immobilized Sodium Alginate Porous Membrane Adsorbent. *Chem. Eng. J.* **2011**, *173*, 511–519.

Chen, J. H.; Ni, J. C.; Liu, Q. L.; Li, S. X. Adsorption Behavior of Cd(II) Ions on Humic Acid-Immobilized Sodium Alginate and Hydroxyl Ethyl Cellulose Blending Porous Composite Membrane Adsorbent. *Desalination* **2012**, *285*, 54–61.

Debrassi, A.; Correa, A. F.; Baccarin, T.; Nedelko, N.; Slawska-Waniewska, A.; Sobczak, K.; Dłużewski, P.; Greneche, J. M.; Rodrigues, C. A. Removal of Cationic Dyes from Aqueous Solutions Using N-Benzyl-O-Carboxymethylchitosan Magnetic Nanoparticles. *Chem. Eng. J.* **2012**, *183*, 284-293.

Deka, J. R.; Liu, C. L.; Wang, T. H.; Chang, W. C.; Kao, H. M. Synthesis of Highly Phosphonic Acid Functionalized Benzene-Bridged Periodic Mesoporous Organosilicas for Use as Efficient Dye Adsorbents. *J. Hazard. Mater.* **2014**, *278*, 539–550.

Deng, S.; Bai, R. B. Aminated Polyacrylonitrile Fibers for Humic Acid Adsorption Behaviors and Mechanisms. *Environ. Sci. Technol.* **2003**, *37*, 5799–5805.

Fan, L.; Luo, C.; Sun, M.; Qiu, H.; Li, X. Synthesis of Magnetic β-Cyclodextrin–Chitosan/Graphene Oxide as Nanoadsorbent and Its Application in Dye Adsorption and Removal. *Colloid Surf. B Biointerf.* **2013**, *103*, 601–607.

Giri, S. K.; Das, N. N.; Pradhan, G. C. Synthesis and Characterization of Magnetite Nanoparticles Using Waste Iron Ore Tailings for Adsorptive Removal of Dyes from Aqueous Solution. *Colloid Surf. A: Physicochem. Eng. Aspects* **2011**, *389*, 43–49.

Gupta, V. K.; Mohan, D.; Sharma, S.; Sharma, M. Removal of Basic Dyes (Rhodamine B and Methylene Blue) from Aqueous Solutions Using Bagasse Fly Ash. *Sep. Sci. Technol.* **2000**, *35*, 2097–2113.

Gurses, A.; Dogar, C.; Yalcin, M.; Acikyildiz, M.; Bayrak, R.; Karaca, S. The Adsorption Kinetics of the Cationic Dye, Methylene Blue, Onto Clay. *J. Hazard. Mater.* **2006**, *131*, 217–228.

Gürses, A.; Hassani, A.; Kıranşan, M.; Açışlı, Ö.; Karaca, S. Removal of Methylene Blue from Aqueous Solution Using by Untreated Lignite as Potential Low-Cost Adsorbent: Kinetic, Thermodynamic and Equilibrium Approach. *J. Water Process Eng.* **2014**, *2*, 10–21.

Jain, N.; Bhargava, A.; Panwar, J. Enhanced Photocatalytic Degradation of Methylene Blue Using Biologically Synthesized "Proteincapped" ZnO Nanoparticles. *Chem. Eng. J.* **2014**, *243*, 549–555.

Kaloti, M.; Kumar, A. Synthesis of Chitosan-Mediated Silver Coated γ-Fe_2O_3 (Ag– γ-Fe_2O_3@ Cs) Superparamagnetic Binary Nanohybrids for Multifunctional Applications. *J. Phys. Chem. C* **2016**, *120*, 17627–17644.

Karim, Z.; Mathew, A. P.; Grahn, M.; Mouzon, J.; Oksman, K. Nanoporous Membranes with Cellulose Nanocrystals as Functional Entity in Chitosan: Removal of Dyes from Water. *Carbohydr. Polym.* **2014**, *112*, 668–676.

Khatoon, N.; Sardar, M. Efficient Removal of Toxic Textile Dyes Using Silver Nanocomposites. *J. Nanosci. Curr. Res.* **2017**, *2*, 113–118.

Kong, L.; Gan, X.; Bin Ahmad, A. L.; Hamed, B. H.; Evarts, E. R.; Ooi, B.; Lim, J. Design and Synthesis of Magnetic Nanoparticles Augmented Microcapsule with Catalytic and Magnetic Bifunctionalities for Dye Removal. *Chem. Eng. J.* **2012**, *197*, 350–358.

Kulkarni, S.; Jadhav, M.; Raikar, P.; Raikar, S.; Raikar, U. Core–Shell Novel Composite Metal Nanoparticles for Hydrogenation and Dye Degradation Applications. *Ind. Eng. Chem. Res.* **2019**, *58*, 3630–3639.

Lee, Y. C.; Kim, E. J.; Yang, J. W.; Shin, H. J. Removal of Malachite Green by Adsorption and Precipitation Using Aminopropyl Functionalized Magnesium Phyllosilicate. *J. Hazard. Mater.* **2011**, *192*, 62–70.

Li, H.; Sheng, G.; Teppen, B. J.; Johnston, C. T.; Boyd, S. A. Sorption and Desorption of Pesticides by Clay Minerals and Humic Acidclay Complexes. *Soil Sci. Soc. Am. J.* **2003**, *67*, 122–131.

Li, Y.; Du, Q.; Liu, T.; Sun, J.; Wang, Y.; Wu, S.; Wang, Z.; Xia, Y.; Xia, L. Methylene Blue Adsorption on Graphene Oxide/Calcium Alginate Composites. *Carbohydr. Polym.* **2013**, *95*, 501–507.

Lian, L.; Cao, X.; Wu, Y.; Lou, D.; Han, D. Synthesis of Organo-Functionalized Magnetic Microspheres and Application for Anionic Dye Removal. *J. Taiwan Inst. Chem. E,* **2013**, *44*, 67–73.

Liu, J. F.; Zhao, Z. S.; Jiang, G. B. Coating Fe_3O_4 Magnetic Nanoparticles with Humic Acid for High Efficient Removal of Heavy Metals in Water. *Environ. Sci. Technol.* **2008**, *42*, 6949–6954.

Mahmoodi, N. M. Magnetic Ferrite Nanoparticle–Alginate Composite: Synthesis, Characterization and Binary System Dye Removal. *J. Taiwan Inst. Chem. Eng.* **2013**, *44*, 322–330.

Mahmoodi, N. M. Synthesis of core–shell magnetic adsorbent nanoparticle and selectivity analysis for binary system dye removal. *J. Ind. Eng. Chem.* **2014**, *20*, 2050–2058.

Mittal, H.; Mishra, S. B. Gum Ghatti and Fe_3O_4 Magnetic Nanoparticles-Based Nanocomposites for the Effective Adsorption of Rhodamine B. *Carbohydr. Polym.* **2014**, *101*, 1255–1264.

Naidu, B. V. K.; Rao, K. K.; Aminabhavi, T. M. Pervaporation Separation of Water + 1, 4-Dioxane and Water + Tetrahydrofuran Mixtures Using Sodium Alginate and Its Blend Membranes with Hydroxyethylcellulose—A Comparative Study. *J. Membr. Sci.* **2005**, *260*, 131–141.

Naresh, G.; Malik, J.; Meena, V.; Mandal, T. K. pH-Mediated Collective and Selective Solar Photocatalysis by a Series of Layered Aurivillius Perovskites. *ACS Omega* **2018**, *3*, 11104–11116.

Panneerselvam, P.; Morad, N.; Tan, K. A.; Mathiyarasi, R. Removal of Rhodamine B Dye Using Activated Carbon Prepared from Palm Kernel Shell and Coated with Iron Oxide Nanoparticles. *Sep. Sci. Technol.* **2012**, *47*, 742–752.

Sadasivam, R. K.; Mohiyuddin, S.; Packirisamy, G. Electrospun Polyacrylonitrile (PAN) Templated 2D Nanofibrous Mats: A Platform toward Practical Applications for Dye Removal and Bacterial Disinfection. *ACS Omega* **2017**, *2*, 6556–6569.

Sedláček, P.; Smilek, J.; Klučáková, M. How the Interactions with Humic Acids Affect the Mobility of Ionic Dyes in Hydrogels-2. Non-Stationary Diffusion Experiments. *React. Funct. Polym.* **2014**, *75*, 41–50.

Shariati, S.; Faraji, M.; Yamini, Y.; Rajabi, A. A. Fe_3O_4 Magnetic Nanoparticles Modified with Sodium Dodecyl Sulfate for Removal of Safranin O Dye from Aqueous Solutions. *Desalination* **2011**, *270*, 160–165.

Sharma, R. K.; Chouryal, Y. N.; Chaudhari, S.; Saravanakumar, J.; Dey, S. R.; Ghosh, P. Adsorption-Driven Catalytic and Photocatalytic Activity of Phase Tuned In_2S_3 Nanocrystals Synthesized via Ionic Liquids. *ACS Appl. Mater. Interf.* **2017**, *9*, 11651–11661.

Tan, K. A.; Morad, N.; Teng, T. T.; Norli, I.; Panneerselvam, P. Removal of Cationic Dye by Magnetic Nanoparticle (Fe_3O_4) Impregnated Onto Activated Maize Cob Powder and Kinetic Study of Dye Waste Adsorption. *APCBEE Procedia* **2012**, *1*, 83–89.

Tan, X. L.; Wang, X. K.; Geckeis, H.; Rabung, T. H. Sorption of Eu(III) on Humic Acid or Fulvic Acid Bound to Hydrous Alumina Studied by SEM-EDS, XPS, TRLFS, and Batch Techniques. *Environ. Sci. Technol.* **2008**, *42*, 6532–6537.

Vinod, V. P.; Anirudhan, T. S. Adsorption Behaviour of Basic Dyes on the Humic Acid Immobilized Pillared Clay. *Water Air Soil Pollut.* **2003,** *150,* 193–217.

Wang, Z.; Liu, S.; Zhang, J.; Yan, J.; Zhao, Y.; Mahoney, C.; Matyjaszewski, K. Photocatalytic Active Mesoporous Carbon/ZnO Hybrid Materials from Block Copolymer Tethered ZnO Nanocrystals. *Langmuir* **2017,** *33,* 12276–12284.

Wu, H. H.; Chang, C. W.; Lu, D.; Maeda, K.; Hu, C. Synergistic Effect of Hydrochloric Acid and Phytic Acid Doping on Polyaniline-Coupled g-C_3N_4 Nanosheets for Photocatalytic Cr (VI) Reduction and Dye Degradation. *ACS Appl. Mater. Interf.* **2019,** *11,* 35702–35712.

Xu, H.; Zhang, Y.; Jiang, Q.; Reddy, N.; Yang, Y. Biodegradable Hollow Zein Nanoparticles for Removal of Reactive Dyes from Wastewater. *J. Environ. Manage.* **2013,** *125,* 33–40.

Yagub, M. T.; Sen, T. K.; Afroze, S.; Ang, H. M. Dye and Its Removal from Aqueous Solution by Adsorption: A Review. *Adv. Colloid Interf. Sci.* **2014,** *209,* 172–84.

Yang, H.; Sheikhi, A.; Van De Ven, T. G. Reusable Green Aerogels from Cross-Linked Hairy Nanocrystalline Cellulose and Modified Chitosan for Dye Removal. *Langmuir* **2016,** *32,* 11771–11779.

Yao, Y.; Xu, C.; Qin, J.; Wei, F.; Rao, M.; Wang, S. Synthesis of Magnetic Cobalt Nanoparticles Anchored on Graphene Nanosheets and Catalytic Decomposition of Orange II. *Ind. Eng. Chem. Res.* **2013,** *52,* 17341–17350.

Zafar, M. N.; Dar, Q.; Nawaz, F.; Zafar, M. N.; Iqbal, M.; Nazar, M. F. Effective Adsorptive Removal of Azo Dyes Over Spherical ZnO Nanoparticles. *J. Mater. Res. Technol.* **2019,** *8,* 713–725.

Zhou, K.; Zhang, Q.; Wang, B.; Liu, J.; Wen, P.; Gui, Z.; Hu, Y. The Integrated Utilization of Typical Clays in Removal of Organic Dyes and Polymer Nanocomposites. *J. Cleaner Prod.* **2014,** *81,* 281–289.

CHAPTER 14

Health Safety and Environment Aspect of Nanotextiles

DEVSUNI SINGH[1*], PRASHANSA SHARMA[2], SUMAN PANT[1], and
VIVEK DAVE[3]

[1]*Department of Clothing & Textile, Banasthali Vidhyapith, Rajasthan, India*

[2]*Department of Home Science, Mahila Mahavidyalaya,
Banaras Hindu University, Varanasi, India*

[3]*Department of Pharmacy, School of Health Sciences,
Central University of South Bihar, India*

[*]Corresponding author. Email: singhdevsuni31@gmail.com

ABSTRACT

Nanotechnology has grown up explosively and has created the attention on large scale in scientific, industrial, and technological fieds in the application of textiles through their advanced properties. Today, nanotechnology rapidly emerges and is utilized to produce various consumer products and a large number of nano-based goods are already available on the market. Along with ample of benefits, it is also an important concern about the implication toward the human health and environment. Every concept has two sides, one is the advantage, and the other one is the disadvantage. There is also raised concern about the potential adverse effects of nanomaterials on the human health, safety, and environment. Concerns are also raised about the safety issues mainly for those workers who are subjected to the exposure to product manufacturing and processing, consumers who consume the products or common public who are affected by nanoparticles. Advances of nanomaterials create doubt in the mind of many people whether these products are safe for use. Here, in this chapter, we discuss all the sides of

Fundamentals of Nano-Textile Science. Prashansa Sharma, Devsuni Singh & Vivek Dave (Eds.)
© 2022 Apple Academic Press, Inc. Co-published with CRC Press (Taylor & Francis)

nanotechnology which are necessary to understand in the context of workers, consumer, and exposed public's health. As nanoparticles incorporate into the textile products that is a direct exposure to the human skin, and dispersion of nanoparticles, agglomerates of nanosized particles, or gas emission in the atmosphere might harm the environment. The designing of product, procedure for production of product, life cycle of the product, and many other factors determine the various health exposure and environmental situations. This chapter provides a well understanding of nanotechnology benefits and impact on environment, human health effects, and safety.

14.1   INTRODUCTION

Nanotechnology is progressively enticing the universal attention, as it is widely apparent that it offers an extensive range of consumer goods. Due to its distinctive and innovative properties, nanoptechnology finds its application in several areas including special finishes, sports textile, medical textile, military or health care textile, automotive textile, filtration or purification, protective textile, environment protective textile, and so on. It improves the functionality of textile materials. Despite all these advantages, many people raise a question about its impact toward environment and human's health by using various chemicals, synthetic compounds, new methods, and techniques. This has raised concern about the potential risk for workers, consumers, and impacts of exposure to these hazardous chemicals on health. Numerous researches have been done on this topic, but still we are far away from concrete evidence. However, this necessitates the research to discover a new way of preparation of innovative material that is safe and will solve the health-related issues of nanotextile.

Nanotechnology is the science which deals with the extremely small substances range from 1 to 100 nm. Application of nanotechnology in various fields has been growing from the past. Of all, textile industry is one of the beneficiary which is greatly trusting on nanotechnologies. The textile industry produces diverse types of greatly traded goods using nanotechnology which can be said as "nanotextiles" goods. Nanotextiles/ nanomaterials could open the way of textile products in various applications as multifunctional textiles. It enables the production of textile with enhanced or novel functionality and becomes ubiquitous in industrial sector as well as in the area of consumer products. In textile, because of the unique nanosized nature of nanomaterials, nanotechnology enhances the textile attributes and improves the functionality of the garment in different ways. . This provides the opportunity to produce outstanding features through modifying the textile fiber- producing technique including electrospinning of nanofiber, processing of nanofiber. Moreover, it can also enhance the textiles by surface modification techniques or by nanocoating, nanotreatments, nanocomposites, or by improving the nanoporous nature of the materials, and it is also intended to improve the existing textile properties with the implementation of new functional properties, such as self-cleaning, water repellent, waterproof, soil release, antibacterial, antimicrobial, antifungal, antistatic, flame-retardant, and several other value-added properties.

Contrary to these qualities, some researchers have expressed that nanotechnology-based textile products work upside down to hazardous products, which may be due to the lack of regulatory measures and consumers are in threat for using the nano-based textile goods and consider it as a risk factor for human health and environment. Many studies and researches have been performed to reach any conclusion but still we demandmore studies to have safe application of nanotextiles yet (Montazer and Harifi, 2018). To create awareness, here we discussed some scientific papers and literature reviews to have a clear outlook about this emerging serious issue. This effort is performing for major concerns on consumer's health and environment safety related to nanotechnology in textile.

Since the reports, researches, innovations, processing and manufacturing of nanotextile products have been enormously increasing, there have been massive public disputes about the toxicological, health, and environmental effects of direct and indirect exposure to nanotextile products, this may be due to the inintentional release of nanoparticles from nanotextile products (Kessler, 2011). This can be determined through the phases of product's life cycle. The design and the various stages of the product life cycle of nanotextiles (raw material, manufacture/production, transport/storage, usage, and disposal/recycling) define in what quantities and forms the nanoparticles are unintentionally released into the environment (Som et al., 2010). The whole product life cycle can only determine the exposure to nanotextiles which might directly affect the human health, other organisms, and environment. Depending upon the life cycle of nanotextile from the start phase till the end of the product, basic properties, procedure, various applications used for incorporating the nanoparticles into the textiles (during the fiber production, during finishing directly applied on the fiber surface or polymeric fiber coating), production/manufacturing of nanotextiles, various nanofinishes applied on the nanotextiles, transport/storage of the nanomaterials, usage of the product, external impacts on the nanomaterials (such as UV radiations, over drying, ironing, heat, laundering, abrasion, strain, bleaching, fastness, detergent, water, sweat, solvents and others), end of the product or disposal/ recycling the product will be understood. The end-of-product will affect the environmental compartments, such as air, water, soil, or on human health (skin, respiratory tract, or GIT) (Fig. 14.1).

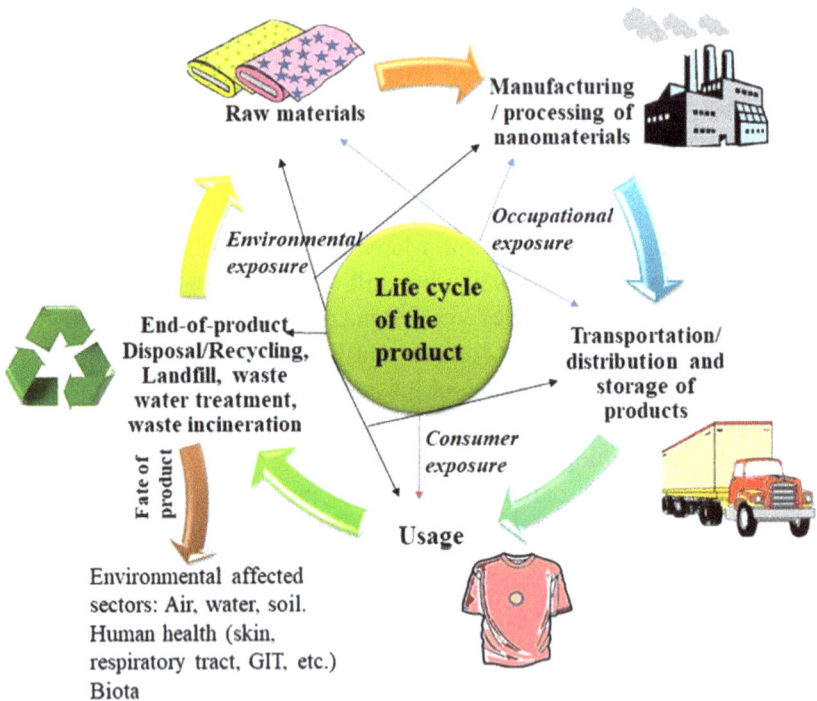

FIGURE 14.1   Life cycle of the product (exposure, fate or health and environment impact).

It is difficult to interpret the comprehensive data, but after performing many case studies on nanotextile applications, we assessed that due to lack of regulatory measures and safety standards, absence of scientific knowledge regarding the toxicity of nanomaterials, mechanism and fate of nanomaterials released during the product's life cycle, and uncertainty in the immense property of nanomaterial substances in terms of particle size, shape, high surface area to volume ratio, crystalline form, surface modification, porosity, purity, density of the nanoparticles, aggregation (strongly bound), and agglomeration (weakly bound).

14.2 NANOTECHNOLOGY

Nanotechnology refers broadly to a field of applied science and technology which deals with the control of matter having one or more dimensions on nanoscale smaller than 1–100 nm, equal to one billionth of a meter. The word nano comes from the Greek word "nanos" meaning "dwarf." Nanotechnology is a broad emerging field having multidisciplinary areas that can bring advantages to different researches and application areas to enhance our way of living in various ways. It firmly has embedded in our society, a number of nanotechnology-embedded materials, products are available in marketplace to serve different purposes. However, the implications of uncontrollable use of nanotechnology make concerns about the exposure and toxic effect toward the human and environmental health.

Nanoparticles occur naturally, incidentally, or intentionally/engineered.

14.2.1 NATURALLY OCCURRING NANOPARTICLES

Naturally occurring nanoparticles can be found in volcanic ashes, mineral springs, fine sand and dust storms, ocean spray, forest fires, and even biological matters (e.g., bacteria, virus). A wide range of nanoparticles occur naturally, in fact, nature itself is providing the nanomaterials in our environment. Nanoashes or soot particles occur due to volcanic ash clouds or other types of combustion that contain polydisperse nanoparticles composed of silicate and iron compounds (range from 100 to 200 nm), they are suspended rapidly in the air and once we inhale these particles, they can cause severe respiratory disorders (Griffin et al., 2017).

14.2.2 INCIDENTALLY OCCURRING NANOPARTICLES

Another type of nanoparticle that occurs unintentionally or accidentally is called incidental nanoparticle compared with the engineered nanoparticles. Incidental nanoparticles differ in shape and size (more variable not well-defined).

Everyday incidences that produce incidental nanoparticles include fire, industries, welding fumes, running diesel engine, transportation activities (vehicle engine exhausts), etc. that contribute to environmental and health hazards.

14.2.3 INTENTIONALLY OCCURRING/ENGINEERED NANOPARTICLES

The well-known nanoparticles are the engineered and man-made nanoparticles that are made intentionally in the laboratories to possess certain properties. These may be (1) metal-based (e.g., Ag, Au, TiO_2) nanoparticles that have silica shell, (2) quantum dots (semiconductor nanocrystal) type, (3) Silica-based, dendrimers (3D polymer structure, 0–7 generation), (4) carbon-based, and (5) carbon black (Ha et al., 2013) (Fig. 14.2).

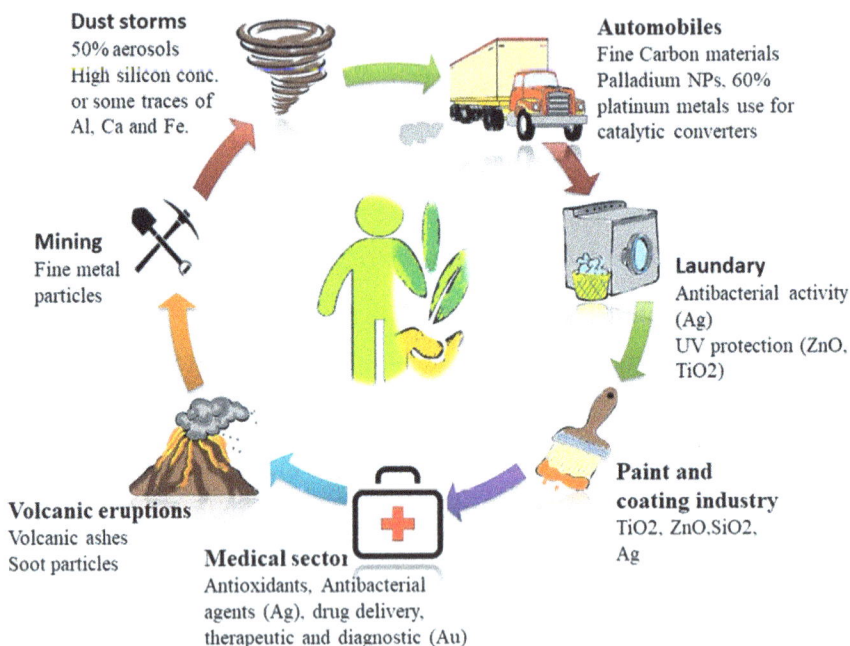

Dust storms
50% aerosols
High silicon conc.
or some traces of
Al, Ca and Fe.

Automobiles
Fine Carbon materials
Palladium NPs, 60%
platinum metals use for
catalytic converters

Mining
Fine metal
particles

Laundary
Antibacterial activity
(Ag)
UV protection (ZnO,
TiO2)

Volcanic eruptions
Volcanic ashes
Soot particles

**Paint and
coating industry**
TiO2, ZnO,SiO2,
Ag

Medical sector
Antioxidants, Antibacterial
agents (Ag), drug delivery,
therapeutic and diagnostic (Au)

FIGURE 14.2   Sources of nanoparticles (naturally, incidentally, or engineered).

14.3 USES OF NANOTECHNOLOGY IN TEXTILES

The success and potential applications of nanotechnology in textile industry have developed the new materials tremendously by improving their functionality using nanotechnology and can provide new or enhanced properties in textile materials.

The application of nanotechnology in textile materials can expand the array of several fabric attributes, such as high durability, abrasion resistance, tear strength, wrinkle-resistance, durable press, fabric softness, and breathability due to their large surface area to volume ratio and have high surface energy, so they provide better affinity for the fabrics. There have been numerous advances of nanotechnology in the textile industry which made the fabric multifunctional, such as water-repellency, soil resistance, fire-retardancy, UV protection, antistatic, antimicrobial properties, anti-odor, improvement of dye ability and the like in fibers, yarns, and fabrics. Because of its boundless potential in consumer-inclinable applications, the textile industry being one of the significant beneficiaries of advances in nanotechnology in terms of conventional process, they impart temporary properties into the fabrics which easily lose their purposes after washing and wearing. In terms of economic potential, nanotechnology application areas include energy saving, control release of substances, packaging, storing, and separating the materials and release under control condition for long duration (David et al., 2002) (Fig. 14.3).

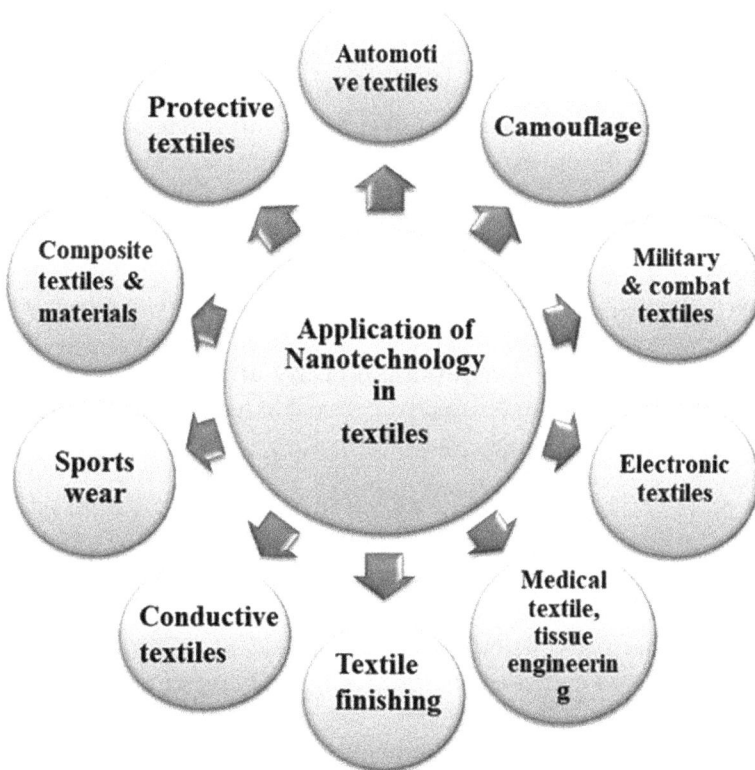

FIGURE 14.3 Applications of nanotechnology in textiles.

Nanomaterials can enhance the property of textiles and make them multi-functional in different ways: during the stage of fiber production (in the core of the fiber) or during the stage of finishing of the fabric (fiber surface as polymeric fiber coating). During the fiber production stage, before spinning the fiber, nanomaterials can be added to the polymer, so that they are uniformly distributed in the whole fiber volume and provide high durability effect. In another way, nanomaterials are incorporated during the stage of fabric finishing as a good polymeric adhesive on the finished fabric surface. Nanomaterials can be coated on the textiles by different techniques, such as plasma polymerization, sol–gel, layer-by-layer, cross-linking, or thin film deposition (Table 14.1).

14.4   RISK ASSESSMENT OF NANOMATERIALS

According to the United States Environmental Protection Agency (EPA) considers "risk to be the chance of harmful effects to human health or to ecological systems results from exposure to an environmental stressor." (US EPA). In toxicology, risk defines the probability of 'exposure" to the "hazards," potential formula: Risk = Hazard × Exposure.

In simple terms, risk involves the probability of uncertainty effects/implications on objectives that may harm the life, human health, animals, property, and/or the environment.

The risk assessment is the overall procedure that helps the identification of hazards and risk factors that cause harmful impact on human health and environment. Identification of hazardous chemical substance is a relevant approach to assess the risks and evaluation of hazards, it is necessary to understand the chemical and physical property of the substance. Chemical risk assessment (CRA) is considered a valuable tool to assess the risk and hazards associated with the human exposure and the environmental to chemical substances (Hristozov and Malsch, 2009) (Fig. 14.4).

14.4.1   HAZARD IDENTIFICATION

Hazard identification is defined as the *"identification of the adverse effects, which a substance has an inherent capacity to cause"* (European Commission JRC et al., 2003). Hazard identification or assessment involves to define the characteristics of a chemical substance, composition, particle size or other properties, and identifies its potential adverse health effects in humans or experimental organisms on exposure to nanomaterials. In the past few years,

TABLE 14.1 Some Nanomaterials Applied on the Textile Finishes

Properties	Nanomaterials	Type of fabric	References
UV protection	ZnO nanoparticles	Cellulosic fabric	Noorian et al. (2020)
	ZnO nanoparticles	Cotton wool fabric	Becheri et al. (2008)
	ZnO nanoparticles	Cotton/cotton polyester fabric	Subramaian et al. (2009)
Flame-retardant	ZnO nanoparticles	Cotton/polyester blend	Alongi et al. (2015)
	SiO_2, colloidal nanoparticles	Polyester fabric	El-Hady et al. (2013)
	TiO_2 nanoparticles	Polyamide and polyester fabric	Apaydin et al. (2015)
	CNT, boroxosiloxane, montmorillonite (nanoclay), Sb_3O_2	Cotton fabric	Zhang et al. (2003)
Antistatic	Silane nanosol	Polyester fabric	Xu et al. (2005)
	Sol–gel coating	Polyester fabric and cotton/polyester blend	Textor and Mahltig (2010)
	Carbon nanotubes	Composite fabric	Sennett and Welsh (2003)
	Nanopolypropylene	Composite fabric	Dong and Huang (2002)
Antibacterial	TiO_2 nanoparticles	Cotton fabric	Burniston et al. (2004)
	ZnO nanoparticles	Cotton/polyester fabric	Farouk et al. (2014)
	SiO_2 nanoparticles	Polyester fabric	El-Gabry et al. (2013)
	Ag nanoparticles	Cellulosic/synthetic fabrics	Lee et al. (2003)
Antimicrobial	Silver nanoparticles	Polyester fabric	Falletta et al. (2008)
	Ag/ZnO composite nanoparticles	Cotton/polyester fabric	Ibanescu et al. (2014)
	Silica sols with silver nanoparticles	Polyamide fabric	Mahltig and Textor (2010)
	ZnO nanoparticles	Cotton fabric	Azam et al. (2012)
Self-cleaning/ superhydrophobic	SiO_2 nanoparticles/ fluoropolymer	Polyester fabric	Xu et al. (2015, 2014)
	TiO_2/SiO_2 coating	Cotton fabric	Yuranova et al. (2006)
	Plasma Polymerization (fluorocarbon chemicals)	Cotton fabric	Zhang et al. (2003)
Wrinkle-resistance	TiO_2 nanoparticles	Cotton fabric	Lam et al. (2010)
	Nano-titanium dioxide and Nanosilica	Cotton/silk fabrics	Wong et al. (2005)

number of experimental studies have shown that exposure to certain nanomaterials can lead to adverse human health effects and other living organisms. Polystyrene nanoparticle deposits in pulmonary system not only cause the lung inflammation but also vascular thrombogenesis (Nemmar et al., 2004). Extrapulmonary translocation after inhalation after exposure to carbon nanoparticles (ultrafine) was reported by Oberdörster et al. (2002). Muller et al. (2005) study shows that multiwall carbon nanotubes caused more inflammatory and fibrogenic reactions, CNTs were still present for 60 days in the rat lungs. Hansen et al. (2007) reported that around 428 studies show toxicity and ecotoxicity of nanomaterials. The potential risk factors involved in the application of nanomaterials in textiles are given in Table 14.2.

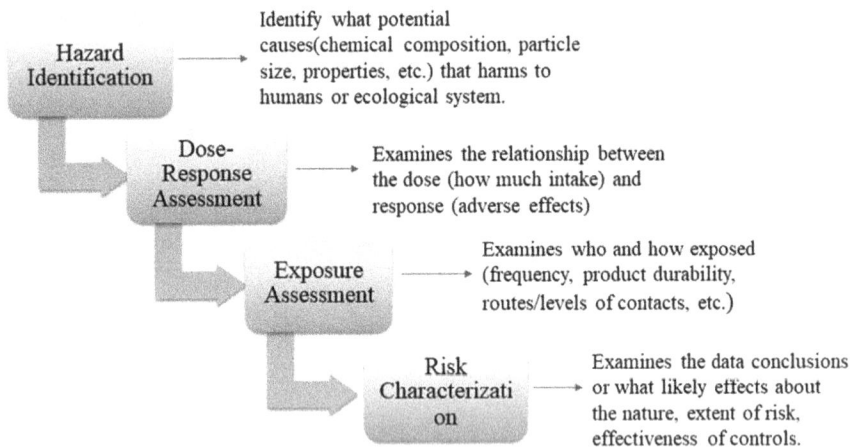

FIGURE 14.4   Risk assessment approach to nanomaterials.

14.4.2   DOSE-RESPONSE ASSESSMENT

Dose–response assessment involves *"an estimation of the relationship between dose, or level of exposure to a substance, and the incidence and severity of an effect"* (European Commission JRC et al., 2003). Dose–response assessment is the second critical step of risk assessment. It determines the relationship between the exposure (dose) and incidence of adverse human health effects (response). The dose is commonly measured in units, such as gram, milligram or per million (g, mg, ppm), respectively. Dose can be classified as three basic types (1) Administered dose (external dose) agent introduced to an organism by inhaled air, eaten food, water intake, (2)

TABLE 14.2 Potential Risk Factors of Nanomaterials.

Nanomaterials	Potential risk factors to human & the environment	References
AgNPs nanoparticles	Citrate-coated silver nanopowders are toxic on human skin HaCaT keratinocytes	Lu et al. (2010)
	AgNPs-coated polyvinyl-pyrrolidine (PVP) can evoke dose-dependent DNA damage in vitro	Nymark et al. (2013)
	AgNPs have genotoxic effects in human normal bronchial epithelial (BEAS-2B) cell due to oxidative stress	Kim et al. (2011) Sharma et al. (2014) and Gliga et al. (2014)
	AgNPs may directly harm the cell membranes, alternate gene expressions, disrupt ATP production, increase the reactive oxygen radicals and replication of DNA. And cytotoxic for human lung at 10 nm.	Choi et al. (2009)
	Ag+ ions are very toxic to microorganism, nanosilver toxicity to nitrifying organism was reduced up to 80% by adding small concentration of sulfide which results no harmful impact on environment	
Metal oxides nanoparticles like TiO₂, ZnO, SiO₂	Human-based 95% ULC of excess risk estimates for lung cancer from TiO₂ exposure in workers. TiO₂ particles induced pulmonary inflammation as well as granuloma formation	NIOSH et al. (2011) Rossi et al. (2009)
	TiO₂: high concentration release in the environment, since changes in the surface properties of nanomaterial will modify their dispersion and fate in aqueous environments	Auffan et al. (2010) Gulson et al. (2010)
	Zn from ZnO particles is absorbed by exposed to sunlight and isotope could be detected in blood and urine	Wang et al. (2006)
	Compared with the microscale zinc powder, nanoscale zinc shows severe symptoms of renal damage, lethargy, anemia, vomiting, anorexia, and diarrhea in the treated mice	Johnston et al. (2000)
	SiO₂: crystalline silica shows mutagenic effects and genotoxic effects, direct or indirect cytotoxicity to target cells. It induced pulmonary inflammation in rats	Chang et al. (2007)
	HSIP "High-tech" wastewater contains numerous nanosized silica particles (2–90 nm), which are not effectively removed from the effluent	

TABLE 14.2 *(Continued)*

Nanomaterials	Potential risk factors to human & the environment	References
Carbon nanotubes	Carbon nanotubes are toxic to humans, could induce dose-dependent pulmonary inflammation and lung fibrosis	Muller et al. (2005)
		Lindberg et al. (2009)
	CNTs and GNFs are potential genotoxic effects, induced DNA damage in human BEAS 2B cells in vitro	Shvedova et al. (2003), Cui et al. (2005), and Kostarelos et al. (2007)
	Produce oxidative stress and cellular toxicity in human epidermal keratinocytes, dose-and time-dependent inhibition of cell proliferation, degradation by enzyme catalyzed reactions	
Quantum dots	DNA: 56% damaged after incubation with dots and exposure to UV light display whereas DNA incubation with quantum dots observed less DNA damaged (29%) in the dark	Green and Howman (2005)
		Chang et al. (2006)
	Cytotoxicity of surface coated QDs (CdSe/CdS) correlates with cell toxicity in intracellular level rather than extracellular level and at 4°C endocytosis of nanoparticles is inhibited	Hoshino et al. (2004)
	Cytotoxicity of fluorescent quantum dot-labeled cells were observed 20% long-lasting traces in mouse (7 days after injection)	
C^{60} fullerenes	$C_{60}(OH)_{24}$ showed cytotoxicity, such as cell density decreased in dose-dependent manner and formation of cytosolic vacuole	Yamawaki and Iwai (2006)
		Rouse et al. (2006)
	Due to maximal dose of lactate dehydrogenase (LDH) release, endothelial toxicity occurs in human umbilical vein	
	Cytotoxicity of fullerene in human epidermal keratinocytes, such as decreased cell viability and pro-inflammatory cytokines IL-6 and IL-8	

the absorbed dose (internal dose) agent comes into contact with the body's internal tissues, (3) cumulative dose (total dose) the total of all individual doses administered to an organism over a given time (Lewandowski and Norman, 2015). Oberdörster et al. (2005b) suggest in toxicity studies that the biological activity of nanoparticles is dependent on physico-chemical properties (particle size, chemical composition, surface area, charge, and porosity) rather than mass-dependent property. Warheit et al. (2007a, 2007b) reported that toxicity was correlated with the surface of nanoparticles (number of functional groups).

14.4.3   *EXPOSURE ASSESSMENT*

Exposure assessment involves *"an estimation of the concentrations/doses to which human populations (i.e., workers, consumers and man exposed indirectly via the environment) or environmental compartments (aquatic environment, terrestrial environment, and air) are or may be exposed"* (European Commission JRC et al., 2003). Exposure assessment is the key element in the process of risk assessment. According to EPA, it is estimating, characterizing, measuring, or modeling the magnitude, frequency, duration of contact with an agent in the environment as well as the number and characteristics of the population exposed. In toxicology, the term exposure assessment is defined as "no exposure = no risk." According to the US Environmental Protection Agency, they divided the tools into three sections for performing exposure assessments to nanoscale materials including (1) monitoring data, (2) exposure models, and (3) the data sets for modeling as shown in Figure 14.5.

Intentionally or unintentionally produced nanomaterials contain number of characteristics, such as particle size, shape, dimensions, agglomeration, or other substances, such as ultrafine particles, airborne particles that may cause toxicity to the human body through various exposure routes, such as inhalation (breathing), ingestion, or oral (eating food, medicines, supplements, or drinking), dermal (direct contact on top layer of skin), and ocular (airborne particles effect the eyes) in Figure 14.6.

14.4.3.1 *SOURCES OF EXPOSURE*

Sources or the fate of nanomaterials can be categorized into three sub-areas: (1) occupational exposure assessment, (2) consumer exposure assessment, and (3) environmental exposure assessment.

Monitoring Data	Biological monitoring, personal sampling, ambient air monitoring, medical surveillance, or worker health monitoring.
Exposure Modeling	Predicts concentration of air toxics through inhalation exposure from external sources
Data Sets for modeling	Collection of data provides underlay for comprehension the exposure and toxicity

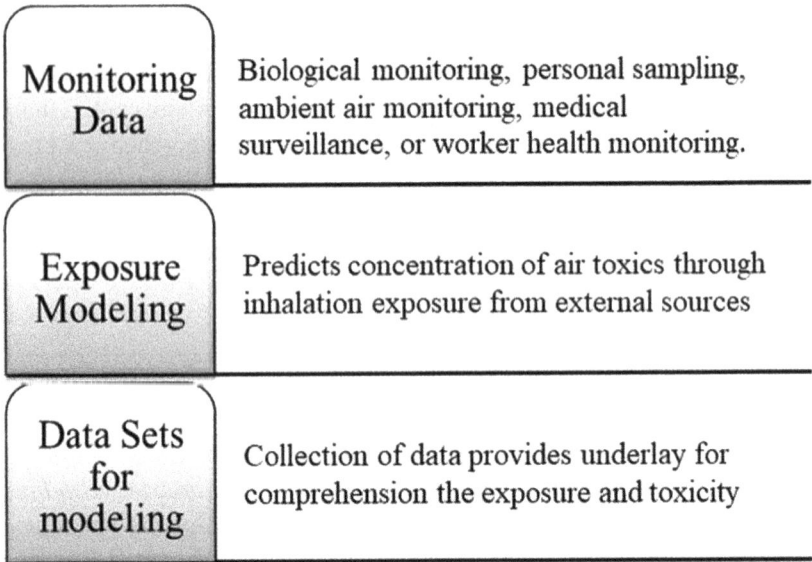

FIGURE 14.5   Tools for exposure assessment.

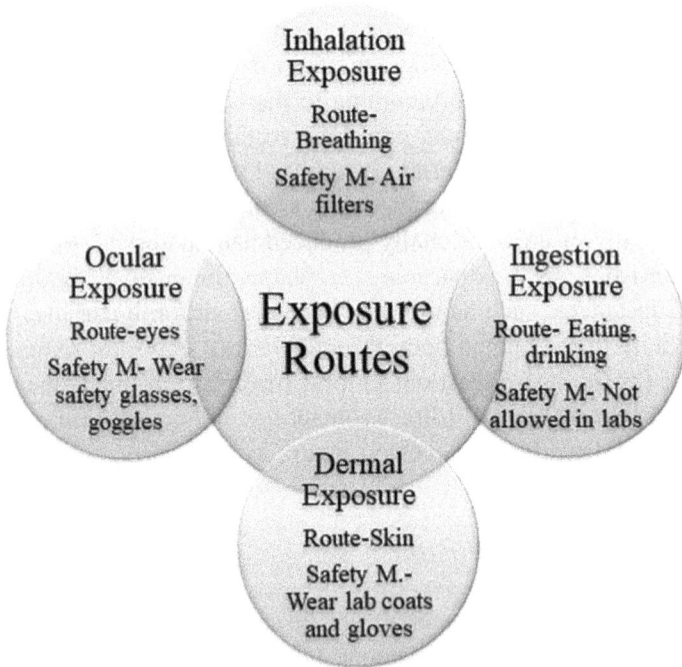

FIGURE 14.6   Potential exposure routes.

14.4.3.1.1 Occupational Exposure Assessment

Occupational exposure is a process for estimating the health risks of workers' exposure to hazardous substance in various levels of workplace (manufacturing/synthesis, production level, industrial applications, distribution/end use of products, or during disposal/recycling). Workers are mainly exposed through the route of inhalation, ingestion, or dermal contact, that is, after the manufacturing process of nanomaterials, when opening the reaction chamber, when the product is drying, at the time or post-processing time of handling the final products (Biswas and Wu, 2005). Workers' exposure more likely occurs during the collection of produced nanomaterials, handling of nanomaterials in filtration system, during cleaning operations, release of nanoparticles in the reaction chamber, nanoconditioning (e.g., coating, compression, or composition of nanoparticles), handling of final products, and these factors pose increased health risks (Biswas and Wu et al., 2005).

Maynard et al. (2004) worked on the potential exposure pathways and the toxicity of raw airborne single-walled CNT. Study was performed to assess the physical nature of aerosol particles formed during agitation while SWCNT material was removed from the production vessels and handled prior to processing. It concluded that there is high risk of airborne exposure and dermal uptake of single-walled CNT concentration while handling the nanomaterial. Gloves deposits were also estimated, suggesting the risk of exposure to nanomaterial during handling and bagging. Aitken et al. (2004) worked on the detection of potential exposures, manufacture and production methods (fullerenes, carbon nanotubes, metal, and oxides) or usage of nanomaterials. They confirmed that mainly there are four groups/phases of risk exposures of nanomaterials in the production processes: vapor deposition/phase (on substrate), gas phase (in air), colloidal phase (liquid suspension), and attrition phase (liquid suspension). All these production processes of nanomaterials can possibly result in the occupational exposure through inhalation, ingestion, or dermal routes.

14.4.3.1.2 Consumer Exposure

EPA estimates the widespread consumer exposure to chemicals in consumer products and materials which can occur by inhalation, oral routes, or dermal contact. Hansen et al. (2008b) used the Technical Guidance Document for consumer exposure assessment of nanoparticles. The concentration is higher in a number of products which fall into several categories, such as appliances, health and fitness, foodstuff and beverages, or home and garden. The consumer exposure was estimated for various fluid products, such as facial lotion, sunscreen lotion (outdoor surface treatment), or spray product (indoor surface treatment). Due to the lack of information about the exposures and the types of nanomaterials used in the products manufacturing, the estimated values of consumer exposure to nanoparticle- containing products can be expressed as 26 µg/kg bw/year for facial lotion, 15 µg/kg bw/year for fluid product, and 44 µg/kg bw/year for spray product. A survey was conducted in the Danish Industry which revealed that in the production of sunscreen lotions, they use 10–20 nm TiO_2 nanoparticles (conc. up to 10%, or surface area 50–200 m^2/g,) as UV absorber (Tønning and Poulsen, 2007).

14.4.3.1.3   Environmental Exposure

As the production of commercial products containing nanomaterials is continuously increasing, there is greater potential of nanomaterial release in the environment.

Potential nanomaterial release in the environment occurs either directly and/or indirectly during all stages of life cycle of the products, such as release during raw material production, manufacturing and processing, release during transport/storage of products, release during consumer usage, release after disposal of products containing nanoparticles (waste landfills, incineration, and wastewater treatment), or during recycling of nanomaterials.

The fate of nanomaterials released in the environment is determined by different routes in the atmosphere, in water, and in soils.

Fate of Nanomaterials in Air: There are various processes and factors influencing the fate of airborne particles with their chemical characteristics and dimension (1) the duration of the particles remaining as airborne, (2) the nature of their interaction with other molecules or particles in the air, and (3) the distance that they may travel in the air before gettingdeposited on to surfaces. The processes which are significant to understand the possible atmospheric transport of nanoparticles include diffusion, aggregation or agglomeration, dry or wet deposition, and gravitational settling. These processes or factors are relatively self-explanatory for studying the ultrafine particles of nanomaterials as well (Wiesner et al., 2006). However, in some cases, there may be significant differences in behavior between intentionally produced nanomaterials and incidental ultrafine particles, for example, when nanoparticles are surface coated to prevent the agglomeration.

With respect to the duration of particles remaining as airborne, particles with diameter (<100 nm) may follow the laws of gaseous diffusion when released to air. The rate of diffusion is inversely proportional to particle diameter, whereas the rate of gravitational settling is proportional to particle diameter (Aitken et al., 2004).

Fate of Nanomaterials in Soil: The fate of nanomaterials released in soil varies depending upon the physiochemical characteristics of the nanomaterial. Soil-sorbed nanoparticles released to soil may be sorbed to soil because of their high surface areas and immobility. Nanoparticles which are smaller do not sorb to soil, but might show greater mobility to travel than larger particles before getting easily trapped into the soil matrix. The strength of the sorption of nanoparticle to soil will be dependent on its chemical composition, particle size and surface, chemistry treatment, and the conditions applied on it.

Fate of Nanomaterials in Water: The fate of nanomaterials in aqueous is determined by (1) aqueous solubility/dispersability, (2) Interactions of nanomaterial with natural or anthropogenic chemicals in the aqueous environment, and (3) biological/abiotic processes. Due to their lower mass, waterborne nanoparticles usually settle down more gradually than large particles of the same material. However, because of their high surface-area-to-mass ratios, nanoparticles are likely to sorb to soil and sediment particles, whereas soil-sorbed nanoparticles can be quickly removed from the aqueous environment (Oberdörster et al., 2005a). Some nanomaterials might be degraded by biotic or abiotic factors (contain hydrolysis/ photocatalysis reaction), resulting in removal from the aqueous column. Nanoparticles in the uppermost surface of aqueous system and water droplet in air are exposed to sunlight. Light-induced photoreactions are essential in the understanding of the environmental fate of chemical matters.

14.4.4 RISK CHARACTERIZATION

Last part of the risk assessment procedure is risk characterization defined as "*estimation of the incidence and severity of the adverse effects likely to occur in a human population or environmental compartment due to actual or predicted exposure to a substance, and may include risk estimation*" (European Commission JRC et al., 2003).

Risk characterization is the last ultimate steps, after gathering the first three steps of information of risk assessment from the first step of hazard identification, dose–response, to the third step, environmental exposure is taken together, weighted, and the actual potential risk is quantified. Muller and Nowack et al. (2008) studied the first completely quantitative environmental risk assessment of nanomaterials. They used threshold conc. of 20 mgL^{-1} and 40 mg^{-1}L of silver nanoparticulate and exposed to *B. subtilis* and *E. coli* bacteria. The results showed that the above concentration of silver nanoparticulate did not affect the integrity of the microorganisms. In addition, they calculated the PNEC values of Ag (0.04 mg^{-1} L), TiO$_2$ (<0.001 mg^{-1}L), and CNTs (<0.0001 mg^{-1}L) of nanoparticulate in water. Merging these PNEC values with the expected exposure, they measured the environmental concentration of the above nanoparticles in Switzerland, stemming from various industries, such as cosmetics, textiles, plastics, coatings, sports gear, and electronics. They observed that the silver nanoparticulate and CNTs were lower than 0.001, and do not cause any risk/harm to aquatic organisms. However, the potential exposure to TiO$_2$ may possibly pose worst-case exposure/risks, ranging from 0.7 to 16.

14.5   MEASURES USED TO REDUCE POTENTIAL EXPOSURE TO NANOMATERIALS

With the continued exploration of nanomaterials, researchers also provide the data about the potential health hazards and exposure limits Appropriate measures must be taken to control the potentialexposure (localized air filters-HEPA, local exhaust ventilation, personal protection equipment's) and to follow paramount practices for providing safe work environment. As per NIOSH, the following engineering controls should be possible to control inhalation exposure risk of nanomaterials (Fig. 14.7):

14.5.1   ENGINEERING CONTROLS

- Use of chemical fume hood/laboratory fume hood for the suspension of aerosols nanoparticles.
- Glove box (HEPA-filtered enclosure) should be used to limit the inhalation of nanomaterials.
- A well-constructed local exhaust ventilation system should be used with a local high-efficiency particulate air (HEPA) filter.
- Use of biological safety cabinet (BSC) class II type A1, A2 (for liquid suspension or bound nanostructure), vented via thimble connection, or B1 or B2 (for bound nanostructures, liquid and dry suspension).

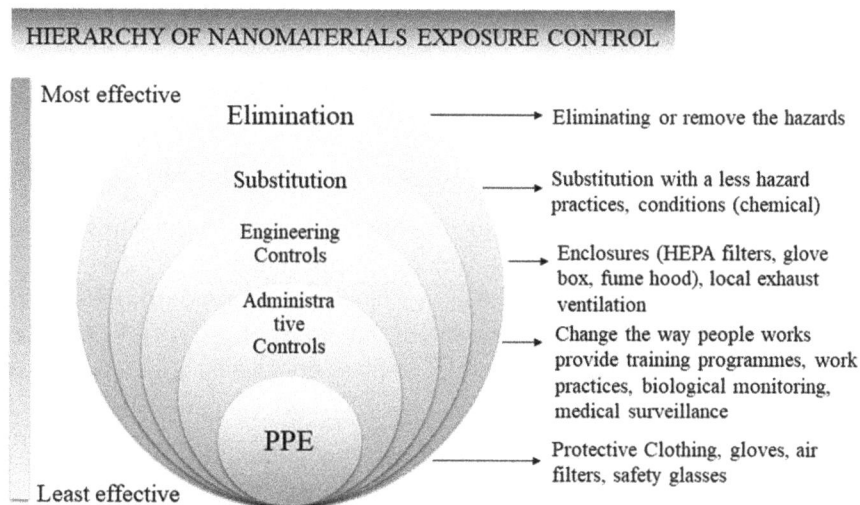

FIGURE 14.7 Hierarchy of nanomaterials exposure control.

- Appropriate equipment and laboratories for monitoring toxic gas (e.g., CO).
- Transport dispersible nanomaterials in sealed containers/air lock containers.

14.5.2 ADMINISTRATIVE CONTROLS

14.5.2.1 EFFECTIVE TRAINING TO THE PERSONNEL

- Frequent hand washing and encouraging good sanitization practices.
- Safe handling of nanomaterials and following standard procedures.
- Laboratory personnel must be appropriately trained and must use extreme caution while performing experiments.
- Laboratory personnel must be well aware about the engineering controls, equipment maintenance, and emergency procedure.

14.5.2.2 GOOD HOUSEKEEPING PRACTICES

- Cleanup the nanomaterial work area with HEPA-vacuum cleaner, use wet wiping method or use compressed air for dust cleaning.

- Properly dispose the used cleaning materials with the hazardous waste procedures.

14.5.2.3   WORK PRACTICES

- Liquid phase–wear gloves before handling, avoid dispersion of liquid droplets, cleanup spills, or working in spill container.
- Gas phase–work in a closed reaction chamber or closed fume hood, use HEPA-filter, and follow safety standards for pressurized vessels.
- Nanomaterials powder–use closed fume hood or an enclosure (glove box), wear class P3, use respiratory filters, cleanup all parts (spills of nanoparticles).

14.5.2.4   MEDICAL SURVEILLANCE

- Make available health monitoring program, medical screening, and surveillance for personnel exposed to nanomaterials.
- Worker health exposure/monitoring assessments and baseline medical evaluations.

14.5.3   PERSONAL PROTECTIVE EQUIPMENT (PPE)

- Protective gloves–wear nitrile gloves to protect the hands when handling nanomaterials and other hazardous substances.
- Eye protection–wear safety glasses or goggles or full-face shield.
- Protective clothing–wear laboratory coats and gowns, it helps to coverage the skin while working with nanomaterials.
- Respiratory protection–wear respiratory air filters (N95 or N100), or full-face piece or air-purifying respirator (P100, N100, R100).

14.5.4   SAFETY AND DISPOSAL PRACTICES OF NANOMATERIAL WASTE

14.5.4.1   SAFETY PRACTICES

- For the identification of the contents, wastes should be labeled on all the chemical vessels.

- For shipment of the nanomaterial, it is recommend that use "Combination package" consists two layers of packaging (1) for inner package, use plastic bags or bottles, and (2) for outer package, use wooden or fiber-board box.
- For transporting the nanomaterial, use sealed or double-contained containers.
- Be certain that test should not be performed on partially oxidized particles.

14.5.4.2 DISPOSAL OF WASTE NANOMATERIALS

- Do not put nanomaterial wastes into the regular trash or down the drain.
- Do not permit nanomaterial wastes to be shifted off-site to home institutions for disposal of waste.
- Follow the list of hazardous property given by UK, LOWR-List of Waste Regulations mentioned the two type of substances, such as H5 "Harmful substances" (worst category case of soluble and insoluble nanoparticles), and H6 "Toxic substances" (CMAR-carcinogenic, mutagenic, asthma-genic, and reproductive toxics).
- Label the waste containers containing nanomaterial waste and dispose the hazardous waste according to the following standard procedures.

14.5.4.3 PREVENTION OF FIRE AND EXPLOSION

- In order to avoid the explosion behavior of nanoparticles, do not put the oxidizable nanomaterials in contact with air and stored safely under inert liquid/gas bath.
- Large samples containing oxidized particles, potentially pyrophoric when exposed in the air.
- Use antistatic shoes and mats in the areas where potentially explosive nanoparticles are handled, they will reduce the static charge.
- Reduction or control the excess release of hazardous substance.
- For a colloid dispersion, use distillation system for vaporizing the solvent within an explosion-proof enclosure.
- Researchers/students should avoid creating large, high-concentration aerosols of combustible or any ignition source nanomaterials to prevent any exposition.

14.6   CONCLUSION

Nanomaterials have the potential to enhance the quality of textile sectors. However, these nano-treated textiles may also pose negative impact against environment and raise human health and safety concerns. Researchers are still finding some scientific data about the safety of nanomaterials for health and environment. The magnitude of the potential risks depends upon the design of the product, circumstances of product usage, end point of the product. Some studies or reports depict the possible risks involved with nano-treated textiles depends upon nanoparticles size, shapes, porosity, density, crystalline form, dosage level, surface area, stability and the end use. Further researchers discussed on the possible negative effects in the application of nanotextile finishing by releasing of nanoparticles into the human skin. There is a lack of regulations and standards regarding restrictions on nanomaterials treated on textiles. Thus increased the concern and need of the regulatory measures and standards for nanotextiles. There are no specific standards for the textile industry, specific standards or preventive measures to nanotechnology should be followed to reduce uncertainty and prevent from hazards. Approaches of risk preventive measures include green chemistry in the life cycle of the product, eco-design, eco-labels, test methods, specification. and innovation in governance regulations are required when nanomaterials are incorporated into the textile materials to ensure the consumers about their products as effective, consistent, less toxic, and safe to use. More encouragement to green nanotechnology, to enhance environmental sustainability and minimize potential risks using environmental friendly steps of manufacturing and processing of nanoproducts are needed. Thus, preventive measures and innovative strategies help to reduce the risks of human health and environment safety.

KEYWORDS

- nanotechnology
- life cycle of the product
- risk assessment approach
- potential risks
- environment and health
- measures to control exposure

REFERENCES

Aitken, R. J.; Creely, K. S.; Tran, C. L. Nanoparticles: An Occupational Hygiene Review. Research Report 274. Prepared by the Institute of Occupational Medicine for the Health and Safety Executive, North Riccarton, Edinburgh, England, 2004.

Alongi, J.; Tata, J.; Carosio, F.; Rosace, G.; Frache, A.; Camino, G. A Comparative Analysis of Nanoparticle Adsorption as Fire-Protection Approach for Fabrics. *Polymers* **2015**, *7*, 47–68, 17.

Apaydin, K.; Laachachi, A.; Ball, V.; Jimenez, M.; Bourbigot, S.; Ruch, D. Layer-by-Layer Deposition of a TiO_2-Filled Intumescent Coating and Its Effect on the Flame Retardancy of Polyamide and Polyester Fabrics. *Colloids Surf. A: Physicochem. Eng. Aspects* **2015**, *469*, 1–10.

Auffan, M.; Pedeutour, M.; Rose, J.; Masion, A.; Ziarelli, F.; Borschneck, D. Structural Degradation at the Surface of a TiO_2-Based Nanomaterial Used in Cosmetics. *Environ. Sci. Technol.* **2010**, *44*, 2689–2694.

Azam, A.; Ahmed, A. S.; Oves, M.; Khan, M. S.; Habib, S. S.; Memic, A. Antimicrobial Activity of Metal Oxide Nanoparticles against Gram-Positive and Gram-Negative Bacteria: A Comparative Study. *Int. J. Nanomed.* **2012**, *7*, 6003–6009.

Becheri, A.; Durr, M.; Lo Nostro, P.; Baglioni, P. Synthesis and Characterization of Zinc Oxide Nanoparticles: Application to Textiles as UV-Absorbers. *J. Nanopart. Res.* **2008**, *10*, 679–689.

Biswas, P.; C.Y, Wu. Critical Review: Nanoparticles and the Environment. *J. Air Waste Manage. Assoc.* **2005**, *55*, 708–746.

Burniston, N.; Bygott, C.; Stratton, J. Nano Technology Meets Titanium Dioxide. *Surf. Coat. Int. Part A: Coat. J.* **2004**, *87* (4), 179–814.

Chang, E.; Thekkek, N.; Yu, W. W.; Colvin, V. L.; Drezek, R. Evaluation of Quantum Dot Cytotoxicity Based on Intracellular Uptake. *Small* **2006**, *2* (12), 1412–1417.

Chang, M. R.; Lee, D. J.; Lai, J. Y. Nanoparticles in Wastewater from a Science-Based Industrial Park-Coagulation Using Polyaluminum Chloride. *J. Environ. Manage.* **2007**, *85*, 1009–1014.

Choi, O.; Cleuenger, T. E.; Deng, B. L.; Surampalli, R. Y.; Ross, L.; Hu, Z. Q. Role of Sulfide and Ligand Strength in Controlling Nanosilver Toxicity. *Water Res.* **2009**, *43*, 1879–1886.

Cui, D.; Tian, F.; Ozkan, C.; Wang, M.; Gao, H. Effect of Single Wall Carbon Nanotubes on Human HEK293 Cells. *Toxicol. Lett.* **2005**, *155*, 73–85.

Dong, W. G.; Huang, G. Research on Properties of Nano Polypropylene/TiO_2 Composite Fiber. *J. Text. Res.* **2002**, *23*, 22–23.

El-Gabry, L. K.; Allam, O. G.; Hakeim, O. A. Surface Functionalization of Viscose and Polyester Fabrics toward Antibacterial and Coloration Properties. *Carbohydr. Polym.* **2013**, *92*, 353–359.

El-Hady, M. M. A.; Farouk, A.; Sharaf, S. Flame Retardancy and UV Protection of Cotton Based Fabrics Using Nano ZnO and Polycarboxylic Acids. *Carbohydr. Polym.* **2013**, *92*, 400–406.

European Commission Technical Guidance Document (TGD) on Risk Assessment; European Commission Joint Research Center (ECJRC): Location, Country, 2003. http://ecb.jrc.ec. europa.eu/tgd/ (accessed 10 Sept 2009).

Falletta, E.; Bonini, M.; Fratini, E.; Lo Nostro, A.; Pesavento, G.; Becheri, A.; Lo Nostro, P.; Canton, P.; Baglioni, P. Clusters of Poly(Acrylates) and Silver Nanoparticles: Structure and Applications for Antimicrobial Fabrics. *J. Phys. Chem C.* **2008**, *112*, 11758–11766.

Farouk, A.; Moussa, S.; Ulbricht, M.; Schollmeyer, E.; Textor T. ZnO-Modified Hybrid Polymers as an Antibacterial Finish for Textiles. *Text. Res. J.* **2014,** *84,* 40–51.

Gliga, A. R.; Skoglund, S.; Wallinder, I. O.; Fadeel, B.; Karlsson, H. Size-Dependent Cytotoxicity of Silver Nanoparticles in Human Lung Cells: The Role of Cellular Uptake, Agglomeration and Ag Release. *Particle Fibre Toxicol.* **2014,** *11* (1), 11.

Green, M.; Howman, E. Semiconductor Quantum Dots and Free Radical Induced DNA Nicking. *Chem. Commun.* **2005,** *1,* 121–123.

Griffin, S.; Masood, M.; Nasim, M.; Sarfraz, M.; Ebokaiwe, A.; Schafer, K.; Keck, C.; Jacob, C. Natural Nanoparticles: A Particular Matter Inspired by Nature. *Antioxidants* **2017,** *7* (1), 3.

Gulson, B.; McCall, M.; Gomez, L.; Korsch, M.; Casey, P.; Kinsley, L. Dermal Absorption of ZnO Particles from Sunscreens Applied to Humans at the Beach. *Int. Conf. Nanosci. Nanotechnol.* **2010,** 22–26 Feb, Sydney.

Ha, S.; Weitzmann, M.; Beck, G. Dental and Skeletal Applications of Silica-Based Nanomaterials. *Nanobiomater. Clin. Dentist.* **2013,** 69–91.

Hansen, S. F.; Michelson, E.; Kamper, A.; Borling, P.; Stuer-Lauridsen, F.; Baun, A. Categorization Framework to Aid Exposure Assessment of Nanomaterials in Consumer Products. *Ecotoxicology* **2008**b, *17* (5), 438–447.

Hansen, S.; Larsen, B.; Olsen, S.; Baun, A. Categorization Framework to Aid Hazard Identification of Nanomaterials. *Nanotoxicology* **2007,** *11,* 243–250.

Hoshino, A.; Hanaki, K.; Suzuki, K.; Yamamoto, K. Applications of T-Lymphoma Labeled with Fluorescent Quantum Dots to Cell Tracing Markers in Mouse Body. *Biochem. Biophys. Res. Commun.* **2004,** *314* (1), 46–53.

Hristozov, D.; Malsch, I. Hazards and Risks of Engineered Nanoparticles for the Environment and Human Health. *Sustainability* **2009,** *1* (4), 1161–1194.

Ibanescu, M.; Musat, V.; Textor, T.; Badilita, V.; Mahltig, B. Photocatalytic and Antimicrobial Ag/ZnO Nanocomposites for Functionalization of Textile Fabrics. *J. Alloys Comp.* **2014,** *610,* 244–249.

Johnston, C. J.; Driscoll, K. E.; Finkelstein, J. N.; Baggs, R.; O'Reilly, M. A.; Carter, J. Pulmonary Chemokine and Mutagenic Responses in Rats after Subchronic Inhalation of Amorphous and Crystalline Silica. *Toxicol. Sci.* **2000,** *56,* 405–413.

Kessler, R. Engineered Nanoparticles in Consumer Products: Understanding a New Ingredient. *Environ. Health Perspect.* **2011,** *119* (3), A120–A125.

Kim, H. R.; Kim, M. J.; Lee, S. Y.; Oh, S. M.; Chung, K. H. Genotoxic Effects of Silver Nanoparticles Stimulated by Oxidative Stress in Human Normal Bronchial Epithelial (BEAS-2B) Cells. *Mutat. Res.* **2011,** *726* (2), 129–135.

Kostarelos, K.; Lacerda, L.; Pastorin, G.; Wu, W.; Wieckowski, S.; Luangsivilay, J. Cellular Uptake of Functionalized Carbon Nanotubes Is Independent of Functional Group and Cell Type. *Nat. Nanotechnol.* **2007,** *2,* 108–113.

Lam, Y. L.; Kan, C. W.; Yuen, C. W. M. Effect of Concentration of Titanium Dioxide Acting as Catalyst or Co-Catalyst on the Wrinkle-Resistant Finishing of Cotton Fabric. *Fibers Polym.* **2010,** *11* (4), 551–558.

Lee, H. J.; Yeo, S. Y.; Jeong, S. H. Antibacterial Effect of Nanosized Silver Colloidal Solution on Textile Fabrics. *J. Mater. Sci.* **2003,** *38,* 2199.

Lewandowski, T.; Norman, J. *Dose-Response Assessment. Toxicological Risk Assessment for Beginners*; Springer: Cham, **2015**; pp 43–66.

Lindberg, H. K.; Falck, G. C.; Suhonen, S.; Vippola, M.; Vanhala, E.; Catalan, J.; Savolainen, K.; Norppa, H. Genotoxicity of Nanomaterials: DNA Damage and Micronuclei Induced by

Carbon Nanotubes and Graphite Nanofibres in Human Bronchial Epithelial Cells in Vitro. *Toxicol. Lett.* **2009**, *186* (3), 166–173.

Lu, W.; Senapati, D.; Wang, S.; Tovmachenko, O.; Singh, A. K.; Yu, H.; Ray, P. C. Effect of Surface Coating on the Toxicity of Silver Nanomaterials on Human Skin Keratinocytes. *Chem. Phys. Lett.* **2010**, *487*, 92–96.

Mahltig, B.; Textor, T. Silver Containing Sol-Gel Coatings on Polyamide Fabrics as Antimicrobial Finish-Description of a Technical Application Process for Wash Permanent Antimicrobial Effect. *Fibers Polym.* **2010**, *11*, 1152–1158.

Maynard, A. D.; Baron, P. A.; Foley, M.; Shvedova, A. A.; Kisin, E. R.; Castranova, V. Exposure to Carbon Nanotube Material: Aerosol Release during the Handling of Unrefined Single-Walled Carbon Nanotube Material. *J. Toxicol. Environ. Health A* **2004**, *67* (1), 87–107.

Muller, J.; Huaux, F.; Moreau, N.; Mission, P.; Heilier, J. F.; Delos, J.; Arras, M.; Fonseca, A.; Nagy, J. B.; Lison, D. Respiratory Toxicity of Multi-Wall Carbon Nanotubes. *Toxicol. Appl. Pharmacol.* **2005**, *207* (3), 221–231

Montazer, M., Harifi, T. *Nanofinishing of Textile Materials: Health Safety, and the Environmental Aspects of Textile Nanofinishing*; Woodhead Publishing Series in Textiles, 2018.

Nemmar, A.; Nemery, B.; HM Hoet, P.; Vermylen, J. Pulmonary Inflammation and Thrombogenicity Caused By Diesel Particles in Hamsters: Role of Histamine. *Am. J. Respirat. Crit. Care Med.* **2004**, *168* (11), 1366–1372.

Noorian, S.; Hemmatinejad, N.; Navarro, J. Ligand Modified Cellulosic Fabrics as Support of Zinc Oxide Nanoparticles for UV Protection and Antimicrobial Activities. *Int. J. Biol. Macromol.* **2020**, *154*, 1215–1226.

Nymark, P.; Catalán, J.; Suhonen, S.; Järventaus, H.; Birkedal, R.; Clausen, P.; Jensen, K.; Vippola, M.; Savolainen, K.; Norppa, H. Genotoxicity of Polyvinylpyrrolidone-Coated Silver Nanoparticles in BEAS 2B Cells. *Toxicology* **2013**, *313* (1), 38–48.

Oberdörster, G.; Maynard, A.; Donaldson, K.; Castranova, V.; Fitzpatrick, J.; Ausman, K.; Carter, J.; Karn, B.; Kreyling, W.; Lai, D.; Olin, S.; Monteiro-Riviere, N.; Warheit, D.; Yang, H. Principles for Characterizing the Potential Human Health Effects from Exposure to Nanomaterials: Elements of a Screening Strategy. *Particles Fibre Toxicol.* **2005b**, *2* (1), 8.

Oberdörster, G.; Sharp, Z.; Atudorei, V.; Elder, A.; Gelein, R.; Lunts, A.; Kreyling, W.; Cox, C. Extrapulmonary Translocation of Ultrafine Carbon Particles Following Whole-Body Inhalation Exposure of Rats. *J. Toxicol. Environ. Health, Part A* **2002**, *65* (20), 1531–1543.

Occupational Exposure to Titanium Dioxide, NIOSH Publisher No. 2011–160, 2011

Rossi, E. M.; Pylkkänen, L.; Koivisto, A. J.; Vippola, M.; Jensen, K. A.; Miettinen, M.; Sirola, K.; Nykasenoja, H. Airway Exposure to Silica-Coated TiO2 Nanoparticles Induces Pulmonary Neutrophilia in Mice. *Toxicol. Sci.* **2009**, *113* (2), 422–433.

Rouse, J.; Yang, J.; Barron, A.; Monteiro-Riviere, N. Fullerene-Based Amino Acid Nanoparticle Interactions with Human Epidermal Keratinocytes. *Toxicol. Vitro* **2006**, *20*, 1313–1320.

Sennett, M.; Welsh, E. Dispersion and Alignment of Carbon Nanotubes in Polycarbonate. *Appl. Phys. A.* **2003**, *76*, 111–113.

Sharma, V. K.; Siskova, K. M.; Zboril, R.; Gardea-Torresdey, J. L. Organic-Coated Silver Nanoparticles in Biological and Environmental Conditions: Fate, Stability and Toxicity. *Adv. Colloid Interf. Sci.* **2014**, *204*, 15–34.

Shvedova, A.; Castranova, V.; Kisin, E.; Schwegler-Berry, D.; Murray, A. R.; Gandelsman, V. Z.; Maynard, A.; Baron, P. Exposure to Carbon Nanotube Material: Assessment of Nanotube Cytotoxicity Using Human Keratinocyte Cells. *J. Toxicol. Environ. Health* **2003**, *66*, 1909–1926.

Som, C.; Berges, M.; Chaudhry, Q.; Dusinska, M.; Fernandes, T. F.; Olsen, S. I. The Importance of Life Cycle Concepts for the Development of Safe Nanoproducts. *Toxicology* **2010,** *269*, 160–169.

Subramaian, K.; D'Souza, L.; Dhurai, B. UV Protection Finishing of Textiles Using ZnO Nanoparticles. *Ind. J. Fibre Text. Res.* **2009,** *34* (3), 267–273.

Textor, T.; Mahltig, B. A Sol-Gel Based Surface Treatment for Preparation of Water Repellent Antistatic Textiles. *Appl. Surf. Sci.* **2010,** *256* (6), 1668–1674.

Tønning, K.; Poulsen, M. Nanotechnology in the Danish Industry—Survey on Production and Application. Environmental Project No. 1206 2007; Danish Ministry of the Environment Danish Environmental Protection Agency: Copenhagen, 2007.

U.S. Environmental Protection Agency Nanotechnology White Paper; U.S. EPA: Washington, DC, USA, 2007. http://www.epa.gov/osa/pdfs/nanotech/epa-nanotechnologywhite-paper-0207.pdf (accessed 10 Sept 2009).

Wang, B., Feng, W.; Wang, T.; Jia, G.; Wang, M.; Shi, J.; Zhang, F.; Zhao, Y.; Chai, Z. Acute Toxicity of Nano- and Micro-Scale Zinc Powder in Healthy Adult Mice. *Toxicol. Lett.* **2006,** *161*, 115–123.

Warheit, D. B.; Webb, T. R.; Colvin, V. L.; Reed, K. L.; Sayes, C. R. Pulmonary Bioassay Studies with Nanoscale and Fine-Quartz Particles in Rats: Toxicity Is Not Dependent Upon Particle Size But on Surface Characteristics. *Toxicol. Sci.* **2007a,** *95* (1), 270–280.

Warheit, D. B.; Webb, T. R.; Reed, K. L.; Frerichs, S.; Sayes, C. M. Pulmonary Toxicity Study in Rats with Three Forms of Ultrafine-TiO$_2$ Particles: Differential Responses Related to Surface Properties. *Toxicology* **2007b,** *230*, 90–104.

Wiesner, M. R.; Lowry, G. V.; Alvarez, P.; Dionysiou, D.; Bisawas, P. Assessing the Risks of Manufactured Nanomaterials. *Environ. Sci. Technol.* **2006,** *40* (14), 4336–4345.

Wong, Y. W. H.; Yuen, C. W. M.; Leung, M. Y. S.; Ku, S. K. A.; Lam, H. L. I. Selected Applications of Nanotechnology in Textiles. *AUTEX Res. J.* **2005,** *6* (1), 1–8.

Xu, L.; Cai, Z.; Shen, Y.; Wang, L.; Ding, Y. Facile Preparation of Superhydrophobic Polyester Surfaces with Fluoropolymer/SiO$_2$ Nanocomposites Based on Vinyl Nanosilica Hydrosols. *J. Appl. Polym. Sci.* **2014,** *131*, 80.

Xu, L.; Shen, Y.; Wang, L.; Ding, Y.; Cai, Z. Preparation of Vinyl Silica-Based Organic/Inorganic Nanocomposites and Superhydrophobic Polyester Surfaces from It. *Colloid Polym. Sci.* **2015,** *293* (8).

Xu, P.; Wang, W.; Chen, S. L. Application of Nanosol on the Antistatic Property of Polyester. *Melliand Int.* **2005,** *11* (1), 56–59.

Yamawaki, H.; Iwai, N. Cytotoxicity of Water-Soluble Fullerene in Vascular Endothelial Cells. *Am. J. Physiol. Cell Physiol.* **2006,** *290* (6), C1495–C1502.

Yuranova, T.; Mosteo, R.; Bandara, J.; Laub, D.; Kiwi, J. Self-Cleaning Cotton Textiles Surfaces Modified by Photoactive SiO$_2$/TiO$_2$ Coating. *J. Mol. Catalys. A: Chem.* **2006,** *244* (1–2), 160–167.

Zhang, J.; France, P.; Radomyselskiy, A.; Datta, S.; Zhao, J.; Ooij, WV. Hydrophobic Cotton Fabric Coated by a Thin Nanoparticulate Plasma Film. *J. Appl. Polym. Sci.* **2003,** *88*, 1473–1481.

Index

A

Actuators, 262–263
Albumin nanospheres, 21–22
Ambient conditions, 14–15
American Association of Textile Chemists
    and Colorists test method (AATCC
    183-2004), 137
Antimicrobial activity, 177
    limitations, 177
    TiO_2, 179–181
    treatment, 177–178
    treatment, necessity, 177–178
Antimicrobial application
    AgNPs, 85
    carboxylic groups, 88
    cell membrane, 84
    copper nanoparticles CuNPs, 87–88
    flame-retardant agent, 89–90
    flasks, 86
    imparting cellulosic textile, 94
    mechanisms of, 83, 84
    self-cleaning, 90–92
    treated fabrics, 85
    UV protection application of, 89
    ZnONPs, 86–87
Antimicrobial nanofinish, 68
Anton Formhalson process, 6
Applied voltage, 38–39
Artificial intelligence-operating system
    (AI-OS), 222–224
Atmospheric-pressure glow discharge
    (APGD), 156
Atmospheric-pressure plasma, 154–155
Australian/New Zealand standard (AS/NZS
    4399, 2017), 136–137

B

Barium titanate (BT) ($BaTiO_3$), 43
Basal cell carcinoma (BCC), 130
Biological oxygen demand, 279–280

C

Cationic polyelectrolytes, 172–173
Cellulosic fabrics
    cellulose polymer, 80–81
    coloration process, 81–82
    finishing of, 82
    pretreatment of wet processing, 81
Chemical oxygen demand, 280
Chitosan-neem nano emulsion (CNNE), 106
CIE (International Commission on
    Illumination), 136
Conductive carbon nanotube (CNT), 44
Co-ZnO/PVDF-HFP nanogenerator, 43

D

Dielectric barrier discharge (DBD), 155

E

Electrospinning, 33
    electrospun web-based piezoelectric
        generators, 40
        Barium titanate (BT) ($BaTiO_3$), 43
        conductive carbon nanotube (CNT), 44
        Co-ZnO/PVDF-HFP nanogenerator, 43
        KNN nanostructures, 41
        microbeads, 45
        nanogenerator, 46
        PVDF-TrFE polymer matrix, 41
        PVDF/ZnO NR-based electrospun
            nanocomposites, 42
        schematic diagram, 44
    parameters and fiber characteristics, 9
        ambient conditions, 14–15
        electro hydrodynamics, 13–14
        process parameter, 10
        solution properties, 12–13
        tip to collector distance, 10–11
    process parameters
        applied voltage, 38–39
        collector and tip, 40

flow rate, 40
  temperature and humidity, 40
set-up, 34
  parts in, 34–35
solution parameters
  concentration of, 36
  molecular mass, 36
  polyvinyl alcohol (PVA), 36
  surface tension, 38
  viscosity of, 35–36

G

Geoffrey Ingram Taylor, 6

H

High-end fibers
  DRYARN, 199
  HYGRA, 198–199
  KILLAT N, 199
  LUMIA, 199
  LYCRA, 199
  ROICA and LEOFEEL, 199
  TRIACTOR, 199
Hydrogen peroxide treatment, 63

I

ISO/EN 13758-1 and EN 13758-2, 137

K

KNN nanostructures, 41

L

Layer-by-layer deposition, 171
  antimicrobial activity, 177
    TiO_2, 179–181
    treatment, necessity, 177–178
    ZnO, 181–182
  modification of fibers, necessity, *175*
  nanoparticles, synthesis of, 175–176
    nano titanium dioxide (TiO_2), 175–176
    nano zinc oxide (ZnO), 176
  PEM formation, preparation, 176–177
  polyelectrolytes, 172–173
  SEM photograph of PEM film, 173–174
  ultraviolet protection, 178
    TiO_2, 185–186

UV absorber, 179
ZnO, 183–185

M

Multijet electrospinning techniques, 15
  multiple needles for, 16
  needle-less, 17–18
  single-needle, 16, 17

N

Nano titanium dioxide (TiO_2), 175–176
Nano zinc oxide (ZnO), 176
Nanobiophotoscouring
  hybrid catalyst system, 63
Nanobleaching, 57
  cotton, 58–59
  wool and silk, 59
    hydrogen peroxide treatment, 63
    melamine, 62
    nanophotobleaching, 61–62
    schematic presentation, 60
    ultrasonic bath sonication set-up, 61
    yellowness index, 61
Nano-encapsulation, 18
  nanofiber applications, 24–25
  nanoparticles methods, production
    electrochemical deposition, 22
    gas-phase condensation methods, 23
    plasma-enhanced chemical vapor
      deposition, 23
    sol-gel method, 22–23
    sputtering, 23–24
  polymerization method
    albumin nanospheres, 21–22
    emulsification/solvent diffusion, 21
    emulsification/solvent evaporation, 20
    emulsion, 19
    interfacial polycondensation, 20
    interfacial polymerization, 19–20
    nanoparticles from preformed
      polymers, 20
    solvent displacement and interfacial
      deposition, 21
Nanofibers, 4
  Anton Formhalson process, 6
  3D-block electrospun fiber work, 7
  electrospinning process, 5, 6, 8

electrospun nanofiber, 9
  setup, 9
experimental works, 5–6
Geoffrey Ingram Taylor, 6
multijet electrospinning techniques, 15
  multiple needles for, 16
  needle-less electrospinning for, 17–18
  single-needle electrospinning, 16, 17
nano-encapsulation, 18
  polymerization method, 19–22
nanotechnology, 4–5
production techniques
  polymers (PLLA poly(L-lactic acid), 8
Nanofinishing and nanosurafce activation,
  63
  antimicrobial nanofinish, 68
  nanocleaning, 64
  nanosoftening
    montmorillonite (MMT), 65
    with nanoclay, 65–66
    with nanosilicones, 66
  UV protection, 66–67
  water repellent and easy care finish,
    67–68
Nanomaterials
  antimicrobial application
    AgNPs, 85
    carboxylic groups, 88
    cell membrane, 84
    copper nanoparticles CuNPs, 87–88
    flame-retardant agent, 89–90
    flasks, 86
    imparting cellulosic textile, 94
    mechanisms of, 83, 84
    self-cleaning, 90–92
    treated fabrics, 85
    UV protection application of, 89
    ZnONPs, 86–87
  biosynthesis, microorganisms used in,
    77–78
  cellulosic fabrics
    cellulose polymer, 80–81
    coloration process, 81–82
    finishing of, 82
    pretreatment of wet processing, 81
  finishing of, 82
  fruit extracts, 79
  green and eco-friendly, 75

microorganisms, synthesis, 75
    plant extracts, 75–76
    plants used in, 79
    traditional synthetic methods, 74
    use, 129
Nanoparticles (NPs)
  textile dyes, removal, 299, 302–304,
    307–310
    effluents, 305–306
  on textiles, application of, 138
    nano-based treatment processes,
      138–140
Nanoscouring, 53
  cotton, 55
    fabrics, 54
  wool and silk
    hydrophilic surface, generation, 56
    photocatalytic, 56
    water absorption, 55
    ZnO nanoparticles, 55
Nanosoftening
  montmorillonite (MMT), 65
  with nanoclay, 65–66
  with nanosilicones, 66
Nanotechnology in textiles, 4–5
  in fiber and textile manufacturing, 7
  industrial applications, 117
    commercializing research, 118
    electrospinning machine, 119
    nano-care, 118–119
    poor breathability, 120
    sonochemical process, 118
Nanotextiles, 317
  measures used, 334
    administrative controls, 335
    engineering controls, 334–335
    fire and explosion, prevention, 337
    good housekeeping practices, 335–336
    medical surveillance, 336
    nanomaterial waste, 336–337
    personal protective equipment (PPE),
      336
    safety and disposal practices, 336–337
    training to personnel, 335
    work practices, 336
  nanomaterials, risk assessment, 324
    dose–response assessment, 326, 329
    hazard identification, 324, 326

potential risk factors, 327–328
on textile finishes, 325
nanotechnology, 321
incidentally occurring nanoparticles, 321
intentionally occurring/engineered nanoparticles, 322
naturally occurring nanoparticles, 321
uses in, 322–324
risk assessment, 329
characterization, 333–334
consumer exposure, 331–332
environmental exposure, 332–333
occupational exposure, 331
sources of, 329–330
Natural fibers, 52
Novel application, nanotechnology new dimension, 237
nanocomposites, functions, 261
actuators, 262–263
defense applications, 262
fire protection, 262
self-cleaning, 263
sensors, 261–262
nanomaterials, 238–239
biomedicine, 241
moisture management, 242
nanofibers, 239–240
nanogenerators, 243–245
nanosols, 240
nanowires, 243
responsive polymers, 241–242
smart textiles, 245–246
personal protection, 246–248
wearable and sensors, 248–250
textile-based PENGS, 257–258
textile 1D yarns/fibers, *258–260*
triboelectric (NANO) generators, *258*
two-dimensional (2D) fabrics, 260–261
yarn and fabric formation, 250–251
conductive nanowires, 253–254
fabrication technology, 251–253
metal nanowires, 254
nanowires, 256
oxide nanowires, 255–256
polymer nanowires, 254–255
semiconducting nanowires, 255
sulfide nanowires, 256

O

Operating-pressure plasma
atmospheric-pressure glow discharge (APGD), 156
atmospheric-pressure plasma, 154–155
corona generating devices, 155
dielectric barrier discharge (DBD), 155
low-pressure plasma, 154
schematic diagram, 156

P

Personal protective equipment (PPE), 336
Piezoelectricity, 32
Plasma in textile field
operating-pressure plasma
atmospheric-pressure glow discharge (APGD), 156
atmospheric-pressure plasma, 154–155
corona generating devices, 155
dielectric barrier discharge (DBD), 155
low-pressure plasma, 154
schematic diagram, 156
plasma in textile field
classification, 153
definition, 151–152
generation, 152–153
principles, 156
activation using, 160
chemical bonds disruption, 157
cleaning and etching, 158–159
composites and laminates, 158
copolymerization and nano coating, 161–164
dyeing and printing, 157
electrical properties, 157
grafting and polymerization, 160–161
mechanical properties, 157
metal-coated organic polymers, 157
species interaction with, 158
wetting, 157
Poly(ethylene imine) (PEI)
dip, 177
fabric strips, 176
Polyelectrolyte
antimicrobial activity
limitations, 177
treatment, 177–178

cationic polyelectrolytes, 172–173
nanoparticles, synthesis
  nano titanium dioxide (TiO$_2$), 175–176
  nano zinc oxide (ZnO), 176
SEM photograph of
  layer-by-layer (LBL) deposition
    technique, 173
  modification, 175
  polyelectrolyte multilayers (PEMs),
    174
Polymers (PLLA poly(L-lactic acid), 8
Polyvinyl alcohol (PVA), 36

S

Smart nanotextiles, 207
in armed forces, intelligent and
  applications, 227–230
designed for armed forces, 218
  artificial intelligence-operating system
    (AI-OS), 222–224
  CBRN protective wearable, 219–220
  extreme environment-resistant,
    220–222
  fitness and alertness model, 218–219
  integration of (IOT), 225–226
  medically active, 222
  physiological changes, management,
    218
management of pandemic
  innovations for, 226–227
material, classifications of, 212
  active smart nanotextiles, 212
  futuristic smart nanotextiles, 212–213
  passive smart nanotextiles, 212
  ultra-smart nanotextiles, 212
in military, application of, 210
  medical applications, 211
  soldiers, 210
textile engineering
  aerogel-incorporated, 215
  biopolymers, 213
  chromic material in, 217
  conductive ink, 215–216
  intrinsically conductive polymers, 214
  metallic fibers, 214
  optical fiber integrated, 216–217
Sportswear, 189
  application of, 191–192

clothing, application, 191
  characteristic required, 192
  utilitarian attire, 192
coated water-repellent breathable fabrics
  filaments, 196
  hydrophilic coatings, 195
  leisurewear, qualities, 195
  miniaturized denier polyester, 195
compactly weaved water-repellent
  fabrics, 194
critical requirements and characteristics,
  190–191
dampness transport, factors influencing
  comfort and relevance, 197
  elements, 196–197
  engineered strands, 197
gloves, 201
high-end fibers for
  DRYARN, 199
  HYGRA, 198–199
  KILLAT N, 199
  LUMIA, 199
  LYCRA, 199
  ROICA and LEOFEEL, 199
  TRIACTOR, 199
laminated breathable, 194–195
latest novelties in leisurewear, 197–198
  high-end fibers, application, 198–199
  sweat and fast-drying property,
    absorption, 198
leisurewear, latest novelties in, 197
  and fast-drying property, 198
  sweat, absorption of, 198
moisture management technology,
  192–193
  crafted strands, 193
  surface pressure, 192
moisture, transport mechanism
  narrow operation, 196
motorbike leather outfits, 201–202
  leather jackets for, 202
  leather pants for, 202
  motocross and motorcycle pants, 203
  printed T-shirts, 203–204, *203–204*
  rain wear, 204
  textile jackets for, 202–203
multilayered structures for, 200–201
  dri-release, 200

Entrant Dermizax EV, 201
mixtures, 200
push-pull fabrics, 200
sports wool, 200
outfit, 201
rainwear and protective gear, 201
requirements and characteristics
cotton T-shirts, 191
Spandex, 190
waterproof fabrics, 194–195
water-repellent breathable fabrics, 193–194
coated water-repellent, 195–196
compactly weaved, 194
factors influencing dampness transport, 196–197
laminated breathable, 194–195
textures, 194
transport mechanism for moisture, 196
waterproof structure, 193
Squamous cell carcinoma (SCC), 130
Sun protection factor (SPF), 135–136

T

Tetraethyl orthosilicate (TEOS), 108
Textile effluent, management, 277
effluent management using nanotechnology, 281–282
adsorption, 282
carbon nanotubes, removal pollutant by, 285–286
electrospun nanofiber adsorbent, pollutant, 286, 287–288
heavy metal, removal, 282–283
nanocomposites, removal pollutant by, 284–285
nanofiltration, 286, 289
nanoparticles, removal pollutants by, 283–284
effluent treatment methods, 280
primary treatment, 280–281
secondary treatment, 281
tertiary treatment, 281
nanotechnology, challenges, 293
parameters used in, 279
biological oxygen demand, 279–280
chemical oxygen demand, 280
total dissolved solids, 280
photocatalysis, 289
chemical oxidation process, 289–290
nano-based ozonation, 293
nanomaterials for, 290–291
pollution and drinking water, 279
zero-liquid discharge, 278
Textile-based PENGS, 257–258
textile 1D yarns/fibers, *258–260*
triboelectric (NANO) generators, *258*
two-dimensional (2D) fabrics, 260–261
Total dissolved solids, 280
Triboelectric (NANO) generators, *258*
Two-dimensional (2D) fabrics, 260–261

U

Ultrasonic bath sonication set-up, 61
Ultraviolet protection factor (UPF), 178
nano-TiO$_2$ particles, 179
antimicrobial property, 181
washing durability, 180
nano-ZnO, 181
fabshield AEM 5700 (Aldrich-Sigma), 182–183
ultraviolet protection, 183–186
UV absorber, 179
UV radiation (UVR)
basal cell carcinoma (BCC), 130
cumulative exposure, 130
evaluation
American Association of Textile Chemists and Colorists test method (AATCC 183-2004), 137
Australian/New Zealand standard (AS/NZS 4399, 2017), 136–137
CIE (International Commission on Illumination), 136
ISO/EN 13758-1 and EN 13758-2, 137
spectrophotometer, 136
sun protection factor (SPF), 135–136
UPF, 136
factors, influencing, 131–135
protective clothing and transmission, 130
protective textiles, 131
squamous cell carcinoma (SCC), 130
ultraviolet spectra, 129

W

Water and wastewater
  biological oxygen demand, 279–280
  chemical oxygen demand, 280
  total dissolved solids, 280
Water-repellent breathable fabrics, 193–194
  coated water-repellent, 195–196
  compactly weaved, 194
  factors influencing dampness transport,
    196–197
  laminated breathable, 194–195
  textures, 194
  transport mechanism for moisture, 196
  waterproof structure, 193
Wet chemical processing
  textile fibers, 146
    bleaching process, 148
    chemical finishing, 150–151
    desizing, 148
    dyeing and printing, 149–150
    mechanical finishing, 150
    mercerization, 149
    scouring process, 148
    singeing process, 147
    treatment of fabrics, 147–149
Wool and silk
  nanobleaching, 59
    hydrogen peroxide treatment, 63
    melamine, 62
    nanophotobleaching, 61–62
    schematic presentation, 60
    ultrasonic bath sonication set-up, 61
    yellowness index, 61

Woolen fabrics, 103
  antimicrobial properties, application
    antibacterial efficiencies, 105
    antibacterial efficiency, 105
    bacterial reduction efficiency, 104
    chitosan-neem nano emulsion (CNNE),
      106
    pad-dry-cure wool fabric, 104
    plasma and enzyme pretreatment, 105
  anti-moth properties, 106
    experiment, 107
    nano-TiO_2-treated samples, 108
  fire-retardant nano finish, 110–113
  multifunctional properties, 113–116
  nanotechnology, 104
  ultraviolet (UV) protection
    nano particles (NPs), 110
    ultraviolet protection factor (UPF), 109
  water-repellant nano finishes
    nano silica coating, 109
    SEM, 108
    tetraethyl orthosilicate (TEOS), 108

Y

Yarn and fabric formation, 250–251
  conductive nanowires, 253–254
  fabrication technology, 251–253
  metal nanowires, 254
  nanowires, 256
  oxide nanowires, 255–256
  polymer nanowires, 254–255
  semiconducting nanowires, 255
  sulfide nanowires, 256
Yellowness index, 61

For Product Safety Concerns and Information please contact our EU
representative GPSR@taylorandfrancis.com
Taylor & Francis Verlag GmbH, Kaufingerstraße 24, 80331 München, Germany

www.ingramcontent.com/pod-product-compliance
Lightning Source LLC
Chambersburg PA
CBHW060800220326
41598CB00022B/2497